Superconductivity of Metals and Alloys

Advanced Book Classics

Anderson: Basic Notions of Condensed Matter Physics, ABC ppbk, ISBN 0-201-32830-5

Atiyah: K-Theory, ABC ppbk, ISBN 0-201-40792-2

Bethe: Intermediate Quantum Mechanics, ABC ppbk, ISBN 0-201-32831-3

Clemmow: Electrodynamics of Particles and Plasmas, ABC ppbk, ISBN 0-20147986-9

Davidson: Physics of Nonneutral Plasmas, ABC ppbk ISBN 0-201-57830-1

DeGennes: Superconductivity of Metals and Alloys, ABC ppbk, ISBN 0-7382-0101-4

d'Espagnat: Conceptual Foundations Quantum Mechanics, ABC ppbk, ISBN 0-7382-0104-9

Feynman: Photon-Hadron Interactions, ABC ppbk, ISBN 0-201-36074-8

Feynman: Quantum Electrodynamics, ABC ppbk, ISBN 0-201-36075-4

Feynman: Statistical Mechanics, ABC ppbk, ISBN 0-201-36076-4

Feynman: Theory of Fundamental Processes, ABC ppbk, ISBN 0-201-36077-2

Forster: Hydrodynamic Fluctuations, Broken Symmetry, and Correlation Functions, ABC ppbk, ISBN 0-201-41049-4

Gell-Mann/Ne'eman: The Eightfold Way, ABC ppbk, ISBN 0-7382-0299-1

Gottfried: Quantum Mechanics, ABC ppbk, ISBN 0-201-40633-0

Kadanoff/Baym: Quantum Statistical Mechanics, ABC ppbk, ISBN 0-201-41046-X

Khalatnikov: An Intro to the Theory of Superfluidity, ABC ppbk, ISBN 0-7382-0300-9

Ma: Modern Theory of Critical Phenomena, ABC ppbk, ISBN 0-7382-0301-7

Migdal: Qualitative Methods in Quantum Theory, ABC ppbk, ISBN 0-7382-0302-5

Negele/Orland: Quantum Many-Particle Systems, ABC ppbk, ISBN 0-7382-0052-2

Nozieres/Pines: Theory of Quantum Liquids, ABC ppbk, ISBN 0-7382-0229-0

Nozieres: Theory of Interacting Fermi Systems, ABC ppbk, ISBN 0-201-32824-0

Parisi: Statistical Field Theory, ABC ppbk, ISBN 0-7382-0051-4

Pines: Elementary Excitations in Solids, ABC ppbk, ISBN 0-7382-0115-4

Pines: The Many-Body Problem, ABC ppbk, ISBN 0-201-32834-8

Quigg: Gauge Theories of the Strong, Weak, and Electromagnetic Interactions, ABC ppbk, ISBN 0-201-32832-1

Richardson: Experimental Techniques in Condensed Matter Physics at Low Temperatures, ABC ppbk ISBN 0-201-36078-0

Rohrlich: Classical Charges Particles, ABC ppbk ISBN 0-201-48300-9

Schrieffer: Theory of Superconductivity, ABC ppbk ISBN 0-7382-0120-0

Schwinger: Particles, Sources, and Fields Vol. 1, ABC ppbk ISBN 0-7382-0053-0

Schwinger: Particles, Sources, and Fields Vol. 2, ABC ppbk ISBN 0-7382-0054-9

Schwinger: Particles, Sources, and Fields Vol. 3, ABC ppbk ISBN 0-7382-0055-7

Schwinger: Quantum Kinematics and Dynamics, ABC ppbk, ISBN 0-7382-0303-3

Thom: Structural Stability and Morphogenesis, ABC ppbk, ISBN 0-201-40685-3

Wyld: Mathematical Methods for Physics, ABC ppbk, ISBN 0-7382-0125-1

SUPERCONDUCTIVITY OF METALS AND ALLOYS

P. G. DE GENNES
Faculté des Sciences
Orsay, France

Translated by P. A. PINCUS
University of California
Los Angeles, California

CRC Press
Taylor & Francis Group
Boca Raton London New York

CRC Press is an imprint of the
Taylor & Francis Group, an **informa** business

Advanced Book Program

First published 1966 by Westview Press

Published 2018 by CRC Press
Taylor & Francis Group
6000 Broken Sound Parkway NW, Suite 300
Boca Raton, FL 33487-2742

CRC Press is an imprint of the Taylor & Francis Group, an informa business

Visit the Taylor & Francis Web site at
http://www.taylorandfrancis.com

and the CRC Press Web site at
http://www.crcpress.com

Library of Congress Catalog Card Number: 99-60033

ISBN 13: 978-0-7382-0101-6 (pbk)

Cover design by Suzanne Heiser

Editor's Foreword

Perseus Books's *Frontiers in Physics* series has, since 1961, made it possible for leading physicists to communicate in coherent fashion their views of recent developments in the most exciting and active fields of physics—without having to devote the time and energy required to prepare a formal review or monograph. Indeed, throughout its nearly forty-year existence, the series has emphasized informality in both style and content, as well as pedagogical clarity. Over time, it was expected that these informal accounts would be replaced by more formal counterparts—textbooks or monographs—as the cutting-edge topics they treated gradually became integrated into the body of physics knowledge and reader interest dwindled. However, this has not proven to be the case for a number of the volumes in the series: Many works have remained in-print on an on-demand basis, while others have such intrinsic value that the physics community has urged us to extend their life span.

The *Advanced Book Classics* series has been designed to meet this demand. It will keep in-print those volumes in *Frontiers in Physics* or its sister series, *Lecture Notes and Supplements in Physics*, that continue to provide a unique account of a topic of lasting interest. And through a sizable printing, these classics will be made available at a comparatively modest cost to the reader.

The lectures transcribed in Nobel Laureate Pierre-Gilles de Gennes's lecture note volume, *Superconductivity of Metals and Alloys*, contain an unusually lucid, physical and original extension of the microscopic theory of Bardeen, Cooper, and Schrieffer (BCS theory) to take into account the role played by inhomogeneity. As a result, they represent a lasting contribution to our understanding of superconductivity, as generation after generation of graduate students and experienced researchers in the field will attest. With the advent of heavy electron and organic superconductors, and especially with the ongoing extraordinarily high level of activity in the field of high temperature superconductivity, it is clear that research on superconductivity will continue to represent a significant portion of research in condensed matter physics for many years to come, and that researchers will continue to turn to de Gennes for enlightenment of the behavior of superconductors in a variety of interesting, real-world situations. I am accordingly very pleased that the publication of *Superconductivity of Metals and Alloys* as part of the *Advanced Book Classics* series will make it readily available to present and future generations of interested readers.

David Pines
Tesuque, NM
January 1999

THESE NOTES ARE DEDICATED

TO THE MEMORY

OF PROFESSOR EDMOND BAUER

Vita

P. G. de Gennes

Professor of Physics at the Collége de France, he is also director of the Ecole de Physique et Chimie in Paris. He studied at the Ecole Normale Supérieure and has served as a research engineer at the Atomic Energy Saclay and as a post-doctoral visitor at the University of California, Berkeley. Professor de Gennes has also been a Professor of Physics at the University of Orsay and is a member of the French Academy of Sciences, the Dutch Academy of Sciences, the American Academy of Sciences, the Royal Society and the National Academy of Science. He is the author of three books and his research now concentrates on magnetism, superconductors, liquid crystals, polymers, and colloids.

Special Preface

Superconductivity of Metals and Alloys was written during the post-BCS period, a time of great enthusiasm. A long period followed where progress in materials science slowed considerably, hampered by the untimely death of B. Matthias. There was a dangerous feeling of comfort on the theoretical side; everything seemed to fit, with the exception of such difficult cases as heavy fermion systems, where many instabilities compete.

Three years ago this feeling of comfort was shaken up, with cuprate systems creating a new wave of enthusiasm from materials scientists. Out of the twenty (or more) theoretical models once proposed, most (including my own) have collapsed at a surprising pace, and those which resist disproving are not necessarily the most exciting.

It may be a good idea, then, that we are again to review the S-state Cooper pair, suitably encouraged by two-dimensional confinement, through the reissue of this book. If this was nature's choice (much to my regret), reprinting this old book may well be justified. Many technical details could be improved, but hopefully the spirit is still there.

I am now too far removed from this field to produce anything which could be considered new. I did accumulate a number of useful corrections suggested by many friends; they have now been incorporated into the text. If we someday get conducting polymers which are liquid, or soluble, and which also superconduct, then I shall rewrite the book. In the meantime, I welcome new readers and offer best wishes to those who boldly attack the cuprates.

P.G. de Gennes
Paris, February, 1989

Preface

The present lecture notes correspond to an *introductory* course given at Orsay during 1962 and 1963. The main purpose of the course was to set up a basic knowledge of superconductivity for both experimentalists and theoreticians in our small group (including the lecturer), and from there on, to plan new experiments.

It is possible (and in fact tempting) to introduce superconductivity as a novel case of long range order, then, from a study of the phase of the order parameter, to deduce the superfluid properties, flux quantization, and the Josephson effect. Later still would come the Landau-Ginsburg equation and a discussion of the magnetic properties of superconductors. Finally with more and more specific assumptions one would reach the Bardeen Cooper Schrieffer theory and its applications.

But in Orsay we wanted to start experiments on superconductivity more urgently than we wanted experiments in teaching; consequently we did not try the above approach. The notes begin with an elementary discussion of magnetic properties of Type I and Type II superconductors. Then the microscopic theory is built up in the Bogolubov language of self-consistent fields: this is powerful enough to cover the amusing situations where the order parameter is modulated in space; it also retains some of the physical insight which we associate with one-particle wave functions. At this stage, the properties of alloys, and in particular of the so-called "dirty" alloys (which, in spite of their name, are often the cleanest systems on which we can experiment) are systematically discussed in parallel with those of the pure metals.

A number of topics have been purposely neglected. The discussion of the electron-electron interactions, for instance, is reduced to a strict minimum, since in most metallic superconductors (just as in ferromagnetic metals) we are unable to compute accurately the fundamental coupling constants. Models which have been historically very useful but which are now of less current application are not mentioned: examples of this class are (1) the Gorter-Casimir theory, and (2) the laminar model for the mixed state of Type II superconductors. Some other problems are not discussed because they require a theoretical level above the one which was chosen: among these, the so-called "strong coupling" effects, the questions of excited states of Cooper pairs, and the effects related to the destruction of the anisotropy of the order parameter in *k* space by alloying.

The notes in their present state represent a sum of many direct and indirect contributions. An early seminar by C. Bloch on the generalized self-consistent field method as applied to finite nuclei was the starting point of Chapter 5. A short note by P. S. Anderson on spin orbit effects was the basis of our approach to dirty superconductor problems. The group at Orsay, J. P. Burger, C. Caroli, G. Deutscher, E. Guyon, A. Martinet, J. Matricon, together with our friends D. Saint James, G. Sarma, M. Tinkham, and P. Pincus, participated in the elaboration and discussion of the more advanced parts, and the writer wishes to express his deep thanks to all of them—with a special mention to P. Pincus, who accepted the task of producing the English version, and to M. Tinkham, for a critical reading of the resulting manuscript. It is hoped that some of the excitement which we felt in these periods of debate and conjecture will finally reach the reader and find him in resonance.

A very small number of useful references are quoted at the end of each chapter. They make no attempt to completeness and take no account of historical priorities: for instance, I always referred the students, not to the original BCS article, but rather to the Les Houches lecture notes by Tinkham, in which the material is presented in a more accessible form.

The technical cooperation of H. Coulle, O. Jancovici, O. Lanouzière (for the French version), and of M. J. Bianquis, Mary Lu Coulter, Kyoto Kaneko, Ann Lampl, Dorothy Pederson, Rachel Shimkin (for the English version) is gratefully acknowledged.

P.G. de Gennes

Orsay
October 1965

Contents

1

FUNDAMENTAL PROPERTIES

1–1 A NEW CONDENSED STATE

We take a piece of tin and cool it down; at a temperature $T_0 = 3.7\,^\circ K$ we find a specific heat anomaly (Fig. 1-1a). Below T_0 the tin is in a new thermodynamical state. What has happened?

It is *not* a change in the crystallographic structure, as far as x rays can tell. It is *not* a ferromagnetic, or antiferromagnetic, transition. (It can be seen by magnetic scattering of neutrons, that tin carries no magnetic moment on an atomic scale.) The striking new property is that the tin has zero electrical resistance. (For instance, a current induced in a tin ring has been observed to persist over times > 1 year.) We say that tin, in this particular phase, is a superconductor, and we call the permanent current a supercurrent.

A large number of metals and alloys are superconductors, with critical temperatures T_0 ranging from less than 1 °K to 18 °K. Even some heavily doped semiconductors have been found to be superconductors.

Historically, the first superconductor (mercury) was discovered by Kammerling Onnes in 1911.

The free energy F_s in the superconducting phase can be derived from the specific heat data and is represented on Fig. 1-1b (solid line). The dotted line gives the corresponding curve F_n for the normal metal. The difference $(F_s - F_n)_{T=0}$ is called the condensation energy. It is *not* of order $k_B T_0$ per electron; it is, in fact, much smaller, of order $(k_B T_0)^2/E_F$ (where E_F is the Fermi energy of the conduction electrons in the normal metal). Typically $E_F \sim 1$ eV and $k_B T_0 \sim 10^{-3}$ eV. Only a fraction $k_B T_0/E_F$ $(\sim 10^{-3})$ of the metallic electrons have their energy significantly modified by the condensation process.

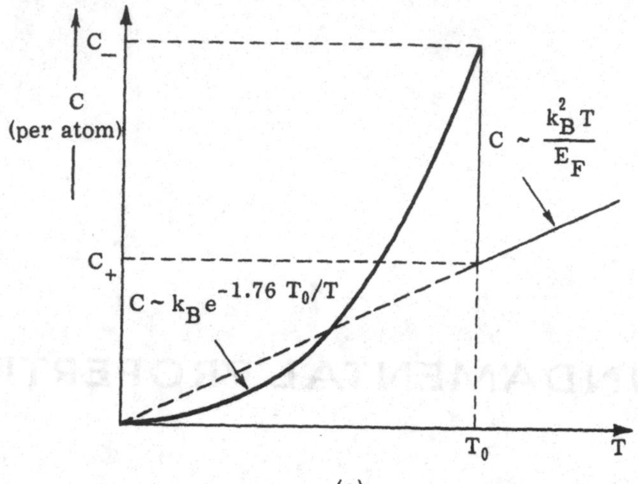

(a)

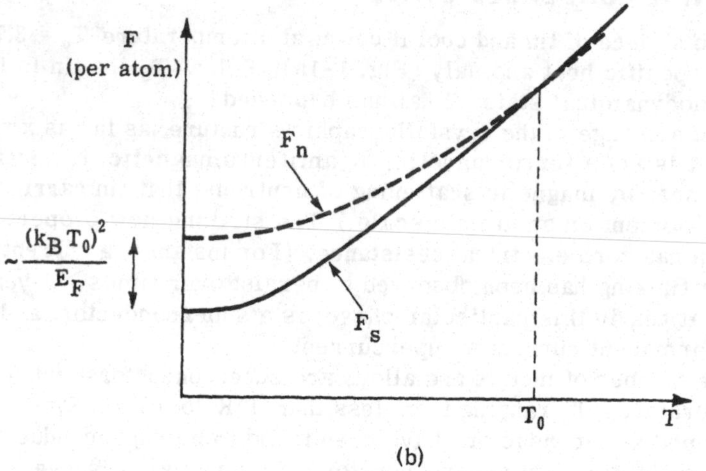

(b)

Figure 1-1

(a) The electronic specific heat C of a superconductor (in zero magnetic field) as a function of temperature (qualitative plot). Above T_0 (in the normal phase) $C_{(\text{per atom})} \sim k_B^2 T/E_F$ where E_F is the Fermi energy. At the transition point T_0, C has a discontinuity. At $T \ll T_0$, C is roughly exponential $C \sim \exp(-1.76\, T_0/T)$.

(b) Free energy of the superconducting phase (F_s) and of the normal phase (F_n) versus temperature. The two curves meet (with the same slope) at the transition point $T = T_0$. At $T = 0$ the difference $F_n - F_s$ is of order $(k_B T_0)^2/E_F$ per atom.

1–2 DIAMAGNETISM

The London Equation

We now extend our energy considerations to situations where there are supercurrents $\mathbf{j}_s(\mathbf{r})$ and associated magnetic fields $\mathbf{h}(\mathbf{r})$ in the sample.[1] We see that in the limit where all fields, currents, and so on, are weak and have a slow variation in space the condition of minimum free energy leads to a simple relation between fields and currents (F. and H. London, 1935).

We consider a pure metal with a parabolic conduction band; the electrons have an effective mass m. The free energy now has the following form:

$$\mathcal{F} = \int F_s \, d\mathbf{r} + E_{kin} + E_{mag} \tag{1-1}$$

where F_s is the energy of the electrons in the condensed state at rest and E_{kin} is the kinetic energy associated with the permanent currents. Let us call $\mathbf{v}(\mathbf{r})$ the drift velocity of the electrons at point $\mathbf{r}$. It is related to the current density $\mathbf{j}_s$ by

$$n_s e\mathbf{v}(\mathbf{r}) = \mathbf{j}_s(\mathbf{r}) \tag{1-2}$$

(where e is the electron charge, and n_s the number of superconducting electrons per cm³). Then we have simply

$$E_{kin} = \int d\mathbf{r} \, \tfrac{1}{2} m v^2 n_s \tag{1-3}$$

the integral being extended over the sample volume. Equation (1-3) would be exact for situations of uniform flow (v = const). It remains approximately correct for our present problem, provided that $\mathbf{v}(\mathbf{r})$ is a *slowly varying function of* $\mathbf{r}$. (We return to this limitation later.)

Finally, E_{mag} is the energy associated with the magnetic field $\mathbf{h}(\mathbf{r})$

$$E_{mag} = \int \frac{h^2}{8\pi} \, d\mathbf{r} \tag{1-4}$$

The field is related to $\mathbf{j}_s$ by Maxwell's equation

$$\operatorname{curl} \mathbf{h} = \frac{4\pi}{c} \mathbf{j}_s \tag{1-5}$$

[1] We use $\mathbf{h}$ to denote a local field value. H will be reserved for the thermodynamic field.

Using (1-3), (1-4), and (1-5) we rewrite the energy E as

$$E = E_0 + \frac{1}{8\pi} \int [\mathbf{h}^2 + \lambda_L^2 \,|\operatorname{curl} \mathbf{h}|^2]\, d\mathbf{r}$$
$$E_0 = \int F_s \, d\mathbf{r} \tag{1-6}$$

where the length λ_L is defined by

$$\lambda_L = \left[\frac{mc^2}{4\pi n_s e^2}\right]^{1/2} \tag{1-7}$$

At T = 0, n_s is equal to n, the total number of conduction electrons per cubic centimeter. We can then compute λ_L explicitly. In simple metals such as Al, Sn, and so on, where m is close to the free electron mass, we find $\lambda_L \sim 500$ Å. For transition metals and compounds with narrow d bands, m is larger and λ_L is also larger (up to 2000Å).

We wish to minimize the free energy (1-6) with respect to the field distribution $\mathbf{h}(\mathbf{r})$. If $\mathbf{h}(\mathbf{r})$ changes by $\delta\mathbf{h}(\mathbf{r})$, E changes by δE

$$\delta E = \frac{1}{4\pi} \int [\mathbf{h} \cdot \delta\mathbf{h} + \lambda_L^2 \operatorname{curl} \mathbf{h} \cdot \operatorname{curl} \delta\mathbf{h}]\, d\mathbf{r}$$

$$= \frac{1}{4\pi} \int [\mathbf{h} + \lambda_L^2 \operatorname{curl} \operatorname{curl} \mathbf{h}] \cdot \delta\mathbf{h}\, d\mathbf{r} \tag{1-8}$$

where we have integrated the second term by parts. The field configuration, in the interior of the specimen, which minimizes the free energy, must therefore satisfy the condition

$$\mathbf{h} + \lambda_L^2 \operatorname{curl} \operatorname{curl} \mathbf{h} = 0 \tag{1-9}$$

Equation (1-9) was first proposed (with a slightly different notation) by F. and H. London. When combined with the Maxwell equation (1-5), it allows us to calculate the distribution of fields and currents.

Meissner Effect

We now apply the London equation and discuss the penetration of a magnetic field $\mathbf{h}$ into a superconductor. We choose the simplest geometry. The surface of the specimen is the xy plane, the region $z < 0$ being empty (Fig. 1-2). The field $\mathbf{h}$ and the current $\mathbf{j}_s$ depend only

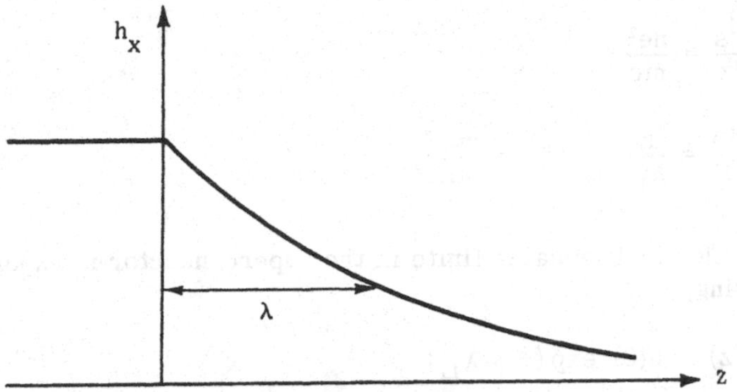

Figure 1-2

Field penetration in a superconductor. The field becomes negligibly small at distances larger than a few penetration depths λ. When the simple London equation (1-9) holds, the penetration is exponential $h = h_0 \exp(-z/\lambda_L)$.

on z. In addition to the relation (1-9), h and j_s are always related by the Maxwell equations

$$\text{curl } \mathbf{h} = \frac{4\pi \mathbf{j}_s}{c} \qquad (1\text{-}10)$$

$$\text{div } \mathbf{h} = 0 \qquad (1\text{-}11)$$

Two cases are possible:

(1) h is parallel to z. Then (1-11) reduces to $\partial h/\partial z = 0$ and h is spatially constant. Therefore curl h = 0 and $j_s = 0$ from Eq. (1-10). Inserting this into Eq. (1-11) we find h = 0. Therefore it is not possible to have a field normal to the surface of the specimen.

(2) h is tangential (and directed along the x axis). Then Eq. (1-9) is automatically satisfied. From Eq. (1-10) j_s is directed along the y axis:

$$\frac{dh}{dz} = \frac{4\pi j_s}{c} \qquad (1\text{-}12)$$

Finally, from Eq. (1-9),

$$\frac{dj_s}{dz} = \frac{ne^2}{mc} h \tag{1-13}$$

$$\frac{d^2h}{dz^2} = \frac{h}{\lambda_L^2} \qquad \lambda_L^2 = \frac{mc^2}{4\pi ne^2} \tag{1-14}$$

The solution that remains finite in the superconductor is exponentially decreasing,

$$h(z) = h(0) \exp(-z/\lambda_L) \tag{1-15}$$

The field h penetrates only to a depth λ_L inside the sample. This result, established here for a semiinfinite slab, is easily generalized to a macroscopic specimen of arbitrary shape. As we have seen, the "penetration depth" λ_L is small. Therefore, in all cases, a *weak* magnetic field practically does not penetrate at all into a macroscopic specimen.[2] The lines of force are excluded as shown on Fig. 1-3.

The superconductor finds an equilibrium state where the sum of kinetic and magnetic energies is minimum, and this state, for macroscopic samples, corresponds to the expulsion of magnetic flux.

Experimentally, the expulsion of lines of force was shown by Meissner and Ochsenfeld in 1933. The Meissner result was particularly important in proving that a true equilibrium state was achieved.

Three remarks concerning the above derivation:

(1) Assuming the existence of permanent currents plus thermodynamic equilibrium, we are led to the diamagnetic properties. It is more usual to go the other way round: Taking the Meissner effect as a starting point, conclude that there exist permanent currents. I chose the first way because I wanted to show you the different contributions to the energy in a superconductor (Eq. 1-6). This list of energies will be useful later (Chapter 3).

(2) We obtained Eq. (1-9) from a minimum condition on the free energy $\mathfrak{F}$. This is the correct thermodynamic potential when the external field sources are permanent magnets. When the source is a coil, with a fixed current I, the correct potential is not $\mathfrak{F}$ but a different function $\mathfrak{G}$ (the "Gibbs potential"). Fortunately, both potentials can be shown to lead to the same local equilibrium condition in the sample. (See Chapter 2 for a discussion of $\mathfrak{F}$ and $\mathfrak{G}$.)

(3) Note that the above calculation is valid only for weak applied fields. In higher fields it may become energetically more favorable

---

[2] In higher field different catastrophes may occur.

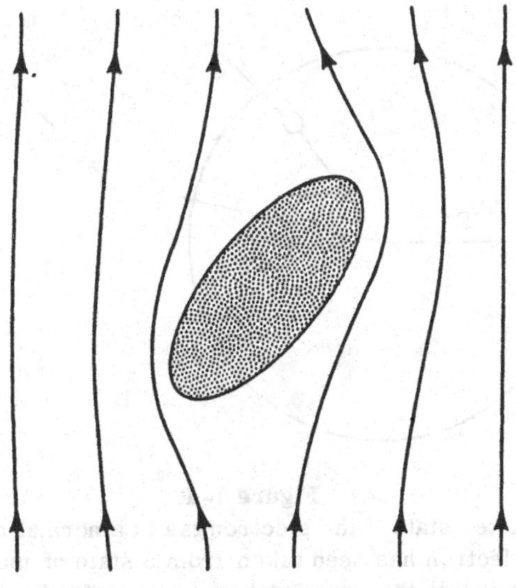

Figure 1-3
Distortion of the magnetic lines of force around a macroscopic superconductor ("macroscopic" means dimensions much larger than the penetration depth). If the fields are not too strong, the superconductor expels the lines completely (Meissner effect).

to destroy superconductivity in some parts of the sample and to allow the flux lines to penetrate. This will be considered in detail in Chapters 2 and 3.

1–3 ABSENCE OF LOW ENERGY EXCITATIONS

Let us begin by considering a free electron gas without interactions. The ground state is obtained by placing an electron into each individual momentum state **p**, of energy $p^2/2m$, until the Fermi energy $E_F = p_F^2/2m$ is reached. Above the Fermi energy E_F, all the levels are empty. (The condition $p = p_F$ defines the Fermi sphere, in momentum space.) In order to construct an excited state of the gas, it suffices to take an electron of momentum **p** from an initially occupied state ($p \leq p_F$) and to place it into a state **p′** initially empty ($p' \geq p_F$) (Fig. 1-4). The excitation energy of this electron-hole pair is

$$E_{pp'} = \frac{p'^2 - p^2}{2m} \geq 0 \tag{1-16}$$

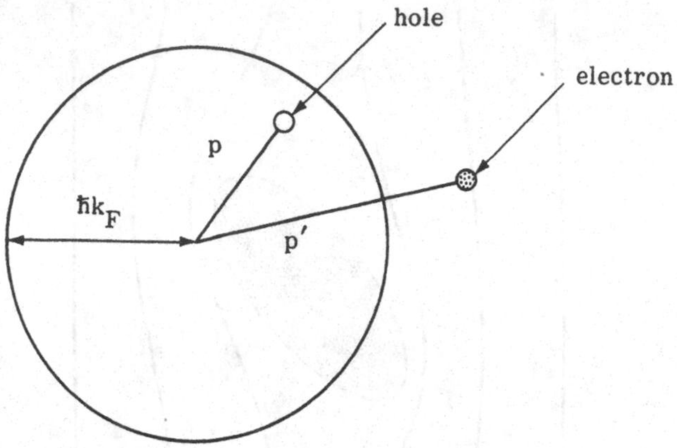

Figure 1-4

An excited state of the electron gas in a normal metal. One electron has been taken from a state of momentum **p** inside the Fermi sphere, to a state of momentum **p**′ outside the Fermi sphere. The excitation energy $(p'^2 - p^2)/2m$ is very low if p and p′ are close to the Fermi momentum $p_F = \hbar k_F$.

If both p and p′ are close to the Fermi momentum, $E_{pp'}$ is very small; in a free electron gas there are numerous low energy excitations. In a normal metal, this free electron picture is not qualitatively modified. The low energy excitations are displayed by the following experiments:

(a) The specific heat is relatively large and proportional to T (of order $k_B \cdot (k_B T/E_F)$ per electron).

(b) Strong dissipative effects appear when the electrons are submitted to low frequency external perturbations (electromagnetic waves, ultrasonic waves, nuclear spin precession, and so on).

In most superconductors, the situation is completely different. The energy $E_{pp'}$ necessary to create a pair of excitations is no longer given by Eq. (1-16). It is necessary to at least furnish a certain "pairing energy" 2Δ:

$$E_{pp'} \geq 2\Delta \tag{1-17}$$

Roughly speaking, this "gap" 2Δ is related to the transition temperature by $2\Delta = 3.5\, k_B T_0$. Thus typically 2Δ is of order 10°K (Table 1-1).

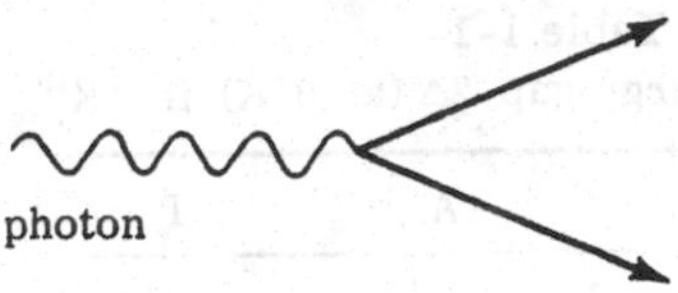

(a)

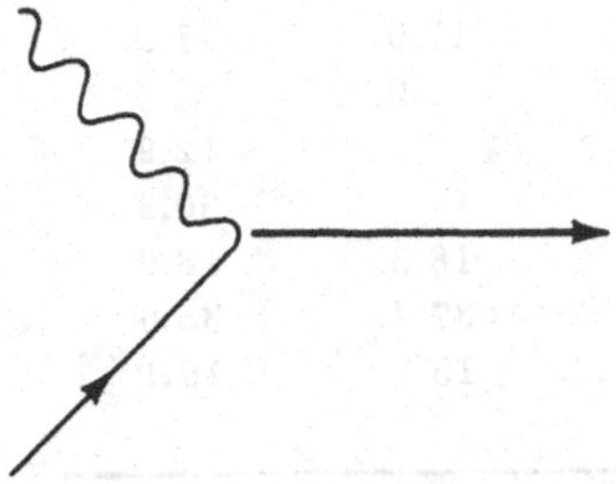

(b)

Figure 1-5
Typical dissipative processes in a superconductor. Fig. 1-5a shows the creation of a pair of excitations by one photon. This process can occur only if $\hbar\omega > 2\Delta$. Fig. 1-5b shows the absorption of the photon by a preexisting excitation. This process can occur even if $\hbar\omega < 2\Delta$, but it is weak at low temperatures, where there are very few thermal excitations. Similar processes are obtained by replacing "photon" by "phonon" in 1-5a and 1-5b.

Note that 2Δ is the energy needed to create two excitations. The energy per excitation is Δ.

Various experiments measure Δ. Here are some of them:

(a) The low temperature specific heat is now exponential and proportional to $\exp(-\Delta/k_B T)$.

(b) Absorption of electromagnetic energy. For $\hbar\omega \geq 2\Delta$ a photon of frequency ω can create an electron-hole pair. [This corresponds to photons in the far infrared; typical wavelengths are in the 1mm range (Fig. 1-5a).]

(c) Ultrasonic attenuation. Here the phonon is of low frequency and cannot decay by creation of a pair of excitations. But it can be absorbed by collision with a preexisting excitation (Fig. 1-5b). This

Table 1-1

Values of the energy gap 2Δ (at $0°K$) in $°K$[a]

	P	A	T
Zn		3.17	
Cd	1.8		
Hg	18.4		18.0
Al	6.01	4.4	4.2
In	13.6	11.9	11.9
Ga		4.03	
Sn	13.0		12.9
Pb	28.7		30.9
V	18.0	18.5	18.0
Nb	27.4	37.4	35.0
Ta		15.7	16.1
La			

[a]For a bibliography on energy gap measurements see D. H. Douglas, Jr., and L. M. Falicov, *Low Temperature Physics*, Vol. IV, edited by C. G. Gorter (Amsterdam: North Holland Publishing Co., 1964). The experiments are classified as follows: P photon absorption (microwave or far infrared photons); A ultrasonic attenuation; T tunneling. The ultrasonic experiments are often performed in single crystals, in which case, 2Δ depends slightly on the direction of the sound wave.

process is proportional to the number of preexisting excitations, thus to $\exp(-\Delta/k_B T)$.

(d) Tunnel effect. A superconductor S and a normal metal N are separated by a thin insulating barrier (typical thickness 25 Å) (Fig. 1-6a). The quantum mechanical tunnel effect allows individual electrons to pass through the barrier. The electron must have been excited from the condensed phase, and this requires an energy Δ. There is no current at low temperatures unless we apply a voltage V across the junction such that the energy gain eV is larger than Δ. The current voltage characteristic has the form shown in Fig. 1-6b.

Question: Is the existence of an energy gap a necessary condition for the existence of permanent current (superfluidity)? The answer is no. A number of situations have been found where superfluidity occurs with no gap in the one particle excitation spectrum. The simplest example is "surface super-

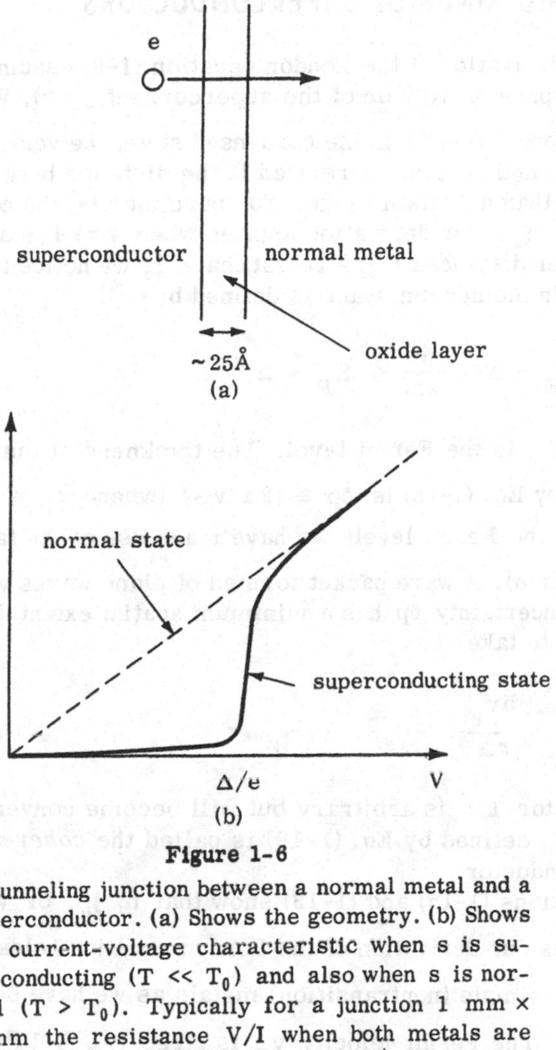

Figure 1-6

A tunneling junction between a normal metal and a superconductor. (a) Shows the geometry. (b) Shows the current-voltage characteristic when s is superconducting ($T \ll T_0$) and also when s is normal ($T > T_0$). Typically for a junction 1 mm × 1 mm the resistance V/I when both metals are normal may range from 10^{-2} to 10^4 ohms. When $T \ll T_0$, to extract one electron from the superconducting condensate requires a minimum energy Δ. Essentially no current flows until $eV = \Delta$.

conductivity''—certain metals or alloys, in a suitable range of field, are superconducting only in a thin sheath (typically 1000Å) near the sample surface. Excitations from the inner (normal) regions can leak up to the surface—there is no gap in the energy spectrum. (This has been checked recently by tunneling experiments.) However the sheath is superconducting! There are other examples, some of which we shall discuss later.

1–4 TWO KINDS OF SUPERCONDUCTORS

Our derivation of the London equation (1-9) assumes a slow variation in space of $\mathbf{v}(\mathbf{r})$ or of the supercurrent $\mathbf{j}_s(\mathbf{r})$. What do we mean by the word "slow"? In the condensed state, the velocities of two electrons (1) and (2) are correlated if the distance between them R_{12} is smaller than a certain range. For pure metals, the correlation length is called ξ_0. Our derivation applies when $\mathbf{v}(\mathbf{r})$ has a negligible variation over distances $\sim \xi_0$. To estimate ξ_0 we notice that the important domain in momentum space is defined by

$$E_F - \Delta < \frac{p^2}{2m} < E_F + \Delta \tag{1-18}$$

where E_F is the Fermi level. The thickness of the shell in p space defined by Eq. (1-18) is $\delta p \cong (2\Delta/v_F)$ (where $v_F = p_F/m$ is the velocity at the Fermi level; we have made use of the fact that $\Delta \ll E_F$ in all cases). A wave packet formed of plane waves whose momentum has an uncertainty δp has a minimum spatial extent $\delta x \sim (\hbar/\delta p)$. This leads us to take

$$\xi_0 = \frac{\hbar v_F}{\pi \Delta} \tag{1-19}$$

(The factor $1/\pi$ is arbitrary but will become convenient later.) The length ξ_0 defined by Eq. (1-19) is called the *coherence length* of the superconductor.

Equations (1-15) and (1-13) show that $\mathbf{h}$, $\mathbf{j}_s$, or $\mathbf{v}$ vary on a scale λ_L. Thus our derivation of the London equation holds only if $\lambda_L \gg \xi_0$.

(1) In simple (nontransition) metals as we have seen, λ_L is small ($\sim$300Å). The Fermi velocity v_F is large ($v_F \lesssim 10^8$ cm/sec) and according to Eq. (1-19) ξ_0 is also large ($\xi_0 \cong 10^4$ Å for aluminum). Thus for these metals the London equation does not apply. In fact, they do exhibit the Meissner effect, but in order to calculate the penetration depth it is necessary to replace Eq. (1-9) by a somewhat more complicated relation, the form of which has been suggested by Pippard. We call these first kind (Type I) or Pippard superconductors and discuss them in Chapter 2.

(2) For transition metals and intermetallic compounds of the type Nb_3Sn, V_3Ga, the effective mass is very large, λ_L is large ($\sim$2000 Å) and the Fermi velocity is small ($\sim 10^6$ cm/sec). Also, in these

compounds the transition temperature T_0 above which superconductivity disappears is found to be high (18°K in Nb_3Sn). As we will see later, Δ is roughly proportional to T_0 and is therefore larger. For all these reasons ξ_0 is very small (~50Å). Therefore for this class of materials Eq. (1-9) is well applicable in weak fields. We call these second kind (Type II) or London superconductors.

In order to complete this discussion, it is necessary to mention the case of superconducting alloys, for which the coherence length and penetration depth are modified by mean free path effects, which we will discuss later. Qualitatively, if the mean free path due to disorder in the structure is short, the coherence length becomes smaller than $\hbar v_F/\pi\Delta$ and λ_L is increased with respect to Eq. (1-7). Therefore it frequently occurs that the addition of impurities into a Pippard superconductor transforms it into a London superconductor.

The distinction here between the two classes is crucial for all experiments made in the presence of external fields. Historically, during a period of 20 years after the discovery of the Meissner effect, experiments were mainly carried out in first kind superconductors. The detailed study of second kind superconductors is much more recent. Paradoxically, the theory has followed the inverse order. Equation (1-9) was introduced by the London brothers in 1935, but the necessary modifications for first type superconductors was only proposed by Pippard in 1953. We now study in detail the magnetic properties of the two types.

REFERENCES

On superfluidity:

F. London, *Superfluids*, Vol. I, 2nd ed. New York: Dover, 1961.

General discussion of experimental data on superconductors:

E. A. Lynton, *Superconductivity*, 2nd ed. London: Methuen and Co., 1965.

More recent advances on the superfluidity concept:

J. Bardeen and R. Schrieffer, *Progress in Low Temperature Physics*, Vol. III, edited by C. G. Gorter. Amsterdam: North Holland, 1961.

J. M. Blatt, *Theory of Superconductivity*. New York: Academic Press, 1964.

Proceedings of the Brighton Symposium on Quantum Fluids (Brighton, 1965) to be published in *Reviews of Modern Physics* 1966.

2
MAGNETIC PROPERTIES OF FIRST KIND SUPERCONDUCTORS

2–1 CRITICAL FIELD OF A LONG CYLINDER

A long superconducting cylinder of radius r_0 is placed in a solenoid of radius $r_1 > r_0$ (Fig. 2-1).

A (weak) current I flows in the coil. The resulting field distribution $\mathbf{h}(\mathbf{r})$ is shown in Fig. 2-2. Outside the sample, $\mathbf{h}(\mathbf{r})$ takes a constant value H. In the sample the field falls rapidly (in a depth $\lambda \sim 500\text{Å}$) to 0. We restrict our attention to a macroscopic specimen ($r_0 \gg \lambda$). Then, on the scale r_0, the field does *not* penetrate into the sample.

This complete flux expulsion is observed for weak external fields H, but, when H reaches a critical value H_c, a radical modification occurs:

(1) The field becomes uniform across the entire section.

(2) The specimen is no longer superconducting (this we can see in principle by measuring the resistivity between the ends of the cylinder). With sufficiently perfect specimens, one can verify that this transition is reversible. If I is now decreased, the superconducting state reappears with total flux expulsion. Typical values of H_c are given in Table 2-1. The critical field decreases with increasing temperature, roughly according to the law

$$H_c(T) = H_c(0)\left[1 - \frac{T^2}{T_0^2}\right] \tag{2-1}$$

(For $T > T_0$ the material is normal even in zero field; T_0 is the

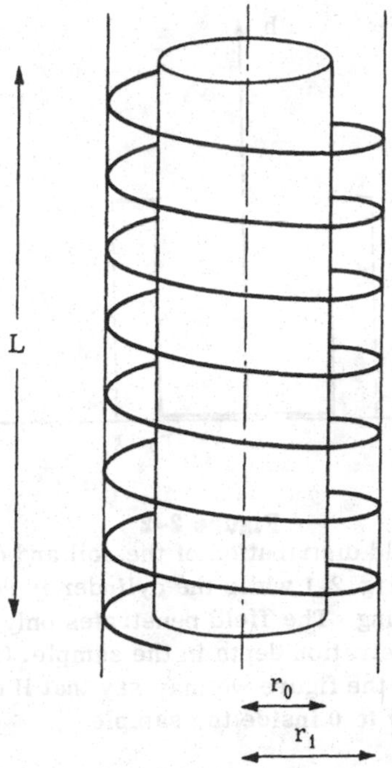

Figure 2-1
Experimental geometry to measure the critical field. The sample is a long oylindcr (of length L and radius r_0). It is placed in a coil of radius r_1.

transition temperature in zero field.) From H_c, we can calculate the difference in free energy between the normal and superconducting states:

(a) When the cylinder is normal the field is uniform across the solenoid

$$h = \frac{4\pi NI}{cL} \tag{2-2}$$

(where N is the number of turns in the solenoid and L is its length, which is also the length of the specimen). The free energy of the system becomes

$$\mathfrak{F}_a = \pi r_0^2 L F_n + \pi r_1^2 L \frac{h^2}{8\pi} \tag{2-3}$$

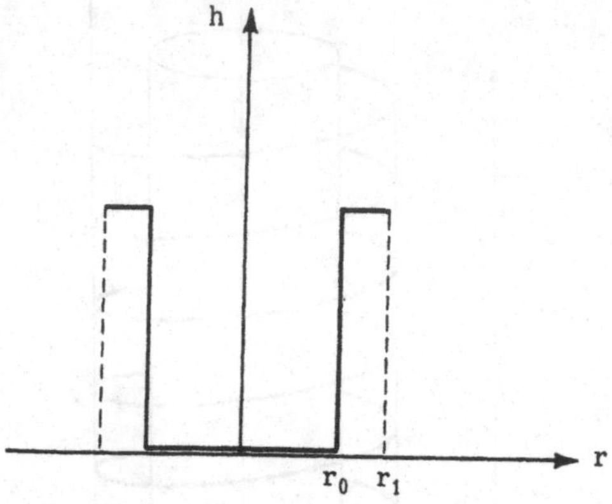

Figure 2-2
The field distribution of the coil and cylinder of Fig. 2-1 while the cylinder is superconducting. The field penetrates only up to one penetration depth in the sample. On the scale of the figure we may say that H drops abruptly to 0 inside the sample.

F_n is the free energy density of the normal sample and the second term is the magnetic energy stored in the coil.

(b) Suppose now that the cylinder becomes superconducting, the current being kept constant in the coil. The field is zero in the sample, but it maintains the value (2-2) in the region $r_0 < r < r_1$.[1] The free energy becomes

$$\mathcal{F}_b = \pi r_0^2 L F_s + \pi (r_1^2 - r_0^2) L \frac{h^2}{8\pi} \tag{2-4}$$

where F_s is the free energy density of the superconducting sample. Equation (2-4) neglects the penetration of the field in the thickness and

---

[1]In order to demonstrate this property, one writes curl $\mathbf{h} = (4\pi/c)\mathbf{j}$, from which

$$\oint \mathbf{h} \cdot d\mathbf{l} = \frac{4\pi}{c} \int \mathbf{j} \cdot d\boldsymbol{\sigma}$$

and one takes for a contour a line of force passing through the coil (in the region $r_0 < r < r_1$). The contributions to $\oint \mathbf{h} \cdot d\mathbf{l}$ from the parts of the line of force enclosing the magnetic current in the exterior of the coil are negligible if the solenoid is long, and one obtains $h = (4\pi NI/cL)$.

Table 2-1
Values of T_c and $H_c(0)$ for some metals

	Zn	Cd	Hg(α)	Al	Ga	In	Tl	Sn	Pb
$H_c(0)$ (G)	53	30		99	51	283	162	306	803
T_c (K)	0.88	0.56	4.15	1.19	1.09	3.41	1.37	3.72	7.18

also the kinetic energy of the surface currents—both these terms are surface effects, and are negligible for a macroscopic cylinder ($r_0 \gg \lambda$).

Note that $\mathfrak{F}_b < \mathfrak{F}_a$, because (1) $F_s < F_n$ and (2) the magnetic term is smaller in state b. What has become of the energy $\mathfrak{F}_a - \mathfrak{F}_b$ when the transition occurs? Answer: The flux ϕ passing through the coil diminishes when we go from (a) → (b). This induces a voltage V in the coil. The work done by this voltage V on the external circuit is

$$\int VI\,dt = \int_a^b -\left(\frac{N}{c}\frac{d\phi}{dt}\right) I\,dt \tag{2-5}$$

The current I is kept constant during the transition. Therefore,

$$\int VI\,dt = \frac{N}{c} I(\phi_a - \phi_b)$$

$$= \frac{NI}{c} \pi r_0^2 h$$

$$= \pi r_0^2 L \frac{h^2}{4\pi} \tag{2-6}$$

At equilibrium between the two states ($h = H_c$), we must have $\mathfrak{F}_a - \mathfrak{F}_b = \int VI\,dt$. Using Eqs. (2-3) and (2-4), we obtain

$$F_n - F_s = \frac{H_c^2}{8\pi} \tag{2-7}$$

From these relations one can deduce a series of thermodynamical properties. Fix the current I and vary the temperature. The entropy (per cubic centimeter) of each phase is determined from the formulas

$$S_n = -\frac{dF_n}{dt} \qquad S_s = -\frac{dF_s}{dT} \tag{2-8}$$

Therefore the difference in entropy between the two phases at equilibrium becomes

$$S_n - S_s = -\frac{1}{4\pi} H_c \frac{dH_c}{dT} \tag{2-9}$$

The latent heat of the transition is

$$L = T(S_n - S_s) = -\frac{T}{4\pi} H_c \frac{dH_c}{dT} \tag{2-10}$$

This heat is positive; it is necessary to furnish this energy to pass from the superconducting state to the normal state.[2]

L vanishes when one considers the transition in zero field: for $T = T_c$, $H_c = 0$, and experiment shows that for all superconductors studied until now dH_c/dT remains finite, therefore $[L]_{H=0} = 0$.

Thus in *zero field* the superconducting transition is of the *second order*. There is a discontinuity in the specific heat

$$C_n - C_s = T \frac{d}{dT}(S_n - S_s)_{T=T_c}$$

$$= -\frac{T}{4\pi}\left(\frac{dH_c}{dT}\right)^2_{T=T_c} \tag{2-11}$$

Formulas (2-10) and (2-11) have been verified to better than 1% in several metals (notably Sn, In, Ta); these experiments demand some care; it is especially necessary to assure that equilibrium has set in, that is, no flux is "frozen" in the superconductor, and so on. The practical interest of the relations (2-10) and (2-11) is to relate the thermodynamic quantities (latent heat and specific heat) to the critical field curve, which is often easier to measure.

---

[2]Consequence: If one passes from the superconducting state to the normal state in a thermally isolated specimen, the temperature of the sample decreases.

2–2 PENETRATION DEPTHS

Relation between Current and Field in a First Kind Superconductor

On the macroscopic scale, a weak field ($h < H_c$) does not penetrate into a first kind superconductor. However, on the microscopic scale, the field does not vary discontinuously at the surface. It penetrates a certain depth λ in the metal. We would like to compute λ.

In Chapter 1 we derived λ from the London equation

$$\operatorname{curl} \mathbf{j} = -\frac{ne^2}{mc}\,\mathbf{h} \tag{2-12}$$

But, in superconductors of the first kind ($\xi_0 > \lambda$), Eq. (2-12) is not valid. We need a more general relation between currents and fields, applicable even when $\mathbf{j}$ and $\mathbf{h}$ have rapid variations in space.

It turns out that the most convenient variable here is not the field $\mathbf{h}$, but rather the vector potential $\mathbf{A}$ such that

$$\operatorname{curl} \mathbf{A} = \mathbf{h} \tag{2-13}$$

This equation does not define $\mathbf{A}$ completely. We often find it convenient to specify $\mathbf{A}$ exactly by adding the supplementary conditions

$$\begin{aligned} \operatorname{div} \mathbf{A} &= 0 \\ A_n &= 0 \qquad \text{on the sample surface} \end{aligned} \tag{2-14}$$

(A_n is the component of $\mathbf{A}$ normal to the surface). When Eqs. (2-14) are satisfied we say that $\mathbf{A}$ is chosen in the London gauge. In the London gauge (2-12) can be rewritten as

$$\mathbf{j} = -\frac{ne^2}{mc}\,\mathbf{A} \tag{2-15}$$

since insertion of (2-13) gives (2-12) and insertion of (2-14) gives

$$\begin{aligned} \operatorname{div} \mathbf{j} &= 0 \qquad \text{(continuity equation)} \\ j_n &= 0 \qquad \text{on the surface} \end{aligned}$$

The latter surface condition is satisfactory if no external current is fed into the sample.

Equation (2-15) applies only when $\mathbf{j}$ and $\mathbf{A}$ vary slowly in space on the scale ξ_0. In general, we may guess that the current $\mathbf{j}(\mathbf{r})$ at one point will depend on the vector potential $\mathbf{A}(\mathbf{r}')$ at all neighboring points $\mathbf{r}'$ such that $|\mathbf{r} - \mathbf{r}'| < \xi_0$. A phenomenological relation constructed to describe this effect, has been proposed by Pippard; in the London gauge it has the form

$$\mathbf{j}(\mathbf{r}) = C \int \frac{(\mathbf{A}(\mathbf{r}) \cdot \mathbf{R})\mathbf{R}}{R^4} e^{-R/\xi_0} \, d\mathbf{r}' \qquad \mathbf{R} = \mathbf{r} - \mathbf{r}' \tag{2-16}$$

The coefficient C is simply determined by noticing that when A varies slowly spatially (over distances $\sim \xi_0$), it can be taken out of the integral, and one must obtain the London value (2-15). This gives

$$C \frac{4\pi}{3} \xi_0 = -\frac{ne^2}{mc} \qquad C = \frac{-\dfrac{3ne^2}{mc}}{4\pi\xi_0} \tag{2-17}$$

The choice of the simple form (2-16) for the relation between current and vector potential was a guess of Pippard. Later, after the development of the microscopic theory, it was verified that the exact relation between field and current is very close to (2-16) if ones takes

$$\xi_0 = \frac{\hbar v_F}{\pi \Delta} \qquad (\text{for } T = 0) \tag{2-18}$$

The exact relation between $\mathbf{j}$ and $\mathbf{A}$ will be discussed in Chapter 4. It is much more complicated algebraically than (2-16) and the approximate Pippard form remains very useful. We now apply it to a study of penetration depths.

The Penetration Depth

Eq. (2-16) is valid for a bulk superconductor. But we now want to study penetration phenomena near the sample surface: in general the relation between $\mathbf{j}$ and $\mathbf{A}$ is then modified. In practice, the following prescription applies: keep Eq. (2-16), but extend the integration $\int d\mathbf{r}'$ only to points $\mathbf{r}'$ in the superconductor, and, more restrictively, to points $\mathbf{r}'$ from which an electron can travel in a straight line to point $\mathbf{r}$. For instance, if there is a small empty cavity in the metal, one must exclude the "shadow" of the cavity, as shown in Fig. 2-3. (This prescription is the correct one for the (usual) case where reflection of individual electrons on the surface is diffuse. We shall derive it later from the microscopic theory.)

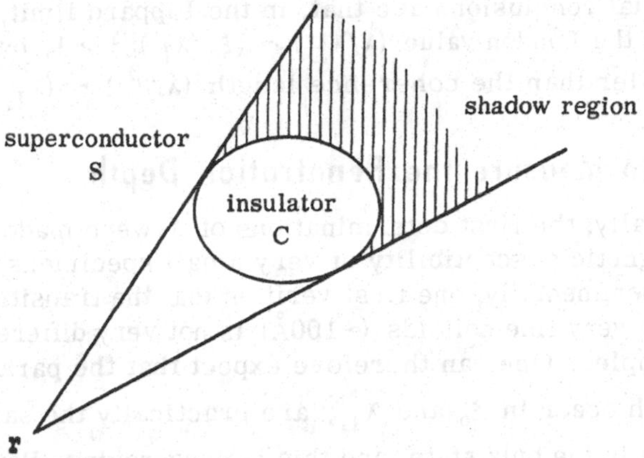

Figure 2-3
"Shadow effect" in the calculation of the current $j(r)$ induced by a magnetic perturbation $A(r')$. A cavity or an insulating precipitate occupies region C inside the superconductor S. Electrons are reflected diffusely at the SC boundary. The magnetic potentials $A(r')$ in C and in the shadow region do not contribute to $j(r)$.

Consider now a specimen with a plane surface xOy, the superconducting side being $z > 0$. A and j are directed along the x axis. The penetration depth λ can then be estimated by another intuitive argument of Pippard's: if $A(z)$ was essentially constant on a thickness $\sim \xi_0$ near the surface, we would have, in this region, from Eq. (2-16): $j = -(ne^2/mc)A$ (the London result). But in fact $A(z)$ is nonzero only in a smaller thickness λ. Thus the integral (2-16) is reduced, roughly by the factor λ/ξ_0:

$$j \simeq -\frac{ne^2}{mc} \cdot \frac{\lambda}{\xi_0} A \qquad (\lambda \ll \xi_0)$$

Using this relation in conjunction with curl $\mathbf{h} = (4\pi/c)\mathbf{j}$, we obtain an (approximate) penetration law of the form $h(z) = h(0)e^{-z/\lambda}$, where λ is defined self-consistently by

$$\frac{1}{\lambda^2} = \frac{4\pi ne^2}{mc^2}\,\frac{\lambda}{\xi_0}$$

$$\lambda^3 = \lambda_L^2\,\xi_0 \qquad (\lambda \ll \xi_0)$$

A (much more complicated) rigorous calculation gives $\lambda^3 = 0.62\lambda_L^2\,\xi_0$.

The essential conclusions are that, in the Pippard limit, λ becomes larger than the London value $(\lambda/\lambda_L) \sim (\xi_0/\lambda_L)^{1/3} > 1$, but λ remains much smaller than the coherence length $(\lambda/\xi_0) \sim (\lambda_L/\xi_0)^{2/3} < 1$.

Methods to Measure the Penetration Depth

Historically, the first determinations of λ were made by measuring the magnetic susceptibility of very small specimens (colloids or films). Experimentally, one first verifies that the transition temperature of even very fine colloids (~100Å) is not very different from that of bulk samples. One can therefore expect that the parameters $[n_s, \Delta(0)]$, which occur in ξ_0 and λ_L, are practically the same in small particles as in the bulk state, and this is encouraging. However, there are several difficulties:

(a) There is a rather large uncertainty in the grain dimensions of the colloid.

(b) In general, one has many lattice defects in such materials and therefore the electronic mean free path ℓ is rather short, and badly known. When ℓ is comparable to ξ_0, the Pippard formula (2-16) must be modified.

The other methods used to determine λ reduce to measuring the value of a self- or mutual inductance in the presence of the superconductor. The principle of the mutual inductance measurement is represented in Fig. 2-4.

The coil 1 creates a field h_e along the exterior of the film and a field $h_i = \rho h_e \ll h_e$ in the interior region. The field h_i is detected by the coil 2. From ρ one can determine λ by suitable theoretical analysis. In practice, one is able to measure coefficients $\rho \lesssim 10^{-9}$. This allows one to work with relatively thick films (~ 10λ) for which the crystallographic state is well defined [and this minimizes objection (b)].

One can also measure the self-inductance of a coil surrounding a superconducting cylinder (method initially suggested by Casimir). If e is the distance between coil and specimen (see Fig. 2-5), the magnetic flux penetrates a surface $\sim \pi R(e + \lambda)$. As $\lambda \sim 10^3$ Å it is necessary to make e as small as possible (typically e ~ 2mm, R being 4mm). In practice e is not known to sufficient precision and the method serves mainly to determine the temperature dependence of λ. A different technique, but similar in principle, is to construct a microwave resonant cavity with the superconducting material.

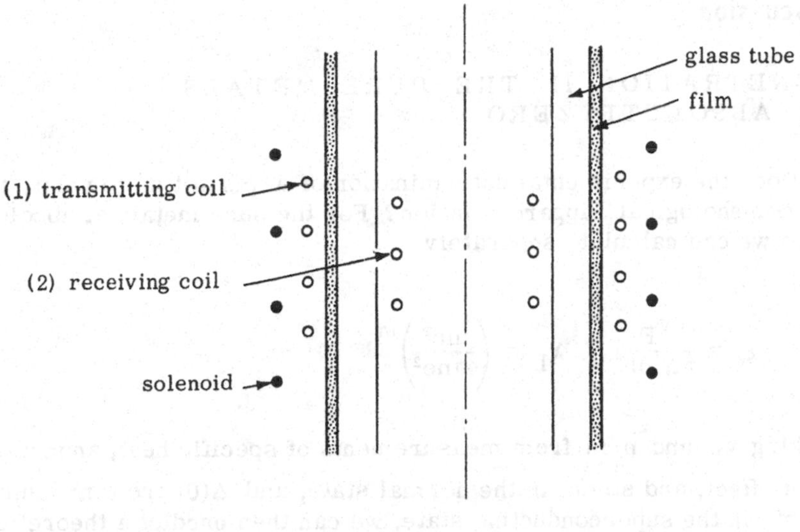

Figure 2-4
Inductive method to measure the penetration depths of thin cylindrical films [after Sarachik, et al., *Phys. Rev. Letters*, 4 52(1960)]. The transmitting coil creates a field attenuated in the film. The receiving coil in the central region detects the field that has leaked through the superconducting film. The solenoid is used to study the field dependence of the penetration depth.

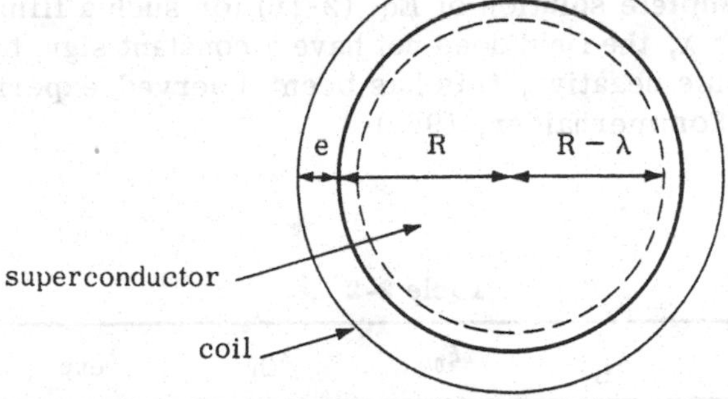

Figure 2-5
Self-induction method to measure penetration depths on a superconducting cylinder of radius R. The cylinder is surrounded by a coil of radius R + e. The flux is contained in a ring of thickness e + λ. In practice e is not known accurately and only the changes of λ with temperature can be measured.

Discussion

PENETRATION IN THE PURE METALS AT ABSOLUTE ZERO

Does the experimental determination of λ permit us to verify the phenomenological Pippard relation? For the pure metals at absolute zero we can calculate separately

$$\xi_0 = \frac{\hbar v_F}{\pi\Delta(0)} \qquad \lambda_L = \left(\frac{mc^2}{4\pi ne^2}\right)^{1/2}$$

Taking v_F and n/m from measurements of specific heat, anomalous skin effect, and so on, in the normal state, and $\Delta(0)$ from measurements in the superconducting state, we can then predict a theoretical value for λ ($\lambda^3 = 0.62\lambda_L^2\xi_0$) for $\lambda_L \ll \xi_0$ (and more complex formulas when $\lambda_L \sim \xi_0$). Table 2-2 gives a comparison between such theoretical values and the experimental data (extrapolated to 0°K). There is satisfactory qualitative agreement, but it would be rather difficult to determine ξ_0 from λ if one had no other information. The detailed form of relation (2-16) cannot be considered established by this type of measurement.

We are now beginning to obtain finer verifications, mainly by measurement of magnetic field transmission through a thin film. In particular, the complete solution of Eq. (2-16) for such a film predicts that when $\xi_0 \gg \lambda$, the field does not have a constant sign, but the ratio ρ can become negative; this has been observed experimentally (Drangeid and Sommerhalder, 1962).

Table 2-2[a]

	λ_L	ξ_0	λ_{th}	λ_{exp}
Al	157	16,000	530	490-515
Sn	355	2,300	560	510
Pb	370	830	480	390

[a]After Bardeen and Schrieffer, *Low Temperature Physics*, edited by C. J. Gorter (Amsterdam: North Holland, 1961), Vol. III, p. 170.

EXTENSION TO ALLOYS

When the electronic mean free path ℓ is limited by the presence of impurities, it is natural to expect that the relation between current (at the point **r**) and vector potential (at the point **r′**) contains an attenuation factor $e^{-|\mathbf{r}-\mathbf{r}'|/\ell}$. One is therefore led to assume, with Pippard,

$$\mathbf{j}(\mathbf{r}) = C \int \frac{[\mathbf{A}(\mathbf{r}')\cdot\mathbf{R}]\mathbf{R}}{R^4}\, e^{-R(1/\xi_0+1/\ell)}\, d\mathbf{r}' \qquad (2\text{-}19)$$

The normalization coefficient C is assumed to be independent of ℓ (therefore equal to its pure metal value $-3ne^2/4\pi mc\,\xi_0$); in other words, one assumes that the contributions to $\mathbf{j}(\mathbf{r})$ coming from points $\mathbf{r}'$ in the neighborhood of $\mathbf{r}$, ($|\mathbf{r}'-\mathbf{r}| \lesssim \ell$) are not modified by the impurities. This hypothesis has permitted Pippard to explain a series of experiments on the dilute SnIn alloy system. One finds that λ increases with the concentration of impurities (Fig. 2-6) and the results are well interpreted by the formula (2-19). One limiting case is particulary remarkable. When $\lambda \gg \ell$ we can neglect the variations of $\mathbf{A}(\mathbf{r}')$ in Eq. (2-19) and perform the integration

$$\mathbf{j}(\mathbf{r}) = C\mathbf{A}(\mathbf{r})\, \frac{4\pi}{3}\, \frac{1}{1/\xi_0 + 1/\ell} \qquad (2\text{-}20)$$

This is a London type equation, but with a coefficient modified with respect to the pure metal. In particular, if $\ell \ll \xi_0$, we have

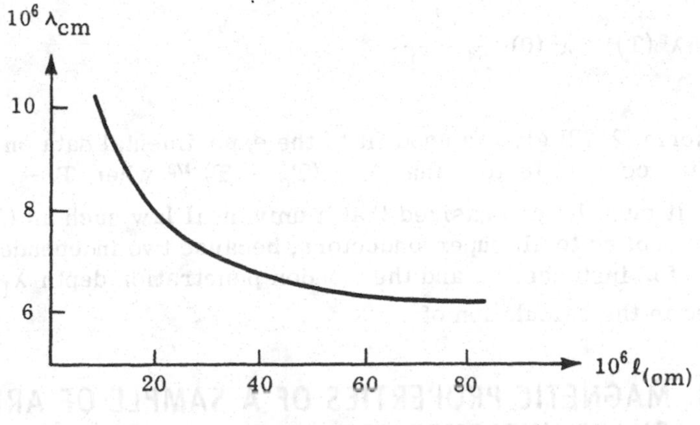

Figure 2-6

The variation of the penetration depth with mean free path in tin-indium alloys. (After A. B. Pippard, *Proc. Roy. Soc. (London)*, **A216**, 547 (1953).) These results are in excellent agreement with Pippard's prediction (2-21).

$$j(r) = -\frac{ne^2}{mc}\frac{\ell}{\xi_0}A(r)$$

which leads to a penetration depth λ

$$\lambda = \lambda_L\left(\frac{\xi_0}{\ell}\right)^{1/2} \qquad (\lambda \gg \ell, \ \xi_0 \gg \ell) \tag{2-21}$$

In this domain λ is proportional to $\ell^{-1/2}$ and therefore to the square root of the impurity concentration.

The agreement between theory and experiment in the SnIn system gives an excellent justification of the Pippard assumptions. In Chapter 4, we see that the microscopic theory confirms them to a large extent.

VARIATIONS OF λ WITH TEMPERATURE

Until now we have essentially limited our discussion to the case $T = 0$. For finite T, the situation can qualitatively be described in the following way: The Pippard equation (2-16) remains valid with a value ξ_0 almost temperature independent. The normalization coefficient C is a function of temperature, and vanishes at the transition point T_c. Very near T_c, C is a linear function of $T_c - T$. A theoretical universal curve for C, derived from the microscopic theory, is given in Chapter 4 (Section 5). Once C is known, one computes λ by solving Eq. (2-20). In an attempt to short circuit this heavy machinery, the experimentalists often refer to the empirical law

$$\lambda^2(T) = \lambda^2(0)\frac{T_c^4}{T_c^4 - T^4} \tag{2-22}$$

The form (2-22) gives a good fit to the experimental data on tin. It has also the correct feature that $\lambda \sim (T_c - T)^{-1/2}$ when $T \to T_c$. However, it must be emphasized that a universal law such as (2-22) *cannot* be applied to all superconductors, because two independent parameters (for instance, ξ_0 and the London penetration depth λ_L) are involved in the calculation of λ.

2—3 MAGNETIC PROPERTIES OF A SAMPLE OF ARBITRARY SHAPE: INTERMEDIATE STATE

Origin of the Intermediate State

In order to define the critical field H_c we chose a cylindrical specimen placed in a field parallel to the axis of the cylinder. This geometry assures that the field h be the same over the entire surface of the specimen ignoring end effects. We now consider a less trivial case, for example, a superconducting sphere of radius a placed in a uniform external field H_0. If H_0 is small, the lines of force are expelled from the specimen and assume the form of Fig. 2-7.

The field configuration external to the sphere is determined by the equations

$$\operatorname{div} \mathbf{h} = \operatorname{curl} \mathbf{h} = 0 \qquad \mathbf{h} \to H_0 \text{ as } r \to \infty \tag{2-23}$$

where r is the distance measured from the center of the sphere. Finally, the Meissner effect imposes the condition that no line of force can penetrate into the sphere. The normal component of h vanishes on the surface of the sphere

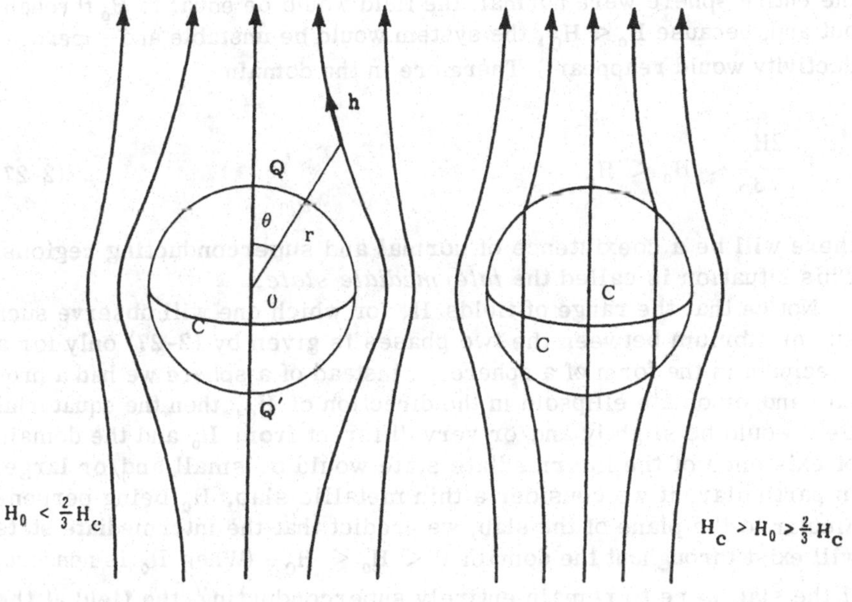

Figure 2-7
The magnetic field distribution about a superconducting sphere of radius a. For applied fields H_0 less than $\frac{2}{3}H_c$, there is a complete Meissner effect and the field at the equator (at any point on circle C) is $\frac{3}{2}H_0$; the field at the poles (Q,Q′) is zero. For $H_c > H_0 > \frac{2}{3}H_c$, the sphere is in the intermediate state.

$$(h_n)_{r=a} = 0 \qquad (2\text{-}24)$$

The appropriate solution in the exterior region is

$$h = H_0 + H_0 \frac{a^3}{2} \nabla \left(\frac{\cos\theta}{r^2} \right) \qquad (2\text{-}25)$$

The component of **h** parallel to the surface of the specimen is

$$|h_\theta|_{r=a} = \tfrac{3}{2} H_0 \sin\theta \qquad (2\text{-}26)$$

At points Q and Q′, h is zero. On the equatorial circle ($\theta = \pi/2$) the tangential component is maximum and is $\frac{3}{2}H_0$. The surface points experience fields between 0 and $\frac{3}{2}H_0$. When H_0 attains the value $\frac{2}{3}H_c$, the field at the equatorial circle becomes equal to H_c. Therefore, for $H_0 > \frac{2}{3}H_c$, certain regions of the sphere pass into the normal state. But there must still exist superconducting regions since $H_0 < H_c$. (If the entire sphere were normal, the field would be equal to H_0 throughout and, because $H_0 < H_c$, the system would be unstable and superconductivity would reappear.) Therefore in the domain

$$\frac{2H_c}{3} < H_0 < H_c \qquad (2\text{-}27)$$

there will be a coexistence of normal and superconducting regions. This situation is called the *intermediate state*.

Notice that the range of fields H_0 for which one will observe such an equilibrium between the two phases is given by (2-27) only for a specimen in the form of a sphere. If instead of a sphere we had a prolate and/or oblate ellipsoid in the direction of H_0, then the equatorial field would be slightly and/or very different from H_0 and the domain of existence of the intermediate state would be small and/or large. In particular, if we consider a thin metallic slab, H_0 being perpendicular to the plane of the slab, we predict that the intermediate state will exist throughout the domain $0 < H_0 < H_c$. (When H_0 is nonzero, if the slab were to remain entirely superconducting, the field at the edge of the slab would be very large, therefore $> H_c$ and it is thus necessary for normal regions to be established.)

Let us consider precisely such a slab of thickness e. The appearance of the normal (N) and superconducting (S) regions is represented in Fig. 2-8. The regions N and S form laminas perpendicular to the plane of the figure. The lines of force only penetrate in the region N. On the contact planes between the N and S regions, the field h must be equal to H_c in order to ensure equilibrium between the two planes.

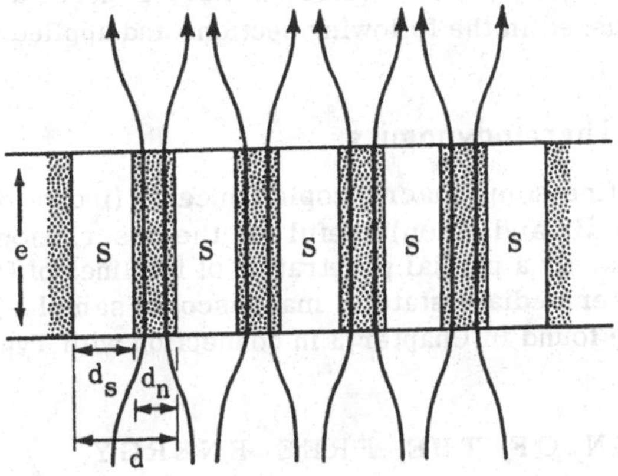

Figure 2-8

Field distribution in a slab in a perpendicular field. For all fields $H < H_c$, the sample is in an intermediate state with a laminar structure of normal and superconducting domains (small distortions of the slabs near the surface are not taken into account).

In the N regions, h is parallel to the z axis, div $\mathbf{h} = 0$ and curl $\mathbf{h} = 0$. Thus h is constant and, therefore, equal to H_c through the N region. In the region S, $h = 0$. For this simple geometry, the fraction $\rho = d_s/(d_s + d_n)$ of S regions is fixed directly by the condition of conservation of flux—far from the film, the field is uniform, $h = H_0$, and the flux becomes SH_0 (where S is the surface area of the film). In the film, the flux is concentrated through a surface $S(1 - \rho)$ and the field is H_c. Therefore

$$SH_0 = S(1 - \rho)H_c \qquad (2\text{-}28)$$

$$\rho = 1 - \frac{H_0}{H_c}$$

An important observation is that, for macroscopic samples ($e \gg \xi_0$) the thicknesses of the layers d_s, d_n are much smaller than the sample dimensions (smaller than e) (this has been verified by detailed experiments to be described later). Thus for many purposes we can ignore the microscopic structure of alternating layers, and we need to know only the relative amount of S regions given by ρ.

The derivation of ρ was very simple for the slab. For more complicated sample shapes, the determination of ρ and of the macroscopic average field of any point requires a more general analysis (due to Peierls), discussed in the following sections and applied to several typical cases.

Preliminary Thermodynamics

We now define some macroscopic concepts (induction **B**, thermodynamic field **H**, and so on) useful for the description of situations where there is only a partial penetration of the lines of force. One such case is the intermediate state of macroscopic samples. Other applications will be found in Chapter 3 in connection with Type II superconductors.

DEFINITION OF THE FREE ENERGY

The system consists of a specimen + external objects (coils, generators). We define a free energy as follows: First, we take the *energy* of the electrons in the sample

$$U = \sum_i \left[\frac{1}{2m}\left(P_i - \frac{eA}{c}\right)^2 + V_i\right] + \sum_{i>j} V_{ij} \tag{2-29}$$

Here **A** is the vector potential, related to the field **h**(**r**) by the relation curl **A** = **h**. V_{ij} represents the electron-electron interactions and V_i the one electron potentials. Second, we add an *entropy* term −TS. Then two contributions appear in practice as an integral over the sample volume.

$$U - TS = \int_{\text{sample}} F_s \, dr \tag{2-30}$$

Third, we include in the free energy the *magnetic field energy* $\int (h^2/8\pi)\,d\mathbf{r}$. The sum of the three contributions we call the free energy

$$\mathfrak{F} = \int_{\text{sample}} F_s \, dr + \int_{\text{all space}} \frac{h^2}{8\pi} \, dr \tag{2-31}$$

DEFINITION OF THE INDUCTION B

We often meet situations where the microscopic field **h**(**r**) in the sample has strong variations in space on a scale Δx much smaller than the sample dimension (Example: the intermediate state pattern of Fig. 2-8). We then find it convenient to introduce a vector **B**(**r**) giving the average of **h** in a region around point **r** of dimensions small compared with the sample, but large compared with Δx.

$$\mathbf{B} = \overline{\mathbf{h}} \tag{2-32}$$

Outside of the sample, we take by definition $\mathbf{B} = \mathbf{h}$. On a microscopic scale, $\mathbf{h}$ satisfies to the equations

$$\operatorname{curl} \mathbf{h} = \frac{4\pi}{c} \mathbf{j}$$
$$\operatorname{div} \mathbf{h} = 0 \tag{2-33}$$

where $\mathbf{j}$ is the local current density. Thus the induction $\mathbf{B}$ satisfies to

$$\operatorname{curl} \mathbf{B} = \frac{4\pi}{c} \bar{\mathbf{j}}$$
$$\operatorname{div} \mathbf{B} = 0 \tag{2-34}$$

$\bar{\mathbf{j}}$ is the macroscopic current density.

DEFINITION OF THE THERMODYNAMIC FIELD H

Suppose that (by changing the currents in the coils) the field distribution $\mathbf{h}(\mathbf{r})$ is changed slightly. At each point $\mathbf{r}$, $\mathbf{B}(\mathbf{r})$ changes by a small amount of $\delta\mathbf{B}(\mathbf{r})$. To first order in $\delta\mathbf{B}$ the change $\delta\mathfrak{F}$ of the free energy is, in its most general form,

$$\delta\mathfrak{F} = \int \frac{\mathbf{H}(\mathbf{r})}{4\pi} \cdot \delta\mathbf{B}(\mathbf{r})\, d\mathbf{r} \tag{2-35}$$

where the factor $(1/4\pi)$ has been inserted for convenience, and $\mathbf{H}(\mathbf{r})$ is some vector function of $\mathbf{r}$. We call $\mathbf{H}(\mathbf{r})$ the thermodynamic field at point $\mathbf{r}$.[3] Outside of the sample we must have

$$\frac{\mathbf{h}\cdot\delta\mathbf{h}}{4\pi} = \frac{\mathbf{H}\cdot\delta\mathbf{B}}{4\pi} \tag{2-36}$$

and $\mathbf{h} = \mathbf{B}$. Thus $\mathbf{h} = \mathbf{H} = \mathbf{B}$ out of the sample. Equation (2-35) is a rather abstract definition of $\mathbf{H}$. To get a physical feeling for $\mathbf{H}$ consider only situations where *no external currents are fed into the sample*. Then we can write for the current

$$\bar{\mathbf{j}} = \bar{\mathbf{j}}_s + \mathbf{j}_{ext} \tag{2-37}$$

---

[3]Note that in the complete Meissner effect region, $\mathbf{B} \equiv 0$ and $\mathbf{H}$ is undetermined.

$\bar{j}_s(\mathbf{r})$ is the average supercurrent at point $\mathbf{r}$ in the sample (by definition j_s is 0 out of the sample). j_{ext} is the current in the coils, generators, and so on. Since no current is fed directly into the sample, the distinction between j_{ext} and $\bar{j}_s$ is clear cut ($j_{ext} = 0$ in the sample). Then we have simply

$$\text{curl}\,\mathbf{H} = \frac{4\pi}{c} j_{ext} \tag{2-38}$$

Proof of Eq.(2-38): When $\mathbf{B}(\mathbf{r})$ varies by $\delta\mathbf{B}(\mathbf{r})$, in a time δt, an electric field $\mathbf{E}$, given by

$$\text{curl}\,\mathbf{E} = -\frac{1}{c}\frac{\delta \mathbf{B}}{\delta t}$$

is generated. The work done by $\mathbf{E}$ on the external currents is $\delta t \int j_{ext} \cdot \mathbf{E}\, dr$ or conversely the work done by the external currents is

$$\delta W = -\delta t \int \mathbf{E} \cdot j_{ext}\, dr \tag{2-39}$$

For a reversible transformation at constant temperature, we know from thermodynamics that $\delta W = \delta \mathfrak{F}$. We want to compare this expression with (2-35). For that purpose we transform (2-35) as follows:

$$\begin{aligned} \delta \mathfrak{F} &= \delta t \frac{1}{4\pi} \int \mathbf{H} \cdot \frac{\delta \mathbf{B}}{\delta t}\, dr \\ &= -\delta t \frac{c}{4\pi} \int \mathbf{H} \cdot \text{curl}\,\mathbf{E}\, dr \qquad (2\text{-}40) \\ &= -\delta t \frac{c}{4\pi} \int \mathbf{E} \cdot \text{curl}\,\mathbf{H}\, dr - \frac{c\delta t}{4\pi} \int (\mathbf{E} \times \mathbf{H}) \cdot d\sigma \end{aligned}$$

The last integral is to be taken over a distant surface surrounding the entire system. It represents the radiated energy and for a slow, reversible transformation it is negligible. Then comparing (2-40) and (2-39) we see that (2-38) is satisfied.

Conclusion. All currents contribute to curl $\mathbf{B}$ (Eq. 2-34), but only j_{ext} contributes to curl $\mathbf{H}$ (Eq. 2-38). We might be tempted to say that $\mathbf{H}$ is the field that would exist in the absence of the specimen (the same j_{ext} being kept on the coils). This is correct for the cylindrical geometry of Fig. 2-1 but is wrong in general. Consider, for instance, the spherical sample of Fig. 2-7. Here the lines of force are distorted in the presence of the sphere and this implies a change of $\mathbf{H}(\mathbf{r})$ (though not of curl $\mathbf{H}$).

Remark on boundary conditions. The equation div $\mathbf{B} = 0$ implies as usual that the normal component of $\mathbf{B}$ be continuous at the surface of the specimen. On the other hand, the tangential component of $\mathbf{B}$ is, in general, not continuous. Curl $\mathbf{B}$ contains the supercurrent. Often we find supercurrents localized on the surface (physically in a thickness λ near the surface). If we let $\mathbf{S}$ be the current per cm on the surface, then by integration of (2-34) we find

$$\mathbf{n} \times (\mathbf{B}_{ext} - \mathbf{B}_{int}) = \frac{4\pi}{c}\mathbf{S} \tag{2-41}$$

($\mathbf{n}$ is a unit vector normal to the surface and directed outward).

Thermodynamic Potential for Fixed T and j_{ext}

When both fields and temperatures are allowed to change, the free energy variation (2-35) becomes

$$\delta \mathcal{F} = \frac{1}{4\pi} \int \mathbf{H} \cdot \delta \mathbf{B} \, dr - S\delta T \tag{2-42}$$

where S is the entropy. When the values of $\mathbf{B}(\mathbf{r})$, at each point $\mathbf{r}$, and the temperature T, are fixed, the equilibrium state corresponds to the minimum of $\mathcal{F}$. Very often, however, this is not the situation of interest. The quantities fixed during the experiment are T and the currents j_{ext} in the coils. We now construct a thermodynamic potential $\mathcal{G}$ adapted to this situation. Define

$$\mathcal{G} = \mathcal{F} - \int \frac{\mathbf{B} \cdot \mathbf{H}}{4\pi} \, dr \tag{2-43}$$

Then

$$\delta \mathcal{G} = - \int \frac{\mathbf{B} \cdot \delta \mathbf{H}}{4\pi} \, dr - S\delta T \tag{2-44}$$

Since div $\mathbf{B} = 0$ we can set $\mathbf{B} = \text{curl } \overline{\mathbf{A}}$ (where $\overline{\mathbf{A}}$ is the macroscopic vector potential) and integrate by parts

$$S\delta T + \delta \mathcal{G} = -\frac{1}{4\pi} \int \text{curl } \overline{\mathbf{A}} \cdot \delta \mathbf{H} \, dr = -\frac{1}{4\pi} \int \overline{\mathbf{A}} \cdot \text{curl } \delta \mathbf{H} \, dr \tag{2-45}$$

$$= -\frac{1}{c} \int \overline{\mathbf{A}} \cdot \delta j_{ext} \, dr$$

Thus $\delta \mathcal{G} = 0$ when T and j_{ext} are fixed. The equilibrium state for fixed T and j_{ext} corresponds to the minimum of $\mathcal{G}$. In practice, we usually construct first $\mathcal{F}$ and B on a microscopic model; then we form $\mathcal{G}$ and finally on minimizing $\mathcal{G}$ we have a condition defining the equilibrium state. This condition plays the same role as the relation $\mathbf{B} = \mu\mathbf{H}$ in a paramagnetic or diamagnetic medium.

Relation between B and H in the Intermediate State

In order to obtain this relation, we calculate the Gibbs potential G d**r** of a volume element d**r** of a specimen when it is in the intermediate state [a fraction ρ of the volume d**r** being superconducting (S) and a fraction $(1 - \rho)$ being normal (N)]. In (N), the microscopic field takes on a certain value $\mathbf{h}_n$; in (S) it is zero. We show that the condition of minimum $\mathcal{G}$ leads to $h_n = H_c$. In terms of ρ

$$B = (1 - \rho)h_n + \rho \cdot 0 = (1 - \rho)h_n \tag{2-46}$$

The free energy per cubic centimeter becomes

$$F = F_n - \frac{\rho H_c^2}{8\pi} + (1 - \rho)\frac{h_n^2}{8\pi} \tag{2-47}$$

The second term represents the condensation energy in the superconducting regions and the third term the magnetic energy. We neglect the (NS) surface energy and also the terms arising from the distortion of the lines of force in the neighborhood of the surface of the specimen; on the macroscopic scale these surface terms are negligible. Expressing F in terms of the variables ρ and **B**, we obtain

$$F = F_n - \frac{\rho H_c^2}{8\pi} + \frac{B^2}{8\pi(1 - \rho)} \tag{2-48}$$

Now forming the thermodynamic potential G, we find

$$G(B, \rho) = F - \frac{BH}{4\pi} = F_n - \frac{\rho H_c^2}{8\pi} + \frac{B^2}{8\pi(1 - \rho)} - \frac{BH}{4\pi} \tag{2-49}$$

If we minimize G: (1) With respect to ρ, we obtain

$$|B| = H_c(1 - \rho) \tag{2-50}$$

On comparing Eqs. (2-50) and (2-46), we see that the field in the normal regions h_n is equal to H_c (this property was previously cited for a simple example on page 26).

(2) With respect to B, we obtain

$$\frac{\partial G}{\partial B} = 0 \qquad B = H(1 - \rho) \tag{2-51}$$

We conclude that (1) the field **H** is parallel to B, (2) the length of **H** must be constant and equal to H_c in the entire specimen. This is the condition equivalent to the relation $B = \mu H$ in a paramagnetic medium. But note that here the relation B(H) is not linear. In order to calculate the field distribution in our sample, it it necessary to solve the equations

$$\text{div}\,\mathbf{B} = 0 \quad (\text{and } \mathbf{B}_{\text{normal}} \text{ continuous at the surface}) \tag{2-52}$$

$$\text{curl}\,\mathbf{H} = 0 \quad (\text{and } \mathbf{H}_{\text{tangential}} \text{ continuous}) \tag{2-53}$$

$$\mathbf{H} = \mathbf{B}\,\frac{H_c}{|\mathbf{B}|} \tag{2-54}$$

Equation 2-53 expresses our assumption that no current is fed into the sample (see Eq. 2-38).

Applications

We still limit ourselves to the (most usual) case where the specimen is not directly connected to the current generator. Then the lines of force, inside the sample, are *straight lines*. In order to show this, we write (proof due to London)

$$H^2 = H_c^2 \qquad 0 = \nabla(H^2) = 2(\mathbf{H}\cdot\nabla)\mathbf{H} + 2\mathbf{H} \times \text{curl}\,\mathbf{H} \tag{2-55}$$

In the sample, curl $\mathbf{H} = 0$, then $(\mathbf{H}\cdot\nabla)\mathbf{H} = 0$. The vector **H** does not vary along a line of forces and therefore the line of force is straight.

Example. Let us return to a spherical sample placed on an exterior field H_0 such that $\frac{2}{3}H_c < H_0 < H_c$. We can construct the solution of the system (2-52), (2-53), (2-54) in the following manner.

In the interior of the sphere ($r < a$) we take **H** and **B** to be parallel to the z axis and constant. In magnitude, $H = H_c$ and B has a certain unknown value B_0. In the exterior, on the other hand,

$$\mathbf{H} = \mathbf{B} = \mathbf{H}_0 - H_1 \frac{a^3}{2} \nabla \frac{\cos\theta}{r^2} \tag{2-56}$$

where H_1 is a constant to be determined. The condition on the continuity of H_{tang} at the surface imposes

$$\left(H_0 + \frac{H_1}{2}\right) \sin\theta = H_c \sin\theta \tag{2-57}$$

The condition of the continuity of B_{norm} gives

$$(H_0 - H_1) \cos\theta = B_0 \cos\theta \tag{2-58}$$

Upon comparing these two equations we find

$$\begin{aligned} B_0 &= 3H_0 - 2H_c \\ H_1 &= 2(H_c - H_0) \end{aligned} \tag{2-59}$$

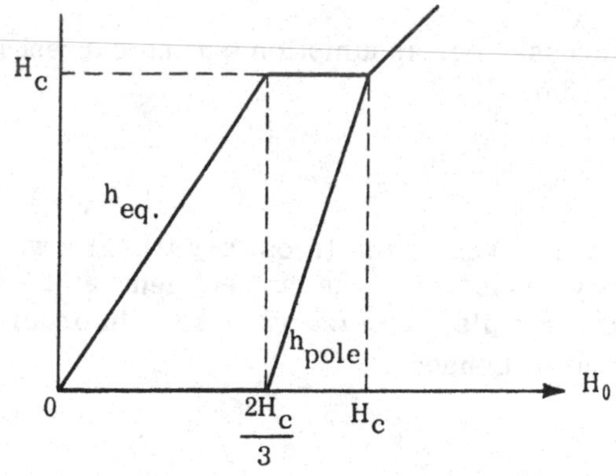

Figure 2-9

The field at the equator and the field at the poles of a superconducting sphere. For $H_0 < \frac{2}{3}H_c$, there is a complete Meissner effect, and the polar field vanishes. For $\frac{2}{3}H_c < H_0 < H_c$, the sphere is in the intermediate state; the equatorial field remains constant at H_c while the polar field increases from zero to H_c.

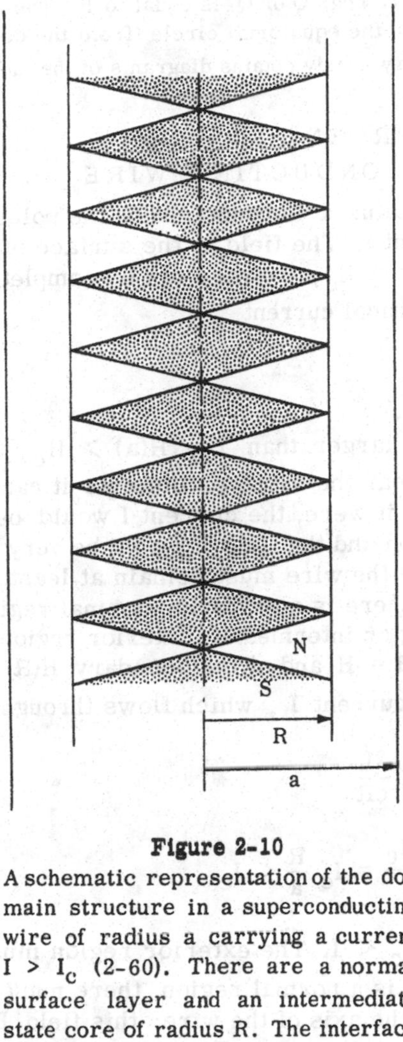

Figure 2-10

A schematic representation of the domain structure in a superconducting wire of radius a carrying a current $I > I_c$ (2-60). There are a normal surface layer and an intermediate state core of radius R. The interface (R) is determined by $H(R) = H_c$.

Conclusions.

(1) There exists a solution such that **B** is constant throughout the sphere. The fraction of superconducting regions $\rho = 1 - B_0/H_c$ is therefore the same throughout the sphere.

(2) The induction B_0 is a linear function of the applied field H_0. In practice, one can measure B_0 by placing a magnetic probe in the neighborhood of the poles Q, Q′ of the sphere. Since B_{norm} is continuous, the field immediately

exterior to the sphere near Q or Q′ is equal to B_0. The field H_c is measured by placing the probe on the equatorial circle (from the continuity of H_{tang}). Experimentally one very nearly obtains diagrams of the theoretical form (Fig. 2-9) in clean samples.

CRITICAL CURRENT OF A SUPERCONDUCTING WIRE

The wire of radius a is connected to the poles of a generator that supplies a current I. The field at the surface of the wire is $H(a) = 2I/ca$. When $H(a) < H_c$, the wire can be completely superconducting. This defines a critical current

$$I_c = H_c \frac{ca}{2} \tag{2-60}$$

When I becomes larger than I_c, $H(a) > H_c$, and the wire must become normal near the surface. However it cannot become entirely normal, since, if it were, the current I would be spread uniformly through the section and the field h would be very small in the central region; therefore the wire must remain at least partially superconducting. Finally there is an exterior normal region $R < r < a$ and a superconducting or intermediate interior region $0 < r < R$. In the exterior region $\mathbf{B} = \mathbf{H}$ and at the boundary $H(R) = H_c$. This allows us to predict the current I_1, which flows through the interior region

$$H(R) = H_c = \frac{2I_1}{cR} \tag{2-61}$$

$$I_1 = \frac{cRH_c}{2} = I_c \frac{R}{a} < I_c \tag{2-62}$$

Therefore $I_1 < I_c < I$. The exterior region must carry a current $I - I_1$. Since this is a normal region, there must be an electric field $\mathbf{E}$ directed along the axis of the wire; this field $\mathbf{E}$ must be the same throughout the wire since curl $\mathbf{E} = 0$.

This shows that the interior region cannot be entirely superconducting since it would short circuit the field $\mathbf{E}$. It must, in fact, be in the intermediate state with N and S laminas perpendicular to the axis of the cylinder (Fig. 2-10). This situation is discussed in detail in the book by London.

PRINCIPLE OF THE CRYOTRON

The cryotron is a control device invented by Buck in 1956; the principle is shown on Fig. 2-11. The "control" current I′ creates a field $H' = 4\pi NI'/cL$ in the solenoid (where N is the number of turns

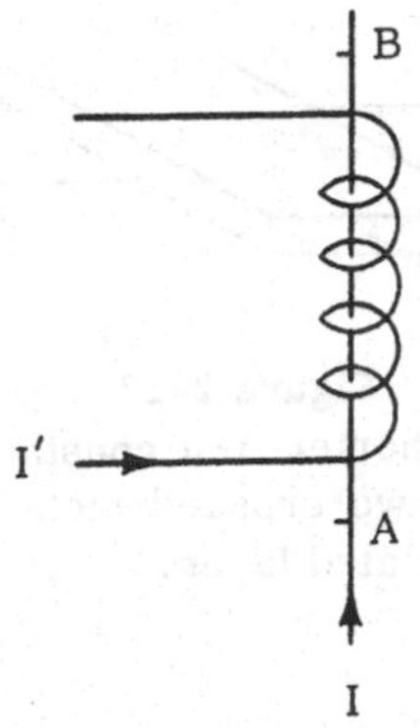

Figure 2-11
The principle of the cryotron. The current passing through the superconducting wire AB is controlled by the current I′ flowing through the solenoid. When the field in the coil reaches H_c, the wire AB becomes normal.

and L is the length of the coil). A superconducting wire AB of diameter 2a and critical field H_c passes through the solenoid. For $H' < H_c$, the wire has zero resistance. For $H' > H_c$, the resistance is finite. It is therefore possible to control a current I passing through AB by the current I′. The minimum current value of I′ required to make AB normal is

$$I'_m = \frac{cL}{4\pi N} H_c \tag{2-63}$$

On the other hand, the current I cannot be too large; otherwise the line AB passes into the intermediate state as we have previously seen

$$I \leq I_c = \frac{caH_c}{2} \tag{2-64}$$

The current gain of the cryotron is

$$G = \frac{I_c}{I'_m} = 2\pi N \frac{a}{L} \tag{2-65}$$

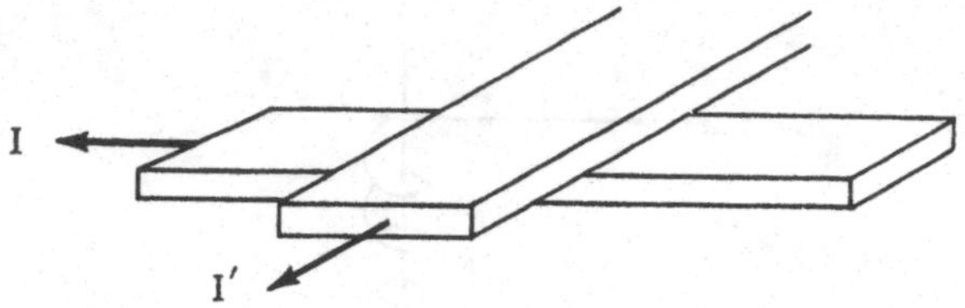

Figure 2-12
A more efficient (shorter time constants) cryotron can be constructed of two crossed metallic films separated by a thin insulated layer.

In Buck's first experiment, the control coil was made of a niobium wire (superconductor with a high critical field ~2000 G) of diameter ~0.1 mm. The wire AB (of tantalum) had $H_c \sim 100$ G at 4.2°K and a diameter $2a = 0.2$ mm. For a coil of only one turn, this corresponds to $Na/L \sim 1$, and $G \sim 6$. In practice it is not useful to increase N for this increases the self-inductance of the coil and finally increases the time constant τ of the cryotron (for only one turn, the τ is not less than 10^{-5} sec). In practice, the time constants are reduced considerably by replacing the set up in Fig. 2-11 by two thin crossed metallic films (Fig. 2-12) separated by an insulating layer approximately 1000Å thick. Since the self-inductance is proportional to this thickness, it is quite small and the time constants fall into the domain 10^{-8}-10^{-9} sec.

Microscopic Structure of the Intermediate State

A slab (Fig. 2-13) of thickness e and lateral dimensions $L_x L_y$ (very large) is placed in a perpendicular field H_0—normal and superconducting domains will occur. We *assume* that these domains are in the form of layers, as shown on the figure. (Experimentally, this arrangement is often found.) We wish to know the thickness d_n or d_s of the normal and of the superconducting regions, and the period $d = d_n + d_s$ of the domain structure.

To do this, we construct the free energy $\mathfrak{F}$, and then minimize it (with respect to d_n, d_s, or equivalent parameters) for fixed H_0 (fixed H_0 means fixed flux through the slab; thus the appropriate potential is $\mathfrak{F}$, not $\mathcal{G}$).

In a zero-order approximation we consider the S and N domains as exactly plane layers and neglect all surface energies. Then the free energy is simply

$$\mathfrak{F}_{\text{macroscopic}} = -\frac{H_c^2}{8\pi} L_x L_y e \frac{d_s}{d} + \frac{H_0^2}{8\pi} \frac{d}{d_n} e L_x L_y \tag{2-66}$$

The first term is the condensation energy in the S regions. The second is the magnetic field energy $h^2/8\pi$, in the N regions, where $h = H_0(d/d_n)$ by flux conservation.

Going now to a first-order approximation, one must add *three* corrections to $\mathfrak{F}$.

(1) To create a wall between an S region and an N region we need a certain surface energy γ per cm² of wall. Dimensionally it is convenient to write this γ in the form

$$\gamma = \frac{H_c^2}{8\pi} \delta \tag{2-67}$$

where the length δ is of order 10^3-10^4 Å. The corresponding contribution to $\mathfrak{F}$ is

$$\gamma e L_y \times (\text{number of walls}) = \frac{H_c^2}{8\pi} \delta e L_y \frac{2L_x}{d} \tag{2-68}$$

(2) The lines of force "open up" near the sample surface and the superconducting domains become thinner near the ends. Thus we lose some condensation energy, in a volume of order $d_s^2 L_y$ for each S domain. This gives a contribution

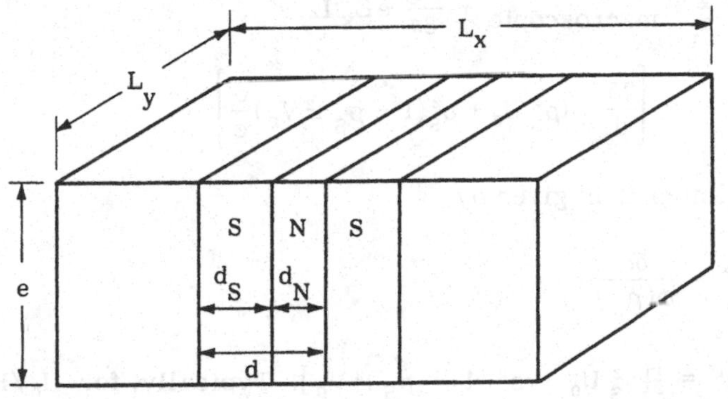

Figure 2-13

Microscopic structure of the normal and superconducting regions in the intermediate state of a slab. The domains form a laminar structure of period $d \sim \sqrt{e\delta}$ (e = slab thickness, $\delta \cong$ thickness of a wall).

$$\frac{H_c^2}{8\pi} d_s^2 L_y \frac{L_x}{d} U_0(\rho_s) \tag{2-69}$$

where $\rho_s = d_s/d$ and $U_0(\rho_s)$ is a dimensionless function that can be computed when the exact shape of the "thinner" regions is known.

(3) The magnetic field energy $\int (h^2/8\pi)\, d\mathbf{r}$ is also modified near the surface, and this gives a term of the form

$$\frac{H_0^2}{8\pi} d_s^2 L_y \frac{L_x}{d} V_0(\rho_s) \tag{2-70}$$

where $V_0(\rho_s)$ is another dimensionless function. (The calculation of U_0 and V_0 is described in Landau and Lifshitz, *Electrodynamics of Continuous Media.*)

We can choose as independent variables ρ_s and d. If we first minimize the free energy with respect to ρ_s, we find that for a macroscopic sample ($e \gg \xi_0$)

$$1 - \rho_s = \frac{H_0}{H_c} \tag{2-71}$$

This, as we know, expresses that the field in the normal regions is equal to H_c. Then we minimize $\mathfrak{F}$ with respect to d. The d-dependent terms are the three correction terms

$$\mathfrak{F} = \mathfrak{F}_{\text{macroscopic}} + \frac{H_c^2}{8\pi} e L_x L_y \times \left[\frac{2\delta}{d} + (\rho_s^2 U_0 + \rho_s^2(1-\rho_s)^2 V_0)\frac{d}{e}\right] \tag{2-72}$$

The optimum d is given by

$$d^2 = \frac{\delta e}{\phi(\rho_s)} \tag{2-73}$$

where $\phi = \frac{1}{2}[\rho_s^2 U_0 + \rho_s^2(1-\rho_s)^2 V_0]$. Typically, for $H_0/H_c \sim 0.7$, $\phi \sim 10^{-2}$ and thus $d \sim 10\sqrt{\delta e}$. Taking e = 1 cm, δ = 3000 Å, we arrive at d ~ 0.6 mm. Near the ends ($\rho_s \to 0$ or $\rho_n \to 1$), ϕ tends toward 0 and d becomes still larger. How can we observe this domain structure? Various methods have been used:

(1) A fine bismuth wire (whose resistance is strongly dependent on

field) is moved close to the surface of the sample. The resistance is large in the N regions and small in the S regions (Meshkovsky and Shalnikov, 1947).

(2) Niobium powder is placed on the sample. Niobium has a high critical field (~2000 G). Thus the grains are always superconducting and diamagnetic, they tend to avoid the lines of force, and they gather on the S regions.

(3) The specimen is covered with a thin layer of cerium glass (typical thickness ~0.1 mm). This glass has a large Faraday rotation. Polarized light traveling along the magnetic field $\mathbf{H}_0$ (normal to the sample surface) has its plane of polarization rotated by an angle θ when it traverses the glass; the total rotation after being reflected from the sample is then 2θ. Since, in general, θ is proportional to the field ($\theta \sim 0.02°$/mm/G), if the specimen is observed between crossed analyzer and polarizer, the N regions, where $h \neq 0$, appear bright. Ultimately, these measurements furnish a determination of the wall energy (characterized by the length δ).

Problem. What are the corrections to the critical field of the plate due to the fact that the thickness e is finite?

Solution. If we minimize the free energy with respect to d, keeping as other independent variables the reduced field $h_r = H_0/H_c$ and $\rho_s = d_s/d$ we find

$$\mathfrak{F} = \frac{H_c^2}{8\pi} e L_x L_y \left[-\rho_s + \frac{h_r^2}{1-\rho_s} + 4\left(\frac{\delta}{e}\right)^{1/2} \phi^{1/2} \right]$$

$$\phi = \tfrac{1}{2}\rho_s^2 [U_0(\rho_s) + h_r^2 V_0(\rho_s)]$$

In the region of interest, h_r is close to 1, ρ_s is close to 0, and we may write for ϕ

$$\phi \cong \theta^2 \rho_s^2$$

where $\theta^2 = \frac{1}{2} | U_0(0) + V_0(0) | = (1/\pi) \ln 2 = 0.22$ (this numerical value is obtained in the previously mentioned calculation of Landau and Lifshitz). Minimizing $\mathfrak{F}$ with respect to ρ_s, we obtain the condition

$$-1 + \left(\frac{h_r}{1-\rho_s}\right)^2 + 4\theta\left(\frac{\delta}{e}\right)^{1/2} = 0$$

The critical field of the plate is reached when $\mathfrak{F}$ is equal to the normal state free energy $(H_0^2/8\pi) e L_y L_x$. This leads to the condition

$$\rho_s\left(\frac{h_r^2}{1-\rho_s} - 1\right) + 4\theta\left(\frac{\delta}{e}\right)^{1/2} \rho_s = 0$$

Both conditions are satisfied when

$$\rho_s = 0$$

$$1 - h_r^2 = 4\theta\left(\frac{\delta}{e}\right)^{1/2}$$

$$H_0 \cong H_c\left(1 - 2\theta\left(\frac{\delta}{e}\right)^{1/2}\right)$$

By taking e = 1 mm and $\delta = 10^4$ Å, we get a 3% decrease of the critical field. Similar deviations from the macroscopic theory occur in all geometries; very often they are important in the discussion of experimental results in type I superconductors.

A PARADOX

If we look carefully at the domain structure of Fig. 2-14a, we notice that at a point such as A, in the N regions, the lines of force have "opened up." Thus the field is significantly lower than at point P on the boundary, that is, lower than H_c. At first sight this is troublesome. We would expect a region near A to become superconducting again. This complication has been considered by Landau. He concluded that the normal regions should "branch" as shown on Fig. 2-14b, possibly up to such a fine scale that the domain structure would become unobservable. In fact, the simple domain structure *is* often observed. Branching does not take place in clean samples with dimensions of order 1 cm. The critical field at point A is reduced below H_c by an effect similar to that discussed in the problem above (but here the dimension of interest is d, not e, thus the effect is large). Thus A can remain normal in rather low fields.

If we had a superconductor of thickness e = 1 mile, d would be of order 1 cm, we could still apply macroscopic considerations in the neighborhood of P, and some branching would take place. The branching model is not wrong; it is simply not adequate for the usual scale of sample dimensions. (Another way to favor branching is not to increase e, but rather to reduce δ—this could be done in suitable alloy systems.)

ORIGIN OF THE SURFACE ENERGY

We qualitatively discuss two extreme cases:

(1) $\xi_0 \gg \lambda$: In our previous macroscopic discussion, there was a sharp boundary between the normal region N and the superconducting region S (to fix ideas, we shall take this boundary as being the yz plane, the fields being along z and the N region corresponding to $x < 0$). On the N side, the thermodynamic potential was lowered by the magnetic

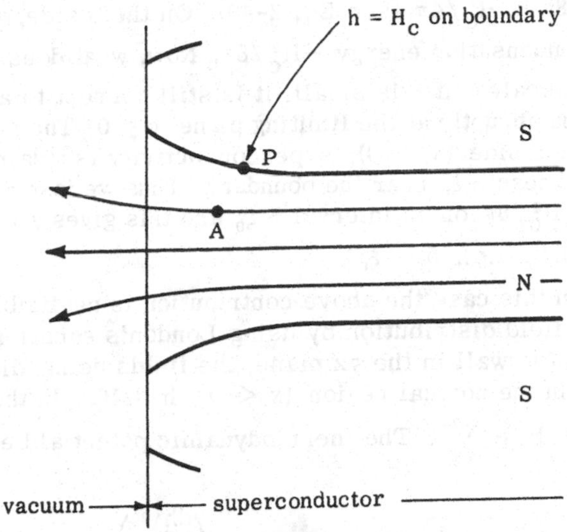

(a)

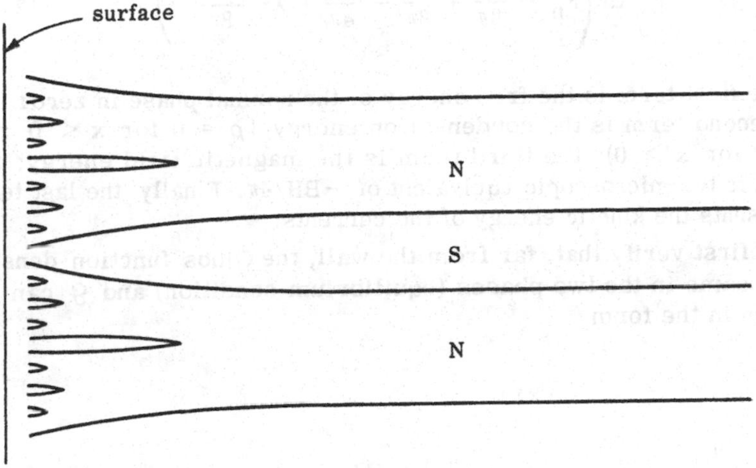

(b)

Figure 2-14

(a) The domain structure near the surface. Note that, at a point such as A, the field is smaller than at point P (as is indicated by the curvature of the lines of flux): $h(a) < h(P) = H_c$. There are some small normal regions near the ends where $h < H_c$! (b) Landau's branching model. Such a branching is required if one wants to keep $h > H_c$ everywhere in the normal regions. In fact branching does *not* take place: The condition $h > H_c$ applies only to *macroscopic* normal regions. Here the small regions of interest near the surface have critical fields significantly lower than H_c; they can remain normal for $h < H_c$ and the simpler model of Fig. 2-14a is the correct one.

field terms $H_c^2/8\pi - H_c^2/4\pi$ (see Eq. 2-49). On the S side, $\mathcal{G}$ was lowered by the condensation energy $-H_c^2/8\pi$. Now, what does happen on a microscopic scale? If λ is small, it is still correct to assume that the field drops abruptly at the limiting plane $x = 0$. The new feature is that on the S side ($x > 0$), superconductivity is "damaged" in a region of thickness $\sim\xi_0$ near the boundary. Thus we lose the condensation energy $H_c^2/8\pi$ on an interval $\sim\xi_0$ and this gives a wall energy $\gamma \sim (H_c^2/8\pi)\,\xi_0$ (i.e., $\delta \sim \xi_0$).

(2) $\xi_0 \ll \lambda$: In this case the above contribution is negligible. We can calculate the field distribution by using London's equation (Fig. 2-15). Let us put the wall in the yz plane, the fields being directed along the z axis. In the normal region ($x < 0$), $h = H_c$. In the S region ($x > 0$), $h = H_c e^{-x/\lambda}$. The thermodynamic potential becomes

$$\mathcal{G} = \int_{x>0} dr \left(F_n - \frac{H_c^2}{8\pi} + \frac{h^2}{8\pi} - \frac{Hh}{4\pi} + \lambda^2 \frac{\left(\frac{dh}{dx}\right)^2}{8\pi} \right) \tag{2-74}$$

The first term is the free energy of the normal phase in zero field; the second term is the condensation energy ($\rho = 0$ for $x < 0$ and $\rho = 1$ for $x > 0$); the third term is the magnetic field energy; the fourth is the microscopic equivalent of $-BH/4\pi$. Finally, the last term represents the kinetic energy of the currents.

We first verify that, far from the wall, the Gibbs function density is the same in the two phases (equilibrium condition) and $\mathcal{G}$ can be written in the form

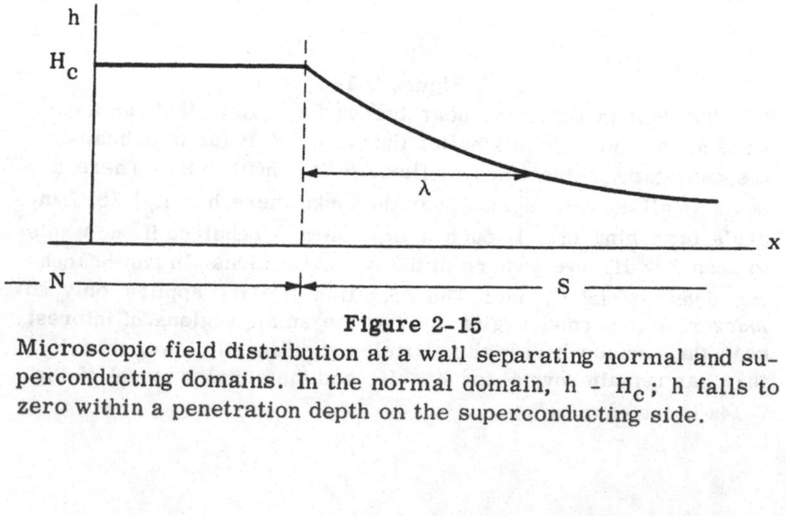

Figure 2-15
Microscopic field distribution at a wall separating normal and superconducting domains. In the normal domain, $h - H_c$; h falls to zero within a penetration depth on the superconducting side.

$$\mathcal{G} = \int dr \left(F_n - \frac{H_c^2}{8\pi}\right) + \gamma S$$

where S is the surface area of the wall and γ is the surface tension given by

$$\gamma = \int_0^\infty dx \left[\frac{h^2 + \lambda^2 \left(\frac{dh}{dx}\right)^2}{8\pi} - \frac{hH_c}{4\pi}\right] = -\frac{H_c^2}{8\pi}\lambda \tag{2-75}$$

In this limit, the surface tension is *negative* and the system decreases its energy by creating new walls; the largest possible number of walls is created, and it is evident that the magnetic properties will be found to be considerably different from those cited previously. It is for this reason that materials having $\xi_0 < \lambda$ are called second kind superconductors. We study them in the next chapter.

REFERENCES

Macroscopic description of 1st type superconductors:

F. London, *Superfluids* (New York: Wiley, 1950), Vol. I.

L. Landau and I. M. Lifschitz, *Electrodynamics of Continuous Media* (Pergamon, 1960), Chap. 6.

Relation between current and vector potential:

A. B. Pippard, *Proc. Roy. Soc. (London)*, **A216**, 547 (1953).

Measurements of penetration depths:

A. L. Schawlow and G. Devlin, *Phys. Rev.*, **113**, 120 (1959).

A. B. Pippard, *Proc. 7th Intern. Conf. Low Temp. Phys.* (Toronto: Toronto Univ. Press, 1960), p. 320.

Intermediate state:

Optical Observation of Domains:

W. De Sorbo, *Phys. Rev. Letters*, **4**, 406 (1960).

W. De Sorbo, *Proc. 7th Intern. Conf. Low Temp. Phys.* (Toronto: Toronto Univ. Press, 1960), p. 370.

Measurement of Wall Energy:

A. L. Schawlow, *Phys. Rev.*, **101**, 573 (1956).

Kinetics of the normal → superconducting phase transition:

T. E. Faber and A. B. Pippard, *Progress in Low Temp. Physics*, edited by C. G. Gorter (Amsterdam: North Holland, 1959), Vol. I, Chap. 9.

3

MAGNETIC PROPERTIES OF SECOND KIND SUPERCONDUCTORS

3–1 MAGNETIZATION CURVES OF A LONG CYLINDER

Type II superconductors are characterized by the following macroscopic properties:

(1) A cylinder placed in a longitudinal field H does not exhibit a "perfect" total flux expulsion (Meissner effect), except for weak field $H < H_{c1}$.[1]

If one calculates the critical field H_c defined by the difference in free energies between the normal and superconducting states in zero field:

$$F_n - F_s = \frac{H_c^2}{8\pi} \tag{3-1}$$

one finds that H_{c1} is clearly smaller than H_c. For example, for the compound V_3Ga, caloric measurements in zero field (giving $F_n - F_s$) indicate $(H_c)_{T=0} \simeq 6000$ G, and magnetic measurements give $(H_{c1})_{T=0} \simeq 200$ G.

(2) For $H > H_{c1}$, lines of force penetrate the cylinder, but even

[1]The notation H_{c1} for the first penetration field has been recommended by the participants of the Colgate Conference on Superconductivity (1963).

at thermal equilibrium this penetration is not complete. The flux ϕ passing through the cylinder remains less than its value when the sample is in the normal state. This implies the existence of permanent currents in the specimen, which is thus still superconducting. This situation exists for fields $H_{c1} \leq H \leq H_{c2}$. H_{c2} is larger than H_c and is sometimes very large—for V_3Ga $(H_{c2})_{T=0} \sim 300{,}000$ G.

(3) For $H > H_{c2}$ a macroscopic sample does not show any expulsion of flux $B \equiv H$. However superconductivity is not completely destroyed. In an interval $H_{c2} < H < H_{c3}$ there remains on the surface of the cylinder a superconducting sheath (of typical thickness 10^3 Å). (In most cases $H_{c3} \sim 1.69 H_{c2}$.) The existence of this sheath can be shown, for instance, by measuring the resistance between two probes on the sample surface. It is found that for low measuring currents the resistance vanishes. Physically the sheath has the following origin: It is easier to nucleate a small superconducting region near the sample surface—just as it is easier to grow bubbles on the bottom of a glass of beer than to grow them from an arbitrary point in the beer. (A somewhat more sophisticated version of this argument will be discussed in Chapter 5.)

The variation with temperature of the fields H_{c1}, H_{c2}, H_{c3} is represented in Fig. 3-1. We now focus our attention on the region $H_{c1} < H < H_{c2}$ where partial flux penetration occurs. The existence of this region of the (H,T) plane was clearly shown for the first time in early experiments on alloys by Schubnikov (1937). We call it the Schubnikov phase, or sometimes the vortex state. (The latter name comes from the microscopic picture to be derived in Section 3-2.)

The partial flux penetration in the Schubnikov phase can be described in terms of a diagram B(H); the aspect of this diagram is shown in Fig. 3-2. Sometimes, instead of the induction B, the experimentalists prefer to plot the "magnetization" M defined by

$$M = \frac{B - H}{4\pi} \tag{3-2}$$

The M(H) curve is shown in Fig. 3-3.

(In practice the observation of these curves is often complicated by difficulties in attaining equilibrium; for example, structural defects oppose the displacement of the lines of force.)

In Fig. 3-3 the dashed line represents the magnetization curve found for a first kind superconductor with the same H_c. These two curves are related by a remarkable property—the areas they subtend are equal.

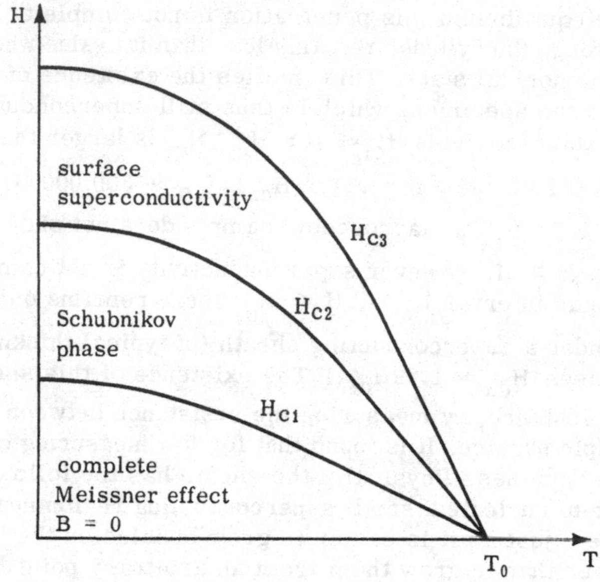

Figure 3-1

Phase diagram for a long cylinder of a Type II superconductor.

Proof: Let G_s be the Gibbs function per unit volume for the superconducting state

$$G_s = F_s(B) - \frac{BH}{4\pi} \tag{3-3}$$

G_s is a minimum for fixed H, that is, at equilibrium

$$\left(\frac{\partial G_s}{\partial B}\right)_H = 0 \tag{3-4}$$

Let G_n be the Gibbs function for the normal state

$$G_n = F_n + \frac{B^2}{8\pi} - \frac{BH}{4\pi} \tag{3-5}$$

At thermodynamic equilibrium in the normal phase $(\partial G_n/\partial B)_H = 0$; therefore, $B = H$ and

$$G_n = F_n - \frac{H^2}{8\pi} \tag{3-6}$$

Let the field vary from H to $H + \delta H$. From (3-3) and (3-4)

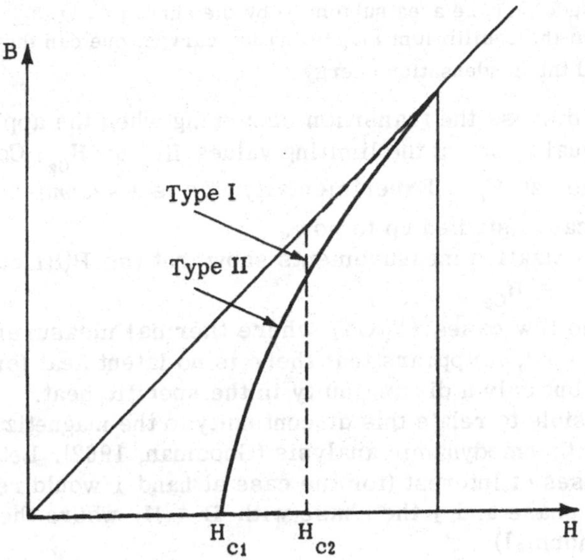

Figure 3-2

The induction (or flux/cm²) in the cylinder as a function of the applied field H. The full curve applies for a Type II superconductor, the dotted curve for a Type I.

$$\frac{\partial G_s}{\partial H} = -\frac{B}{4\pi} \tag{3-7}$$

and from (3-6)

$$\frac{\partial G_n}{\partial H} = -\frac{H}{4\pi} \tag{3-8}$$

$$\frac{\partial}{\partial H}(G_n - G_s) = \frac{B - H}{4\pi} = M \tag{3-9}$$

We now integrate this relation between $H = 0$ and $H = H_{c2}$. At $H = H_{c2}$, the two phases are in equilibrium and $G_n = G_s$. For $H = 0$, $B = 0$ we have $G_n = F_n$, $G_s = F_s$, and by definition

$$(F_n - F_s)_{B=0} = \frac{H_c^2}{8\pi}$$

The result is therefore

$$\int_0^{H_{c2}} M\,dH = \frac{-H_c^2}{8\pi} \tag{3-10}$$

and we conclude that the area subtended by the curve of Fig. 3-3 depends only on H_c. From the equilibrium magnetization curves, one can therefore determine H_c and the condensation energy.

We now discuss the transition occurring when the applied field H becomes equal to one of the limiting values H_{c1} or H_{c2}. Consider first the transition at H_{c2}. Experimentally this is a *second-order* transition in all cases studied up to now.

(1) Magnetization measurements show that the B(H) curve is continuous at $H = H_{c2}$.

(2) In the few cases (V_3Ga) where thermal measurements have been performed, it appears that there is no latent heat for the transformation, but only a discontinuity in the specific heat.

It is possible to relate this discontinuity to the magnetization curves by a purely thermodynamic analysis (Goodman, 1962). Let i and j be the two phases of interest (for the case at hand i would represent the Schubnikov phase and j the phase with $B \equiv H$, where the bulk of the sample is normal)

$$G_i = F_i(T, B_i) - \frac{B_i H}{4\pi} \tag{3-11}$$

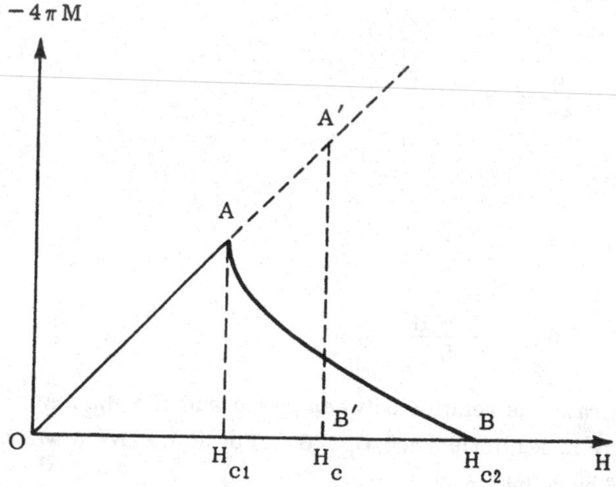

Figure 3-3

The reversible magnetization curve of a long cylinder of Type I (dotted line) or Type II (solid line) superconductor. If the two materials have the same thermodynamic field H_c, the areas OAB and OA′B′ are equal.

is the Gibbs function per cm³ of the ith phase. The relation between the field H and the induction B in this phase is obtained by minimizing G for fixed H and T

$$\frac{\partial}{\partial B_i} F_i(T, B_i) = \frac{H}{4\pi} \tag{3-12}$$

the entropy S_i is deduced from the relations

$$S_i = -\left(\frac{\partial G_i}{\partial T}\right)_H = -\frac{\partial F_i}{\partial T} \tag{3-13}$$

When the field satisfies a certain condition $H = H^*(T)$, there is an equilibrium between the phases i and j (i.e., for the present case $H^* = H_{c2}$). On this curve, one has

$$G_i = G_j \tag{3-14}$$

Suppose that there is no latent heat associated with the transformation. Then, along the curve $H = H^*(T)$, the two phases have the same entropy

$$S_i = S_j \tag{3-15}$$

We first show that this excludes any discontinuity in B at the transition. To show this, we calculate the variation of F_i when one moves along the equilibrium curve $[dH = (dH^*/dT)\,dT]$

$$\frac{dF_i}{dT} = \frac{\partial F_i}{\partial T} + \frac{\partial F_i}{\partial B_i}\frac{dB_i}{dT} \tag{3-16}$$

From this equation and using (3-12) and (3-13), we obtain the variation of G along the equilibrium curve

$$\frac{dG_i}{dT} = -S_i - \frac{B_i}{4\pi}\frac{dH^*}{dT} \tag{3-17}$$

Along the equilibrium curve we have constantly $G_i = G_j$, therefore $dG_i/dT = dG_j/dT$; if also $S_i = S_j$ we necessarily have $B_i = B_j$; B is continuous at the transition.

We now calculate the specific heat in constant field

$$C_i = T\left(\frac{\partial S_i}{\partial T}\right)_H \tag{3-18}$$

We can transform the second factor by calculating the total derivative of the entropy along the equilibrium curve

$$\frac{dS_i}{dT} = \left(\frac{\partial S_i}{\partial T}\right)_H + \left(\frac{\partial S_i}{\partial H}\right)_T \frac{dH^*}{dT} \tag{3-19}$$

From (3-15), $dS_i/dT = dS_j/dT$ along the equilibrium curve and

$$C_j - C_i = T\frac{dH^*}{dT}\left[\left(\frac{\partial S_i}{\partial H}\right)_T - \left(\frac{\partial S_j}{\partial H}\right)_T\right] \tag{3-20}$$

We transform $(\partial S_i/\partial H)_T$ by using (3-13) and (3-12)

$$\left(\frac{\partial S_i}{\partial H}\right)_T = \left(\frac{\partial S_i}{\partial B_i}\right)_T \left(\frac{\partial B_i}{\partial H}\right)_T = -\frac{\partial^2 F_i}{\partial B_i \partial T}\left(\frac{\partial B_i}{\partial H}\right)_T$$

$$= -\frac{1}{4\pi}\frac{\partial H(B_i\, T)}{\partial T}\left(\frac{\partial B_i}{\partial H}\right)_T \tag{3-21}$$

Finally, we write the variation of H^* with respect to T in the form

$$\frac{dH^*}{dT} = \left(\frac{\partial H}{\partial T}\right)_{B_i} + \left(\frac{\partial H}{\partial B_i}\right)_T \frac{dB}{dT} \tag{3-22}$$

where $dB/dT = dB_i/dT = dB_j/dT$ represents the variation of B along the equilibrium curve. On inserting $(\partial H/\partial T)_B$ from (3-22) into (3-21) we find

$$\left(\frac{\partial S_i}{\partial H}\right)_T = -\frac{1}{4\pi}\frac{dH^*}{dT}\left(\frac{\partial B_i}{\partial H}\right)_T + \frac{1}{4\pi}\frac{dB}{dT} \tag{3-23}$$

$$C_j - C_i = \frac{T}{4\pi}\left(\frac{dH^*}{dT}\right)^2\left[\left(\frac{\partial B_j}{\partial H}\right)_T - \left(\frac{\partial B_i}{\partial H}\right)_T\right] \tag{3-24}$$

Therefore, if one knows $H^*(T)$ and the permeabilities $(\partial B_i/\partial H)_T$ for each of the phases, one can predict the discontinuity in the specific heat. For the transition (i → j) we have a finite permeability (> 1) in the vortex state (i) from Fig. 3-2 and a permeability equal to 1 in the normal state (j). Therefore $C_i > C_j$. At this time all the necessary information is not available to compare (3-24) to experiment. However, for V_3Ga, $C_i - C_j$ and dH_{C2}/dT are known and if one makes a reasonable extrapolation for $\partial B/\partial H$ in order to predict its value at H_{C2} one finds an agreement to within about 10% from magnetic and calorimetric measurements. A similar analysis can be carried out in principle for the transition at $H = H_{C1}$. Here, however, the permeability $(\partial B/\partial H)_{H=H_{C1}}$ in the Schubnikov phase is probably infinite, as shown by the theoretical calculations of Section 3-2. From (3-24) this leads to an infinite peak in the specific heat at the transition. The singularity is, in fact, weak and easily masked by hysteresis effects. It has been observed recently (on Niobium) by the Rutgers group.

3–2 VORTEX STATE: MICROSCOPIC DESCRIPTION

Negative Surface Energy

We have previously seen that in a London ($\xi < \lambda$) superconductor the surface tension of a wall separating normal and superconducting regions becomes negative. Under these conditions, we guess that in the presence of a field a state is created where the N and S regions are finely divided and where the wall energy gives an important contribution to the thermodynamic potential. This situation is very different from that encountered for a Pippard superconductor where the walls are less numerous and where their energy can be neglected in a macroscopic treatment.

Consider, for example, the limit where B is small (that is, few lines of force penetrate the specimen and only a small fraction of the sample is normal). There are essentially two possibilities to maximize the surface to volume ratio for the N regions:

We can form lamina of very small thickness ($\geq \xi$) or filaments of small diameter ($\sim \xi$). In the case of $\lambda \gg \xi$, theoretical calculations show that the second solution is lowest in energy.[2] We therefore find filaments. They are represented in Fig. 3-4a.

Each filament has a hard core of radius ξ where the superconducting electron density n_s falls as is shown in Fig. 3-4c. The lines of force are not confined to the hard core; the field is maximum at the center of the filament but extends a distance λ (Fig. 3-4b). Annular

[2]See problem, page 71.

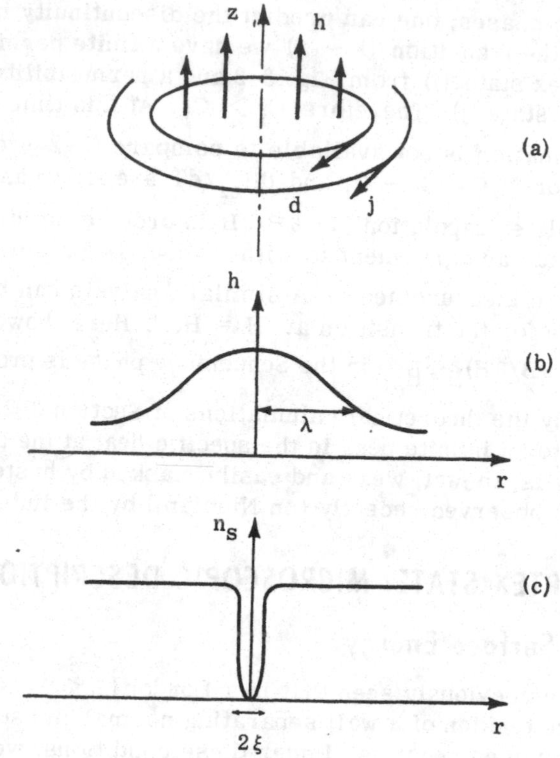

Figure 3-4

Structure of one vortex line in a Type II superconductor. The magnetic field is maximum near the center of the line. Going outwards, h decreases because of the screening in an "electromagnetic region" of radius ~λ (the penetration depth) (Fig. 3-4b). On the other hand, the number of superconducting electrons per cm^3 n_s is reduced only in a small "core region" of radius ξ (Fig. 3-4c).

currents j encircle the filament and screen out the field for $r \gtrsim \lambda$. For $r > \xi$, that is, in most of the region of interest, the currents and fields can be simply calculated by London's equations. It will be shown that the radius and exact form of the hard core only appear in the argument of a logarithm; therefore it will not be necessary to know them accurately. In the limit $\xi \ll \lambda$ we shall, in fact, see that the properties of a filament are very easy to calculate.

What is the flux $\phi = \int h d\sigma$ carried by a filament?

Experiment and theory show us that if a bulk superconducting annulus surrounds some lines of force the enclosed flux can only take on discrete values

$$\phi = k\phi_0 \qquad (\text{k is an integer})$$
$$\phi_0 = \frac{ch}{2e} = 2 \times 10^{-7}\ \text{G cm}^2 \tag{3-25}$$

The explanation of this effect will be given at the end of Chapter 4. The same result applies here. In order to attain the state of maximum subdivision, each filament carries one quantum of flux ϕ_0. This condition fixes the field scale in Fig. 3-4b, and the structure of the filament is completely defined.

Such a quantized filament, formed of a very thin hard core surrounded by currents rotating about the axis, is very analagous to the vortex lines found in superfluid He^4 when the helium container is rotated. The only important difference is that the helium atoms are not charged, making $e = 0$ in Eq. (1-13). We see that the penetration depth λ in He^4 is infinite and the particle currents j, instead of decreasing exponentially for $r > \lambda$, decrease very slowly (as in $1/r$) far from the filament. Historically, vortex lines were first discussed for the He^4 problem by Onsager and Feynman; the generalization to superconductivity is due to Abrikosov (1956). When the superconducting metal contains a finite density of these lines, we say that it is in the vortex state.

Properties of One Isolated Vortex Line

We now study in detail the structure of one vortex line in the limit $\lambda \gg \xi$. The "hard core" of radius ξ is very small and we shall, for the moment, neglect completely its contribution to the energy.

Then the line energy is given by the formula

$$\mathfrak{I} = \int_{(r > \xi)} d\mathbf{r}\ \frac{1}{8\pi}\left[\mathbf{h}^2 + \lambda^2 (\text{curl}\ \mathbf{h})^2\right] \tag{3-26}$$

Equation (3-26) has been derived in Chapter 1 (assuming $\lambda \gg \xi$). For a pure superconductor, the penetration depth λ has the London value

$$\lambda_L = \left[\frac{mc^2}{4\pi n_s e^2}\right]^{1/2} \tag{3-27}$$

For a superconducting alloy, with $\lambda \gg \xi$, Eq. (3-26) still applies, but with a modified (larger) value of λ, as explained in Chapter 2 (see the discussion after Eq. 2-20). In Eq. (3-26) the integration is carried out in all space outside of the "hard core" ($r > \xi$). We also choose to compute the energy per unit length of line; the resulting energy per cm $\mathfrak{J}$ is called the line tension. Demanding that $\mathfrak{J}$ be a minimum leads as usual to the London equation

$$\mathbf{h} + \lambda^2 \text{ curl curl } \mathbf{h} = 0 \qquad |\mathbf{r}| > \xi \tag{3-28}$$

In the interior of the hard core Eq. (3-28) must be replaced by something more complicated. But, since the hard core has a very small radius, we can try to replace the corresponding singularity, simply by a two-dimensional delta function $\delta_2(\mathbf{r})$, and write

$$\mathbf{h} + \lambda^2 \text{ curl curl } \mathbf{h} = \boldsymbol{\phi}_0 \delta_2(\mathbf{r}) \tag{3-29}$$

where $\boldsymbol{\phi}_0$ is a vector along the line direction. We now show that the strength ϕ_0 in (3-29) represents the total flux carried by the line.

Integrate (3-29) over the interior surface of a circle C of radius r encircling the axis of the cylinder and use the curl formula:

$$\int \mathbf{h} \cdot d\boldsymbol{\sigma} + \lambda^2 \oint \text{curl } \mathbf{h} \cdot d\boldsymbol{\ell} = \phi_0 \tag{3-30}$$

if the circle has a radius $r \gg \lambda$, the currents $\mathbf{j}(\mathbf{r}) = c/4\pi$ curl $\mathbf{h}$ are negligible and the line integral along the perimeter of the circle vanishes. Thus the total flux carried by the filament has the value ϕ_0.

We now pass on to the explicit solution of (3-30) to which is added the Maxwell equation

$$\text{div } \mathbf{h} = 0 \tag{3-31}$$

The field h is directed along the z axis: the current lines are circles in the xy plane. It is easy to predict the value of curl **h**, that is, the current, in the region $\xi \ll r \ll \lambda$. In fact, if we reconsider (3-30) with a circle C whose radius is in this domain, the term $\int \mathbf{h} \cdot d\boldsymbol{\sigma}$ is negligible (only a fraction r^2/λ^2 of the flux ϕ_0 passes through the circle C) and we have

$$\lambda^2 \, 2\pi r \, |\text{curl } \mathbf{h}| = \phi_0 \tag{3-32}$$

and

$$|\text{curl } \mathbf{h}| = \frac{\phi_0}{2\pi\lambda^2} \frac{1}{r} \qquad (\xi < r \ll \lambda) \tag{3-33}$$

Since h is directed along the z axis, we have $|\text{curl}\,\mathbf{h}| = -dh/dr$ and upon integrating this relation

$$h = \frac{\phi_0}{2\pi\lambda^2}\left[\ln\left(\frac{\lambda}{r}\right) + \text{const}\right] \qquad (\xi < r \ll \lambda) \tag{3-34}$$

In order to derive the constant of integration in (3-34), it is necessary to write the complete solution to (3-30) and (3-32), which is

$$h = \frac{\phi_0}{2\pi\lambda^2} K_0\left(\frac{r}{\lambda}\right) \tag{3-35}$$

where K_0 is the zero-order Bessel function of an imaginary argument defined as in Morse and Feshbach.[3] The important properties of this solution are the asymptotic form (3-34) for $r \ll \lambda$ (it is found that the constant vanishes) and the asymptotic form for large distances

$$h = \frac{\phi_0}{2\pi\lambda^2}\sqrt{\frac{\pi\lambda}{2r}}\, e^{-r/\lambda} \qquad (r \gg \lambda) \tag{3-36}$$

Once the fields are determined, it is easy to calculate the energy $\mathfrak{J}$. On integrating the second term in (3-28) by parts,

$$\mathfrak{J} = \frac{\lambda^2}{8\pi}\int d\sigma \cdot \mathbf{h} \times \text{curl}\,\mathbf{h} \tag{3-37}$$

where the integral $\int d\sigma$ is to be taken over the surface of the hard core (cylinder of radius $\sim\xi$). It is convenient to calculate $\mathfrak{J}$ per cm of length along the filament. Then

$$\mathfrak{J} = \frac{\lambda^2}{8\pi}\, 2\pi\xi h(\xi)\,|\text{curl}\, h(\xi)| \tag{3-38}$$

which is, from (3-34) and (3-33),

$$\mathfrak{J} = \left(\frac{\phi_0}{4\pi\lambda}\right)^2 \ln\left(\frac{\lambda}{\xi}\right) \tag{3-39}$$

DISCUSSION OF THIS FORMULA

(1) $\mathfrak{J}$ only depends upon ξ logarithmically.

(2) $\mathfrak{J}$ is a quadratic function of the flux. Upon going to a situation where the flux is $2\phi_0$, it is preferable to have two filaments of flux ϕ_0 (total energy $2\mathfrak{J}$) than a filament of double flux (energy $4\mathfrak{J}$). This

---

[3] *Methods of Theoretical Physics*, (New York: McGraw-Hill, 1953), Chap. 10, p. 1321.

justifies the choice for ϕ_0 of the minimum flux value, that is, the quantum of flux.

(3) It is possible to rewrite $\mathfrak{J}$ at $T = 0$ in a rather different form by using the relations

$$\phi_0 = \frac{ch}{2e}, \quad \xi = \xi_0 = \frac{\hbar v_F}{\pi\Delta(0)} \tag{3-40}$$

and a relation (which will be proven by the microscopic theory) between the condensation energy and the energy gap $\Delta(0)$:

$$\frac{H_c^2}{8\pi} = \tfrac{1}{2} N(0)\Delta^2(0) \tag{3-41}$$

where $N(0) = m^2 v_F / 2\pi^2\hbar^3$ is the density of states (for one direction of spin) at the Fermi energy in the normal state per unit energy and per cm^3. Upon regrouping these formulas, we obtain

$$\mathfrak{J} = \frac{\pi^3}{3} \frac{H_c^2}{8\pi} \xi^2 \ln \frac{\lambda}{\xi} \qquad (T = 0) \tag{3-42}$$

This formula is interesting for the following reason:

Until now we have neglected the contribution of the hard core to the line energy. In fact, superconductivity is more or less destroyed in the hard core section and this takes an extra energy $\mathfrak{J}_{int} \sim (H_c^2/8\pi)\xi^2$.

Dimensionally from (3-42) this energy is comparable to $\mathfrak{J}$. Numerically, however, it is much smaller. A more detailed calculation gives for the total energy

$$\mathfrak{J} = \left(\frac{\phi_0}{4\pi\lambda}\right)^2 \left(\ln \frac{\lambda}{\xi} + \epsilon\right) \qquad (\lambda \gg \xi) \tag{3-43}$$

The numerical constant ϵ includes the effect of the hard core and is of the order of 0.1.

Problem. Discuss the structure of vortices in a thin film, the applied magnetic field being normal to the film surface (J. Pearl, 1964).

Solution. Again there is a "hard core" of radius ξ (which we assume to be small) surrounded by current rings. But since the currents are restricted to the thickness d of the film, their screening capacity is weak and the "electromagnetic region" is more spread out than in a long vortex line.

Inside the film, we apply Eq. (3-29)

$$h + \frac{4\pi\lambda^2}{c} \text{ curl } \mathbf{j} = \phi_0 \delta_2(\mathbf{r}) \mathbf{n}_z$$

where $\mathbf{j}$ is the current density and $\mathbf{n}_z$ is a unit vector normal to the film. It is convenient to use the vector potential **A** rather than the field $\mathbf{h} = \text{curl } \mathbf{A}$. In the London gauge, we find

$$\mathbf{A} + \frac{4\pi\lambda^2}{c} \mathbf{j} = \mathbf{\Phi}$$

where $\Phi_r = \Phi_z = 0$ and $\Phi_\theta = \phi_0/2\pi r$.

Now average over the thickness d of the film. If $d \ll \lambda$, **A** and $\mathbf{j}$ are nearly constant in the thickness. Call **J** the total current $\mathbf{J} = \mathbf{j}d$. Then

$$\mathbf{J} = \frac{c}{4\pi} \frac{1}{\lambda_{\text{eff}}} (\mathbf{\Phi} - \mathbf{A})$$

$$\lambda_{\text{eff}} = \frac{\lambda^2}{d}$$

Now replace the film by a infinitesimally small current carrying sheet in the plane $z = 0$, the current density being $\mathbf{J}\delta(z)$. This will be valid when d is much smaller than the range of the electromagnetic region.

In terms of the current sheet the equation valid for all space is

$$\text{curl curl } \mathbf{A} = \text{curl } \mathbf{h} = \frac{4\pi}{c} \mathbf{j} = \frac{1}{\lambda_{\text{eff}}} \delta(z)(\mathbf{\Phi} - \mathbf{A})$$

or (since curl curl $\mathbf{A} = -\nabla^2 \mathbf{A}$ in the London gauge)

$$-\nabla^2 \mathbf{A} + \mathbf{A} \frac{1}{\lambda_{\text{eff}}} \delta(z) = \mathbf{\Phi} \frac{1}{\lambda_{\text{eff}}} \delta(z)$$

This result was derived here from the London Eq. (3-29). In actual thin films, such a simple equation does not usually hold. But it is still correct to assume a linear current response of the form $\mathbf{J} = (c/4\pi\lambda_{\text{eff}})(\mathbf{\Phi} - \mathbf{A})$, where λ_{eff} is some unknown constant, which can be obtained from another experiment (λ_{eff} is, in fact, the effective penetration depth in parallel fields, which could be measured on a hollow cylinder made with the same film).

To solve the equation for **A**, introduce the three-dimensional Fourier transform

$$\mathbf{A}_{qk} = \int \mathbf{A}(xyz) \exp i(q_x x + q_y y + kz) \, dx \, dy \, dz$$

and the two-dimensional transforms

$$\mathbf{A}_q = \frac{1}{2\pi} \int dk\, \mathbf{A}_{qk} = \int \mathbf{A}\delta(z) \exp i(q_x x + q_y y)\, dx\, dy\, dz$$

$$\mathbf{\Phi}_q = \int \mathbf{\Phi}(xy) \exp i(q_x x + q_y y)\, dx\, dy = i \frac{\phi_0}{q^2} \mathbf{n}_z \times \mathbf{q}$$

Then

$$(q^2 + k^2)\mathbf{A}_{qk} + \frac{1}{\lambda_{eff}} \mathbf{A}_q = \frac{1}{\lambda_{eff}} \mathbf{\Phi}_q$$

Solve for $\mathbf{A}_{qk}$ and integrate over k:

$$\mathbf{A}_q = -\frac{1}{2\pi} \int dk \frac{1}{q^2 + k^2} (\mathbf{A}_q - \mathbf{\Phi}_q) \frac{1}{\lambda_{eff}} = -\frac{1}{2q\lambda_{eff}} (\mathbf{A}_q - \mathbf{\Phi}_q)$$

$$\mathbf{A}_q = \mathbf{\Phi}_q \frac{1}{1 + 2q\lambda_{eff}}$$

From this, all required information can be extracted:

(a) The current has components

$$\mathbf{J}_q = \frac{c}{4\pi\lambda_{eff}} (\mathbf{\Phi}_q - \mathbf{A}_q) = \frac{c}{4\pi\lambda_{eff}} \mathbf{\Phi}_q \frac{2q\lambda_{eff}}{1 + 2q\lambda_{eff}}$$

When $q \gg \lambda_{eff}^{-1}$, $\mathbf{J}_q$ is proportional to $\mathbf{\Phi}_q$. Thus at small distances r from the center of the vortex

$$\mathbf{J}(r) = \frac{c}{4\pi\lambda_{eff}} \mathbf{\Phi}(r)$$

$$J = \frac{c\phi_0}{8\pi^2 \lambda_{eff} r} \qquad (\xi \ll r \ll \lambda_{eff})$$

When $q \ll \lambda_{eff}^{-1}$

$$\mathbf{J}_q \cong \frac{c}{4\pi\lambda_{eff}} 2q\lambda_{eff} \mathbf{\Phi}_q = \frac{c\phi_0}{2\pi} \frac{i\mathbf{n}_z \times \mathbf{q}}{q}$$

$$J = \frac{c\phi_0}{4\pi^2 r^2} \qquad (r \gg \lambda_{eff})$$

The size of the screening region is λ_{eff}. But even beyond λ_{eff}, J decreases only slowly with distance.

(b) The normal field component h_z in the film is derived from

$$h_{zq} = \left(-i\mathbf{q} \times \mathbf{A}_q\right)_z = \frac{\phi_0}{1 + 2q\lambda_{eff}}$$

When $q \gg \lambda_{eff}^{-1}$

$$h_{zq} \sim \frac{\phi_0}{2q\lambda_{eff}}$$

$$h_z(r) \sim \frac{\phi_0}{4\pi\lambda_{eff} r} \qquad (\xi < r \ll \lambda_{eff})$$

At large r we derive h_z most easily from the current J

$$h_z = -\frac{4\pi\lambda_{eff}}{c}\frac{1}{r}\frac{d}{dr}(Jr)$$

$$\cong \frac{2}{\pi}\frac{\phi_0\lambda_{eff}}{r^3}$$

(c) The self-energy of the vortex is derived from Eq. (3-26); the required components of **h** and curl **h** at the core surface ($\xi \ll \lambda_{eff}$) are quoted above. The result is

$$E = \left(\frac{\phi_0}{4\pi}\right)^2 \frac{1}{\lambda_{eff}} \log\frac{\lambda_{eff}}{\xi} = \frac{137}{16}\frac{\hbar c}{\lambda_{eff}} \log\frac{\lambda_{eff}}{\xi}$$

Typically $\lambda_{eff} \sim 1000$Å and $E \sim 30$eV.

(d) The force between two vortices is

$$\mathbf{F}_{12} = \frac{\phi_0}{c}\mathbf{n}_z \times \mathbf{J}(R_{12})$$

Note that at long distances $J \sim 1/R^2$ and the repulsion energy decreases only like $1/R$. This long range is due to the fact that most of the interaction takes place not through the superconductor, but through the empty space above and below.

Interactions between Vortex Lines

TWO VORTEX LINES

Consider two parallel lines directed along the z axis with positions $\mathbf{r}_1 = (x_1, y_1)$, $\mathbf{r}_2 = (x_2 y_2)$. The magnetic field distribution is determined by the equations

$$\mathbf{h} + \lambda^2 \text{ curl curl } \mathbf{h} = \phi_0[\delta(\mathbf{r} - \mathbf{r}_1) + \delta(\mathbf{r} - \mathbf{r}_2)] \qquad (3\text{-}44)$$

which is the generalization of (3-29). The solution **h** is the superposition of the fields $\mathbf{h}_1$ and $\mathbf{h}_2$ due to the filaments (1) and (2).

$$h(r) = h_1(r) + h_2(r)$$

$$h_1(r) = \frac{\phi_0}{2\pi\lambda^2} K_0 \left(\frac{r - r_1}{\lambda}\right) \tag{3-45}$$

The energy of the system is still written

$$F = \int \frac{h^2 + \lambda^2 (\text{curl } h)^2}{8\pi} \, dr = \frac{\lambda^2}{8\pi} \int \mathbf{h} \times \text{curl } \mathbf{h} \cdot d\sigma \tag{3-46}$$

Here the integral $\int d\sigma$ is to be taken over the surfaces of the two hard cores ($|r - r_i| = \xi$); writing explicitly the two contributions to h we obtain

$$F = \frac{\lambda^2}{8\pi} \int (d\sigma_1 + d\sigma_2) \cdot (\mathbf{h}_1 + \mathbf{h}_2) \times (\text{curl } \mathbf{h}_1 + \text{curl } \mathbf{h}_2) \tag{3-47}$$

There are 8 terms that we regroup as follows: First the individual energy of each filament

$$\frac{\lambda^2}{8\pi} \left[\int d\sigma_1 \cdot \mathbf{h}_1 \times \text{curl } \mathbf{h}_1 + \int d\sigma_2 \cdot \mathbf{h}_2 \times \text{curl } \mathbf{h}_2 \right] = 2\mathfrak{J}$$

then the terms

$$\int (\mathbf{h}_1 + \mathbf{h}_2) \cdot (\text{curl } \mathbf{h}_1 \times d\sigma_2 + \text{curl } \mathbf{h}_2 \times d\sigma_1)$$

which tend toward 0 in the limit where $\xi \ll \lambda$ because $\mathbf{h}_1 + \mathbf{h}_2$ and curl $\mathbf{h}_1$ are finite in the domain of integration $\int d\sigma_2$. There remains an important contribution

$$U_{12} = \frac{\lambda^2}{8\pi} \int (\mathbf{h}_1 \times \text{curl } \mathbf{h}_2 \cdot d\sigma_2 + \mathbf{h}_2 \times \text{curl } \mathbf{h}_1 \cdot d\sigma_1) \tag{3-48}$$

In effect, curl $\mathbf{h}_2$ is proportional to $1/|r - r_2|$ for $|r - r_2| \ll \lambda$ from (3-33) and after integration one obtains a finite result as $\xi \to 0$. If we set

$$h_{12} = h_1(r_2) = h_2(r_1) = \frac{\phi_0}{2\pi\lambda^2} K_0 \left(\frac{r_1 - r_2}{\lambda}\right) \tag{3-49}$$

then using (3-33) we obtain

$$U_{12} = \frac{\phi_0 h_{12}}{4\pi} \tag{3-50}$$

U_{12} represents the interaction energy (per cm) of two filaments. This is a *repulsive* energy which decreases as $(1/\sqrt{r_{12}})\, e^{-r_{12}/\lambda}$ at large distances, and which diverges as $\ln(|\lambda/r_{12}|)$ at short distances.

A remark about forces. Let us compute the force f_2 experienced by line 2, as due to the interaction U_{12}

$$f_{2x} = -\frac{\partial U_{12}}{\partial x_2} = -\frac{\phi_0}{4\pi}\frac{\partial h_{12}}{\partial x_2} \tag{3-51}$$

Introduce now the current $\mathbf{j} = n_s e\mathbf{v}$, which would exist, in the presence of line 1 alone, at the point $x_2 y_2$. Then $j_y = -(c/4\pi)(\partial h_{12}/\partial x_2)$ by Maxwell's equation and we have

$$f_{2x} = \frac{\phi_0}{c}\, j_y = \tfrac{1}{2} hnv_y \tag{3-52}$$

When more than one line is acting on line 2, Eq. (3-52) remains valid provided we interpret $\mathbf{v}$ as the total superfluid velocity at point $(x_2 y_2)$.

Conclusion. A line is in static equilibrium when the superfluid velocity at any point on the line is 0.

MAGNETIZATION CURVES

We now form the Gibbs function, minimize it, and deduce the density of vortices existing in the sample in thermal equilibrium.

$$G = n_L \mathfrak{I} + \sum_{ij} U_{ij} - \frac{BH}{4\pi} \tag{3-53}$$

The first term represents the individual energies of the lines, n_L is the number of lines per cm², related to the induction B by

$$B = n_L \phi_0 \tag{3-54}$$

(This expresses the fact that each vortex carries a flux ϕ_0.) The second term in (3-53) describes the repulsive interactions between vortices, the explicit form of U_{ij} is given by (3-49) and (3-50). Finally, the last term gives the effect of the field H and favors large values of B. It plays the role of a pressure that tends to increase the density of vortices.

In order to numerically evaluate the interaction term, it is useful to distinguish several regions:

(1) In the small induction region ($n_L \lambda^2 \ll 1$), only the interaction

between nearest neighbor vortices is important and the sum ΣU_{ij} converges rapidly.

(2) When B is larger ($n_L \lambda^2 \gg 1$), the range λ of the interaction becomes large compared to the spacing of the filament lattice, and other methods are preferable to evaluate ΣU_{ij}.

(3) Finally, when n_L becomes comparable to $1/\xi^2$, the hard cores begin to overlap and the elementary methods used in this section are no longer valid. But, qualitatively, we may guess that when the hard cores do overlap, superconductivity is destroyed in the bulk. This corresponds to inductions $B \sim \phi_0/\xi^2$.

THE FIRST PENETRATION FIELD H_{c1}

At very low line densities (low B), the interaction term in Eq. (3-53) is small and we shall first neglect it completely. Then, using Eq. (3-54) we get

$$G \cong B\left(\frac{\mathfrak{J}}{\phi_0} - \frac{H}{4\pi}\right) \tag{3-55}$$

When $H < 4\pi\mathfrak{J}/\phi_0$, G is an increasing function of B. The lowest G is obtained for $B = 0$ (complete Meissner effect).

When $H > 4\pi\mathfrak{J}/\phi_0$, we can lower G by choosing $B \neq 0$. There is some flux penetration.

We conclude that the first penetration field is given by

$$H_{c1} = \frac{4\pi\mathfrak{J}}{\phi_0} = \frac{\phi_0}{4\pi\lambda^2}\log\frac{\lambda}{\xi} \tag{3-56}$$

H_{c1} is often much smaller than the "thermodynamic" field H_c defined by (3-1). For $T = 0$, for example, on using (3-40) and (3-41), we find

$$\frac{H_{c1}}{H_c} = \frac{\pi}{\sqrt{24}}\frac{\xi}{\lambda}\ln\left(\frac{\lambda}{\xi}\right) \tag{3-57}$$

The result is therefore $H_{c1}/H_c \sim \xi/\lambda$ and this may be much smaller than 1. A measurement of H_{c1} and H_c, in principle, allows a determination of ξ and λ. For example, for V_3Ga, if at $T = 0$, $H_c \sim 6000$ G and $H_{c1} \sim 200$ G, one finds from (3-57) that $\lambda/\xi \sim 80$ and then, from (3-56), $\lambda \sim 2000$Å and $\xi \sim 25$Å. These orders of magnitude are still rather inaccurate because of current uncertainties in H_c and H_{c1}, but it is hoped that the situation will improve in the near future.

FIELDS SLIGHTLY LARGER THAN H_{c_1}

For finite line densities (finite B's) we must take into account the interaction term in Eq. (3-53). To minimize this repulsive energy, the lines will take a regular arrangement. Detailed calculations based on (3-53) show that, at all B's, the most favorable arrangement is triangular, as shown on Fig. 3-5 (J. Matricon, 1964).

If H is only slightly larger than H_{c_1}, we can guess that the equilibrium density of lines n will be small, and thus the distance between neighboring lines d will be large. If $d > \lambda$, we may keep only the nearest neighbor contributions to the interaction term in (3-53) and write

$$G \cong \frac{B}{4\pi}\left[H_{c_1} - H + \tfrac{1}{2} z \frac{\phi_0}{2\pi\lambda^2} K_0\left(\frac{d}{\lambda}\right)\right] \qquad (3\text{-}58)$$

where z is the number of nearest neighbors of one line (z = 6 for the triangular lattice), d is related to the induction B through the relation

$$B \equiv \phi_0 n_L = \frac{2}{\sqrt{3}} \frac{\phi_0}{d^2} \quad \text{(triangular lattice)} \qquad (3\text{-}59)$$

Equation (3-59) can be easily verified on Fig. 3-5. The function G(B) is represented in Fig. 3-6. Since $H > H_{c_1}$, the initial slope $(\partial G/\partial B)_{B=0}$

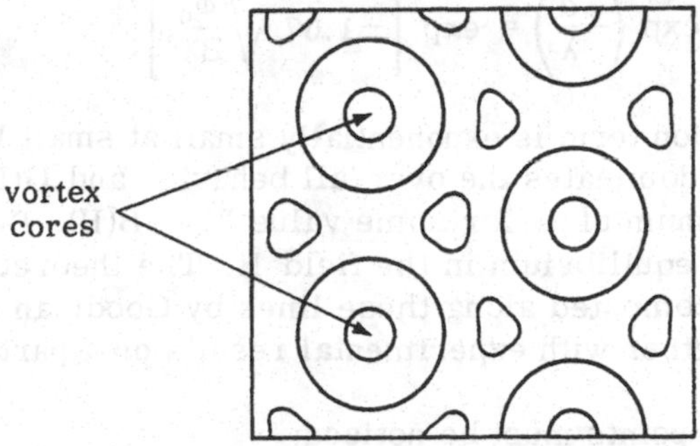

Figure 3-5

A triangular lattice of vortex line (after Kleiner, Roth, and Autler, *Phys. Rev.*, 133A, 1226 (1964). The plane of the figure is normal to the field direction. The contours give the lines of constant n_s. This figure describes the situation at high fields (nearly overlapping cores).

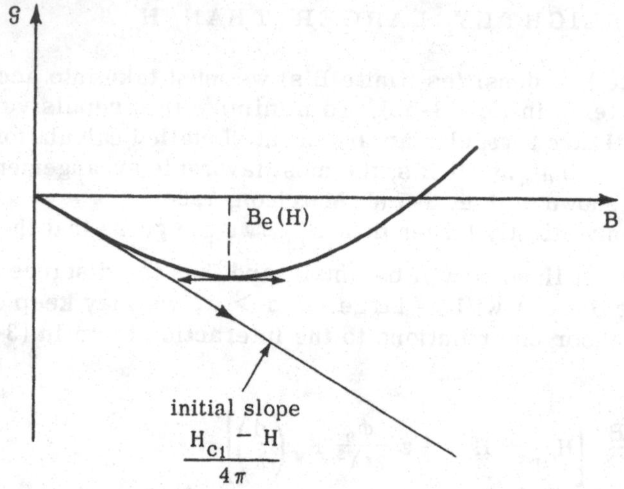

Figure 3-6
The thermodynamic potential $\mathcal{G}$ as a function of the induction B ($B = n_L \phi_0$ measures the number of vortices per cm² n_L). The equilibrium value of B [$B_e(H)$] corresponds to the minimum of $\mathcal{G}$.

is negative. As B increases, the interaction term begins to contribute but rather slowly, since it is proportional to $K_0(d/\lambda)$. When $d > \lambda$ we may write, according to (3-36)

$$K_0\left(\frac{d}{\lambda}\right) \sim \exp\left(-\frac{d}{\lambda}\right) = \exp\left[-1.07\sqrt{\frac{\phi_0}{B\lambda^2}}\right] \qquad (3\text{-}60)$$

Thus the interaction term is exponentially small at small B's. At larger B's however, it dominates the over-all behavior and G(B) increases. There is a minimum of G for some value B = B(H). B(H) is the induction found at equilibrium in the field H. The theoretical B(H) or M(H) has been computed along these lines by Goodman and is shown on Fig. 3-7, together with experimental results on a particularly good MoRe alloy.

The following points must be noticed:

The theoretical curve has an infinite slope $(\partial M/\partial H)_{H=H_{c1}} = \infty$ at the first penetration field. Physically, this reflects the fact that the lines repel each other like $e^{-d/\lambda}$, that is, we may think of their interaction as having a finite range λ. At field slightly larger than H_{c1} it is thus possible to form many lines in the sample without competing against the interaction energy. The experimental curve does not show

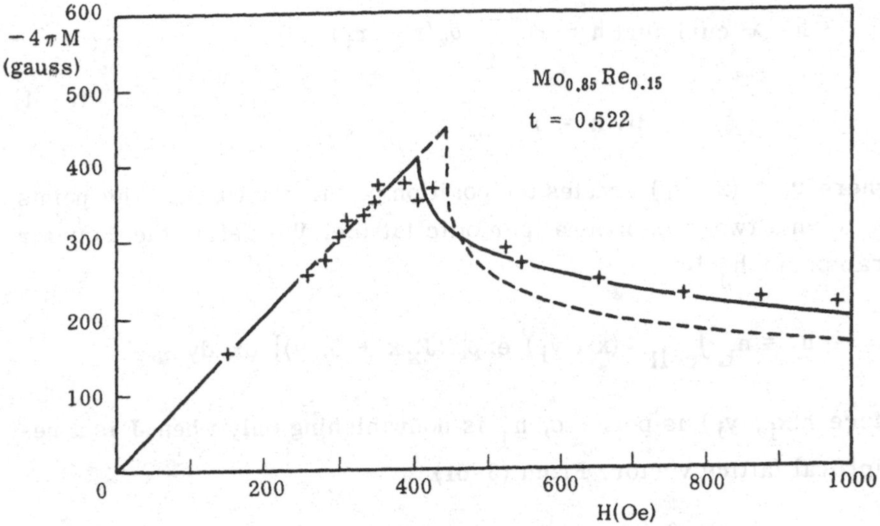

Figure 3-7
Experimental magnetization of a molybdenum-rhenium alloy at $T = 0.52\,T_0$ (after Joiner and Blaugher, *Rev. Mod. Phys.*, **36**, 67 (1964). Also shown are two theoretical magnetization curves (after B. B. Goodman). The broken curve is for a laminar model, the continuous one for vortex lines.

a very large slope $(\partial M/\partial H)_{H=H_{c_1}}$; this is not very surprising since in the region of interest the interactions between lines are very weak and the lines can easily be pinned by structural defects. However, as we depart from H_{c_1} by more than 10%, we get good agreement between theory and experiment.

A similar theoretical curve can be drawn for another model where the flux-carrying units are not vortex lines, but laminas (see problem, p. 71). If the distance between laminas is d, we again find a repulsion between units proportional to $e^{-d/\lambda}$. However, in this case, the induction B is proportional to d^{-1}, while in the line case it is proportional to d^{-2} as shown by Eq. (3-59). Thus the fall of M(H) for $H > H_{c_1}$ is more rapid in the laminar model than in the vortex line model. The two theoretical curves are compared on Fig. 3-7; it is apparent that the vortex line gives a better fit, as emphasized by Goodman.

DOMAIN $\frac{1}{\lambda^2} \ll n_L \ll \frac{1}{\xi^2}$

In this region the vortices form a rather dense lattice and the interactions extend to distant neighbors. The interaction energy can then be calculated by the following method: The field h(r) directed along the z axis is the solution of

$$\mathbf{h} + \lambda^2 \text{ curl curl } \mathbf{h} = \phi_0 \sum_i \delta_2(\mathbf{r} - \mathbf{r}_i)$$
$$\text{div } \mathbf{h} = 0 \tag{3-61}$$

where $\mathbf{r}_i = (x_i, y_i)$ denotes the position of the ith vortex. The points $\mathbf{r}_i$ form a two-dimensional periodic lattice. We define the Fourier transform h_J by

$$h_J = n_L \int_{\text{cell}} h(x_i, y_i) \exp[i(J_x x + J_y y)] \, dx \, dy$$

Since $h(x_i, y_i)$ is periodic, h_J is nonvanishing only when $\mathbf{J}$ is a reciprocal lattice vector. From (3-61),

$$h_J = \frac{n_L \varphi_0}{1 + \lambda^2 J^2} \tag{3-62}$$

Finally the free energy becomes

$$\mathfrak{F} = \frac{1}{8\pi} \int (h^2 + \lambda^2 \text{ curl}^2 \, h) \, dr$$
$$= \frac{1}{8\pi} \sum_J h_J^2 (1 + \lambda^2 J^2) = \frac{B^2}{8\pi} \sum_J \frac{1}{1 + \lambda^2 J^2}$$
$$= \frac{B^2}{8\pi} + \frac{B^2}{8\pi} \sum_{J \neq 0} \frac{1}{1 + \lambda^2 J^2} \tag{3-63}$$

In the sum $\Sigma_{J \neq 0}$, the minimum magnitude of the vectors $\mathbf{J}$ is of the order $1/d \sim \sqrt{n_L}$ and $\lambda^2 J^2 \sim n_L \lambda^2 \gg 1$ in the domain of interest. Therefore $1/(1 + \lambda^2 J^2)$ can be replaced by $1/\lambda^2 J^2$. Finally we must perform the sum $\Sigma_{J \neq 0} 1/J^2$, which depends on the particular lattice considered. Here, we will simplify the calculation by replacing the sum by an integral

$$\sum \frac{1}{J^2} \rightarrow \frac{1}{(2\pi)^2} \frac{1}{n_L} \int \frac{dJ_x \, dJ_y}{J^2} \rightarrow \frac{1}{2\pi n_L} \int_{J \min}^{J \max} \frac{J \, dJ}{J^2}$$
$$= \frac{1}{2\pi n_L} \ln \left| \frac{J \max}{J \min} \right|$$

with $J \min \sim 1/d$ and $J \max \sim 1/\xi$ (the Fourier components relative to the interior of the hard core must be excluded). We finally find

$$F = \frac{B^2}{8\pi} + \frac{B}{4\pi} H_{c_1} \frac{\ln \beta d/\xi}{\ln \lambda/\xi}$$

$$G = F - \frac{BH}{4\pi} \tag{3-64}$$

In (3-64) β is a numerical constant of the order unity (for the triangular lattice, Matricon has calculated $\beta = 0.381$). The B(H) relation is obtained as usual by imposing $\partial G/\partial B = 0$. This gives

$$H = B + H_{c_1} \frac{\ln\left(\beta' \frac{d}{\xi}\right)}{\ln \frac{\lambda}{\xi}} \tag{3-65}$$

where $\beta' = \beta\, e^{-1/2}$ and where d is always related to B by Eq. (3-59). The logarithmic dependences predicted by (3-65) are in rather good agreement with the experimental data on reversible magnetization curves in materials with $\lambda \gg \xi$.

DOMAIN $n_L \sim \xi^{-2}$

Here, as already pointed out, our simple model breaks down, and we shall need a more elaborate approach based on the Landau-Ginsburg equations (Chapter 6). The upper critical field H_{c_2} is of order φ_0/ξ^2. This, physically, corresponds to the onset of overlap between the hard cores.

Problem. Compare the Gibbs function in the filamentary structure described above with that of a possible laminar structure.

Solution. As before, we shall limit our considerations to the case $\lambda \gg \xi$. The laminar structure will be formed of planes, for example, perpendicular to the x axis, and equidistant (spacing d)(Fig. 3-8). In the neighborhood of each of these planes, over a thickness $\sim 2\xi$, the superconductivity is strongly perturbed (N regions). In the remainder (S regions), the density of superconducting electrons has the value n_s. Such a model has been discussed in detail by Goodman (1961). The fields h(x) (parallel to the z axis) are determined by the London equation

$$h = \lambda^2 \frac{d^2 h}{dx^2}$$

except in the thin (N) regions. The solution is of the form

$$h = H_m \cosh (x/\lambda)/\cosh P$$

where $P = d/2\lambda$ and H_m is the field in the N regions. The free energy of the (S) regions becomes, from (3-26),

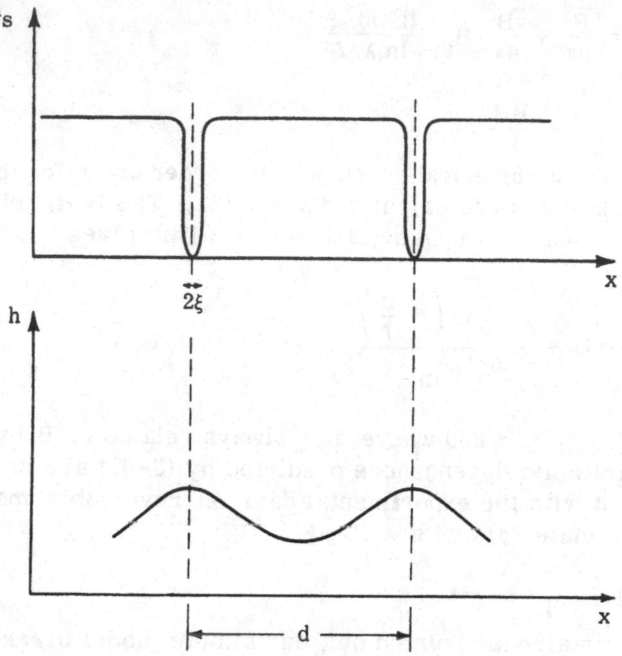

Figure 3-8

The laminar model for the Schubnikov phase. Thin normal sheets N of thickness $\sim 2\xi$ alternate with superconducting sheets S. The N sheets repel each other. The range of the repulsive forces is the penetration depth λ.

$$F_1 = \frac{2}{d}\int_0^{d/2} dx\, \frac{h^2 + \lambda^2\left(\frac{dh}{dx}\right)^2}{8\pi} = \frac{H_m^2}{8\pi}\,\frac{\tanh P}{P}$$

It is necessary to add the formation energy of the N regions

$$F_2 \cong \frac{H_c^2}{8\pi}\,\frac{2\xi}{d} = \frac{H_c^2}{8\pi}\,\frac{1}{P\kappa}$$

where $\kappa = \lambda/\xi$. Finally, to obtain the Gibbs function, we must add a term

$$-\frac{BH}{4\pi} = -H\,\frac{H_m}{4\pi}\,\frac{\tanh P}{P}$$

$$G_{\text{laminar}} = \frac{1}{8\pi}\left[H_m^2\,\frac{\tanh P}{P} + \frac{H_c^2}{P\kappa} - 2HH_m\,\frac{\tanh P}{P}\right]$$

On minimizing G with respect to H_m we obtain $H = H_m$.

$$G_{\text{laminar}} = \frac{1}{8\pi P}\left[-H^2 \tanh P + \frac{H_c^2}{\kappa}\right]$$

For $H < H_c/\sqrt{\kappa}$ the minimum G is obtained for infinite P, which corresponds to a complete Meissner effect. For $H > H_c/\sqrt{\kappa}$ the minimum occurs for finite P. The initial field for penetration is therefore $H_c/\sqrt{\kappa}$ for the laminar model. This is to be compared with the result for the vortex line model, Eq. (3-56).

$$H_{c1} = \frac{\pi}{\sqrt{24}} \frac{H_c}{\kappa} \ln\kappa \quad (\kappa \gg 1)$$

for $\kappa \gg 1$, $H_{c1} < H_c/\kappa$. For $H_{c1} < H < H_c/\kappa^{1/2}$, we have

$$G_{\text{vortex}} < G_{\text{Meissner}}$$

$$G_{\text{laminar}} = G_{\text{Meissner}}$$

therefore, $G_{\text{vortex}} < G_{\text{laminar}}$; that is, the vortex state is more favorable in the weak induction domain.

It is also possible to make the comparison in the region where H is larger ($H \sim H_c$, for example). We are then in the region $P \ll 1$ for the laminar model By expanding $\tanh P \cong P - P^3/3$ and minimizing G, we obtain

$$G_{\text{laminar}} = -\frac{H^2}{8\pi} + \left(\frac{3}{2\kappa}\right)^{2/3} \frac{H_c^{4/3} H^{2/3}}{8\pi}$$

In the vortex line model, the potential is determined from (3-64) and (3-65)

$$G_{\text{vortex}} = -\frac{1}{8\pi}(H - H')^2$$

$$H' = H_{c1}\frac{\ln d/\xi}{\ln \kappa} = \nu\frac{H_c}{\kappa}$$

where ν is a constant of order unity. For $T = 0$

$$\nu \sim \frac{\pi}{\sqrt{24}} \ln \frac{d}{\xi}$$

In the region of interest $H' \ll H$ and

$$G_{\text{vortex}} = -\frac{H^2}{8\pi} + \nu\frac{HH_c}{4\pi\kappa}$$

therefore

$$\frac{G_{\text{laminar}} + \frac{H^2}{8\pi}}{G_{\text{vortex}} + \frac{H^2}{8\pi}} = \text{const}\left(\frac{\kappa H_c}{H}\right)^{1/3}$$

when $H < \kappa H_c$ (which roughly corresponds to the upper critical field H_{c2}), $G_{\text{laminar}} > G_{\text{vortex}}$. Thus the vortex state is still most stable in the intermediate and high field regions.[4]

Problem. Discuss the equilibrium magnetization curves for a second kind superconductor in the form of an ellipsoid of revolution, the field being applied along the ellipsoidal axis.

Solution. The equations div $\mathbf{B} = 0$, curl $\mathbf{H} = 0$, and $\mathbf{B} = (\mathbf{H}/|\mathbf{H}|)B_e(H)$, where $B_e(H)$ is the equilibrium induction in the presence of a field H measured for a long cylinder, allow a solution where B and H are constant in the ellipsoid with $H = H_0 - NM = H_0 - N(B - H)/4\pi$ where N is the demagnetizing coefficient of the ellipsoid. The relation between B and the applied field is therefore given by the implicit formula

$$B = B_e\left(\frac{H_0 - \frac{NB}{4\pi}}{1 - \frac{N}{4\pi}}\right)$$

B is nonzero for $H_0 > H_{c1}(1 - N/4\pi)$. The slope $(dB/dH_0)_{B=0}$ is finite and equal to $4\pi/N$. The upper critical field remains equal to H_{c2} since $B(H_{c2}) = H_{c2}$ when the transition is second order.

Problem. Discuss the scattering of slow neutrons by a regular lattice of vortex lines in a superconductor.

Solution. The interaction between neutron and lines is $\mu_n h(\mathbf{r})$ where $\mu_n = 1.91\,e\hbar/2Mc$ is the neutron moment and M the neutron mass. Consider a scattering event where the neutron momentum changes from $\hbar\mathbf{k}_0$ to $\hbar(\mathbf{k}_0 + \mathbf{q})$. The corresponding scattering amplitude is given by the Born approximation formula

$$a = \frac{M}{2\pi\hbar^2}\int \mu_n h(\mathbf{r})\, e^{i\mathbf{q}\cdot\mathbf{r}}\, d\mathbf{r}$$

This is nonzero only if $\mathbf{q} = \mathbf{J}$, where $\mathbf{J}$ is a reciprocal lattice vector associated with the two-dimensional "line lattice." From (3-62) we find

---

[4]These elementary calculations of G are not sufficient in the neighborhood of H_{c2}. We return later to a study of the region using the Landau-Ginsburg equations.

$$\int h(r)\, e^{iJ\cdot r}\, dr = \frac{BV}{1 + \lambda^2 J^2} = \frac{n_L \phi_0 V}{1 + \lambda^2 J^2}$$

where V is the sample volume and n_L the number of lines per cm^2. Thus

$$a_J = \tfrac{1}{2}\, 1.91\, \frac{n_L V}{1 + \lambda^2 J^2}$$

For a triangular lattice of lines with nearest neighbor distance d, we have $n_L = (2/\sqrt{3})(1/d^2)$ and, for the first reflection, $J = (4\pi/\sqrt{3})\, d^{-1}$. Taking B = 2000 G ($n_L = 10^{10}$), we get d ~ 10^3 Å and J ~ 6.7 10^5 cm^{-1}. For λ = 1000Å this gives $(\lambda J)^2 \cong 45 >> 1$. We compute the amplitude a per atom (since this is the quantity familiar to experimentalists). Inserting for V an atomic volume of 30$Å^3$ we get a ≅ 0.7 10^{-13} cm. The corresponding "coherent scattering cross section" is $4\pi a^2 \sim 5 \times 10^{-28}$ cm^2 = 0.5 millibarns—a small, but measurable value.

The scattering angle θ for this first reflection is very small $\theta \cong (J/k_0) = (2/\sqrt{3})(\lambda_n/d)$ where we have introduced the neutron wavelength $\lambda_n = 2\pi/k_0$. At best, with subthermal neutrons we can make λ_n as large as ~5Å. For the above example this leads to angles $\theta \sim 6.10^{-3}$ rad (or 20' of arc). The experiment has been performed on Nb metal (Cribier, Jacrot, et al., 1964). Due to the $1/(1 + \lambda^2 J^2)$ dependence of a, it has been possible only to observe the first reflection (with the smallest J). (See Fig 3-9.)

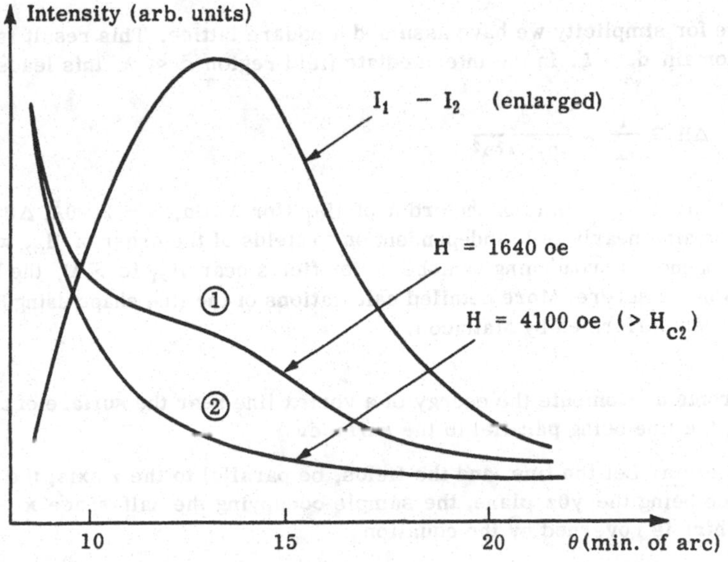

Figure 3-9
Neutron scattering by an array of vortex lines in niobium metal at T = 4.2° K. (Courtesy of D. Cribier.)

Problem. Calculate the broadening of a nuclear magnetic resonance line arising from the field inhomogeneities associated with the vortex state (P. Pincus, 1964).

Solution. In intermediate fields $H_{c1} \ll H \ll H_{c2}$ where the vortices form a dense lattice, the field distribution is given by the Fourier transform of Eq. (3-62). A knowledge of all the moments of the field distribution would completely determine the nuclear magnetic resonance line shape (if this were the only source of broadening). However if the line does not have anomalous wings, as is often the case (Jaccarino and Gossard, 1964), the second moment is a good measure of the line width (see A. Abragam, ***Principles of Nuclear Magnetism***, Clarendon Press, 1961). Thus the line width is given by

$$\Delta H = [\langle h^2 \rangle - \langle h \rangle^2]^{1/2}$$

where $\langle \rangle$ denotes a spatial average. The term $\langle h \rangle$ is just $n_L \phi_0 = B$. $\langle h^2 \rangle$ can easily be calculated from Eq. (3-62):

$$\langle h^2 \rangle = S^{-1} \int h^2(\mathbf{r})\, d\sigma = \sum_J h_J h_{-J} = n_L^2 \phi_0^2 \sum_J [1 + (\lambda J)^2]^{-2}$$

where S is the surface area of the sample perpendicular to the field. On replacing the sum over reciprocal lattice vectors by an integral in a similar manner as is done in the text after Eq. (3-63), we obtain

$$\Delta H = \frac{B}{\sqrt{4\pi}} \frac{d}{\lambda} \left[1 + \left(\frac{2\pi\lambda}{d}\right)^2\right]^{-1/2}$$

where for simplicity we have assumed a square lattice. This result is valid in the domain $d \gg \xi$. In the intermediate field region $d \ll \lambda$, this leads to

$$\Delta H \cong \frac{1}{\sqrt{2}} \times \frac{\phi_0}{(2\pi)^{3/2} \lambda^2}$$

Notice that this width is of the order of H_{c1} (for V_3Ga, $\lambda \sim 2000$Å, $\Delta H \approx 20$ Oe) and remains nearly field independent on to fields of the order of H_{c2} where the inhomogeneous broadening vanishes. For fields near H_{c1} ($d \gtrsim \lambda$), the broadening is more severe. More detailed calculations of the line shape using Eq. (3-62) have been performed by Matricon.

Problem. Compute the energy of a vortex line near the surface of the specimen, the line being parallel to the surface.

Solution. Let the line, and the fields, be parallel to the z axis, the limiting surface being the yOz plane, the sample occupying the half-space $x > 0$. The field $\mathbf{h}(\mathbf{r})$ is governed by the equation

$$\mathbf{h} + \lambda^2 \operatorname{curl} \operatorname{curl} \mathbf{h} = \phi_0 \delta_2(\mathbf{r} - \mathbf{r}_L)$$

where $\mathbf{r}_L$ represents the two-dimensional coordinate of the line [we shall take

$\mathbf{r}_L = (x_L, 0)]$ and ϕ_0 is a vector, of length ϕ_0, along z. The boundary conditions on the surface are

$$\mathbf{h} = \mathbf{H} \qquad (\text{curl } \mathbf{h})_x = 0 \qquad (0 \text{ normal current})$$

where H is the applied field. The solution $\mathbf{h}(\mathbf{r})$ will be written as

$$\mathbf{h} = \mathbf{h}_1 + \mathbf{h}_2$$

where $\mathbf{h}_1 = \mathbf{H} \exp(-x/\lambda_L)$ represents the field penetration in the absence of any line, while $\mathbf{h}_2$ is due to the line, and can be obtained by a method of images. To the line $(x_L, 0)$ we add an image of opposite sign located at $(-x_L, 0)$ and take for $\mathbf{h}_2$ the algebraic sum of the field due to the line and image. Thus $\mathbf{h}_2$ automatically vanishes on the limiting surface $x = 0$ and the boundary condition is satisfied.

Having constructed $\mathbf{h}(\mathbf{r})$, we now compute the thermodynamic potential

$$\mathcal{G} = \int d\mathbf{r} \left\{ \frac{\mathbf{h}^2 + \lambda^2 (\text{curl } \mathbf{h})^2}{8\pi} - \frac{\mathbf{H} \cdot \mathbf{h}}{4\pi} \right\}$$

The integral is taken in the sample volume $(x > 0)$ except for the core region of the line, which is excluded. The last term is the microscopic analog of the standard $\mathbf{B} \cdot \mathbf{H}/4\pi$ term for macroscopic systems. We transform $\mathcal{G}$ into a surface integral, using the London equation for $\mathbf{h}$, and we obtain

$$\mathcal{G} = \frac{\lambda^2}{4\pi} \int_{(\text{core and plane})} d\boldsymbol{\sigma} \cdot (\tfrac{1}{2}\mathbf{h} - \mathbf{H}) \times \text{curl } \mathbf{h}$$

The surface integral $\int d\sigma$ includes the surface of the hard core (giving a contribution $\mathcal{G}'$) and the surface of the specimen (giving a contribution $\mathcal{G}''$). As usual the only important term in $\int_{\text{core}} d\sigma$ (in the limit $\xi \to 0$) comes from the singular term in curl h, and the result is

$$\mathcal{G}' = \frac{\phi_0}{4\pi} (\tfrac{1}{2}h(\mathbf{r}_L) - H)$$

The second term $\mathcal{G}''$ may be written as

$$\mathcal{G}'' = -\frac{\lambda^2}{8\pi} \int_{\text{plane}} d\boldsymbol{\sigma} \cdot \mathbf{h}_1 \times \text{curl } \mathbf{h}$$

since on the sample surface $\mathbf{h} = \mathbf{h}_1 = \mathbf{H}$. Writing curl $\mathbf{h}$ = curl $\mathbf{h}_1$ + curl $\mathbf{h}_2$ we can separate in $\mathcal{G}''$ a term involving $\mathbf{h}_1 \times \text{curl } \mathbf{h}_1$, which is the energy in the absence of the line, an additive constant we drop from now on. We are left with

$$\mathcal{G}'' = -\frac{\lambda^2}{8\pi}\int_{\text{plane}} d\sigma\cdot\mathbf{h}_1 \times \text{curl}\,\mathbf{h}_2$$

We rewrite this integral as

$$\int_{\text{plane}} = \int_{\text{core+plane}} - \int_{\text{core}}$$

By making use of London's equation for $\mathbf{h}$, in the region outside the core, we have

$$\int_{\text{core+plane}} d\sigma\cdot\mathbf{h}_1 \times \text{curl}\,\mathbf{h}_2 = \int_{\text{core+plane}} d\sigma\cdot\mathbf{h}_2 \times \text{curl}\,\mathbf{h}_1$$

$\mathbf{h}_1$ is not singular near the line axis; thus the core contribution to the right-hand side vanishes when $\xi \to 0$. The integral on the plane also vanishes since $(\mathbf{h}_2)_{x=0} = 0$. Finally

$$\mathcal{G}'' = \frac{\lambda^2}{8\pi}\int_{\text{core}} d\sigma\cdot\mathbf{h}_1 \times \text{curl}\,\mathbf{h}_2 = \frac{\phi_0 h_1(\mathbf{r}_L)}{8\pi}$$

$$\mathcal{G} = \frac{\phi_0}{4\pi}\left[H \exp(-x_L/\lambda) + \tfrac{1}{2} h_2(\mathbf{r}_L) - H\right]$$

[Note incidentally that $\mathcal{G} = 0$ when $x_L = 0$, that is, when the line is just on the surface, since $h_2(x = 0) = 0$.] If we analyze $h_2(\mathbf{r}_L)$ into a direct term and an image term, the direct term gives as a contribution to $\mathcal{G}$ the line self-energy $\mathcal{E} = \phi_0 H_{c1}/4\pi$. The image term describes an attraction between line and image, of value $-(\phi_0/8\pi)\,h(2x_L)$ where $h(r)$ is the function giving the field at distance r of a single line (Eq. 3-35). Finally

$$\mathcal{G} = \frac{\phi_0}{4\pi}\left[H \exp(-x_L/\lambda) - \tfrac{1}{2} h(2x_L) + H_{c1} - H\right]$$

Discussion

(1) The term $(\phi_0 H/4\pi)\exp(-x_L/\lambda)$ describes the interaction of the line with the external field and the associated screening currents. It has the same form as Eq. (3-50). It is a repulsive term.

(2) The term $-\phi_0 h(2x_L)/8\pi$ represents the attraction between the line and its image. The magnitude of this energy differs from Eq. (3-50) by a factor $\frac{1}{2}$. But the force derived from it has the conventional magnitude $\phi_0 j/c$ [when differentiating $h(2x_L)$ with respect to x_L, we get a factor 2].

(3) The aspect of $\mathcal{G}(x_L)$ for various values of the applied field H is shown on Fig. 3-10. When $H \sim H_{c1}$ there is a strong barrier opposing the entry of a line. We can understand this barrier as follows: When $H = H_{c1}$, $\mathcal{G}(x_L = 0) = \mathcal{G}(x_L = \infty) = 0$. if we start from x_L large and bring the line closer to the surface,

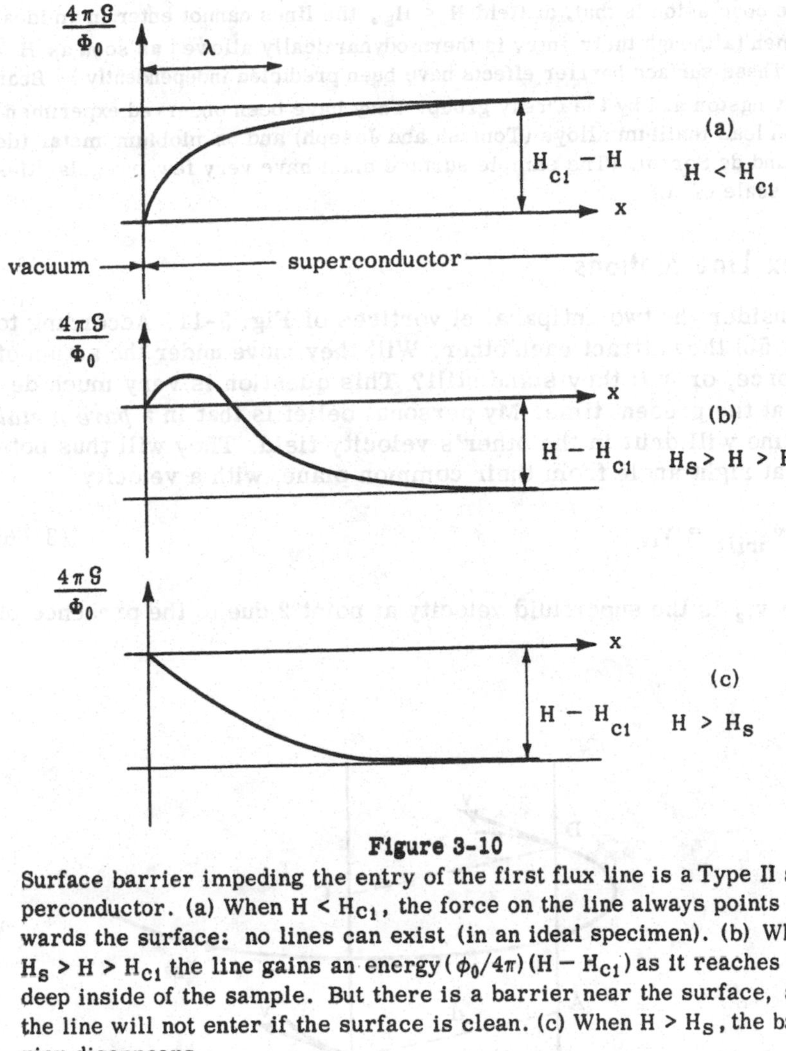

Figure 3-10

Surface barrier impeding the entry of the first flux line is a Type II superconductor. (a) When $H < H_{c1}$, the force on the line always points towards the surface: no lines can exist (in an ideal specimen). (b) When $H_s > H > H_{c1}$ the line gains an energy $(\phi_0/4\pi)(H - H_{c1})$ as it reaches the deep inside of the sample. But there is a barrier near the surface, and the line will not enter if the surface is clean. (c) When $H > H_s$, the barrier disappears.

the repulsive term ($\sim \exp(-x_L/\lambda)$) dominates the image term ($\sim \exp(-2x_L/\lambda)$). Thus $\mathcal{G}$ becomes positive and we have a barrier. The barrier disappears, however, in high fields as is clear in Fig. 3-10(c). When $H > H_s = \phi_0/4\pi\lambda\xi$, it can be seen from the equation for $\mathcal{G}$ that the slope $(\partial\mathcal{G}/\partial x_L)_{x_L=\xi}$ becomes negative.[5]

---

[5]We shall see later from the microscopic analysis that the field H_s thus defined is of the order of the thermodynamic critical field H_c.

The conclusion is that, at field $H < H_S$, the lines cannot enter in an ideal specimen (although their entry is thermodynamically allowed as soon as $H > H_{c1}$). These surface barrier effects have been predicted independently by Bean and Livingston and by the Orsay group. They have been observed experimentally on lead thallium alloys (Tomash and Joseph) and on niobium metal (de Blois and de Sorbo). (The sample surface must have very few irregularities on the scale of λ.)

Vortex Line Motions

Consider the two antiparallel vortices of Fig. 3-11. According to Eq. (3-50) they attract each other. Will they move under the action of this force, or will they stand still? This question is very much debated at the present time. My personal belief is that in a *pure metal* each line will drift in the other's velocity field. They will thus both move at right angle from their common plane, with a velocity

$$v_{drift} = v_{12} \tag{3-66}$$

where v_{12} is the superfluid velocity at point 2 due to the presence of line 1.

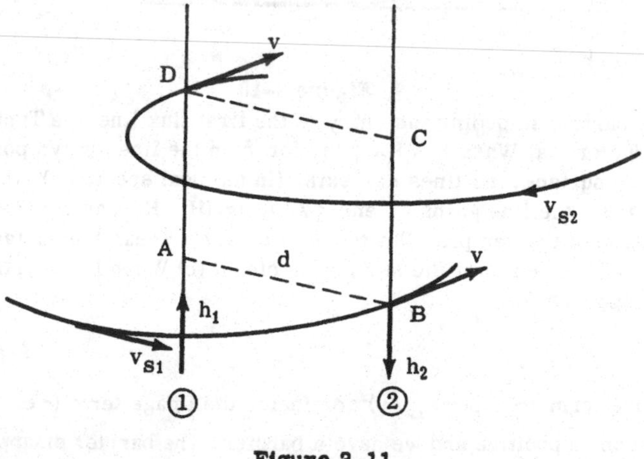

Figure 3-11

Two antiparallel vortex lines in a pure superconductor of Type II. $\mathbf{v}_{S1}$ ($\mathbf{v}_{S2}$) is the superfluid velocity induced by line 1 (2). Each line drifts with the local superfluid velocity **v**. For that particular geometry both lines go with the same velocity. Note that **v** is normal to the plane ABCD of the lines.

Such drift motions should lead to amusing collective modes for an assembly of vortex lines in a very pure metal of Type II. (P. G. de Gennes, J. Matricon, 1962.)

In *dirty* superconductors, on the other hand, friction between the lines and the lattice will dominate the motion. The two antiparallel vortex lines AD and CB will then move toward one another, as shown on Fig. 3-12, with a velocity

$$v_{\text{drift}}^{(2)} = -v_{\text{drift}}^{(1)} = \frac{f}{\eta} \tag{3-67}$$

where f is the attractive force between the lines, as given by Eq. (3-52) and η is a viscosity coefficient. We can estimate η with the following assumptions: suppose that the currents due to line 1 are not distorted near the core of line 2. This core then carries a current density $j = nev_{12}$. But this core is essentially normal. Thus we expect a loss (per unit length of line 2)

$$W = \frac{j^2}{\sigma\pi\xi^2}$$

where $\sigma = ne^2\tau/m$ is the normal state conductivity and ξ the core radius. This power dissipation must also be equal to $fv_{\text{drift}}^{(2)} = 1/\eta f^2$. Recalling from Eq. (3-52) that $f = \frac{1}{2}nhv_{12}$. we obtain

$$\eta = \frac{n\tau h^2}{4\pi m\xi^2} \tag{3-68}$$

Viscous motions which are reasonably well described by this type of damping have been observed in dirty materials by Kim and coworkers.

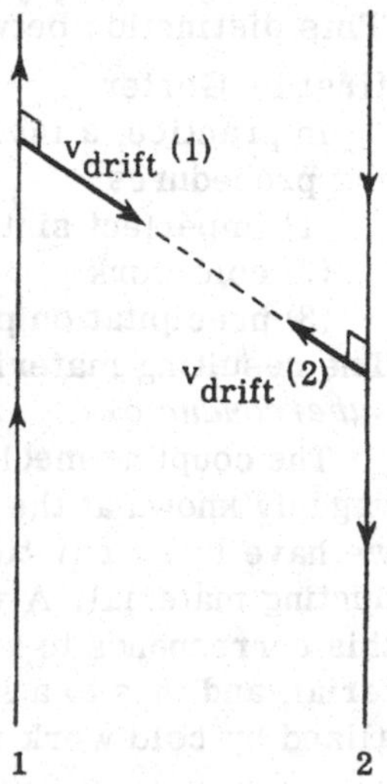

Figure 3-12

Two antiparallel vortex lines in a *dirty* superconductor of Type II: the lines move toward each other with a drift velocity controlled by friction with the lattice.

3–3 NONEQUILIBRIUM PROPERTIES

Up to now, we have restricted our attention to the *reversible* behavior of second type superconductors. We have seen that when the coherence length ξ is small, they can remain superconducting up to very high fields, of order $H_{C2} \sim \phi_0/\xi^2$.

From a technical point of view, however, what is most interesting is to obtain superconducting wires that can carry high currents. But this condition cannot be realized at thermal equilibrium, as shown by the following argument: Consider a cylindrical wire of radius a carrying a total current I. (When I is weak, this current is, in fact, entirely carried by a surface sheet, of thickness λ, around the cylinder.) The field at the surface of the wire is

$$H = \frac{2I}{ca} \tag{3-69}$$

The situation is stable when $H < H_{C1}$. When $H > H_{C1}$, vortex lines begin to appear. They are bent in circles (following the lines of force). Once created at the surface, with radius a, they tend to shrink (to decrease their line energy) and finally annihilate near the axis of the wire. This process dissipates energy. Thus in an ideal specimen we have 0 resistance only if $H < H_{C1}$ or $I < (ca/2)\, H_{C1}$. If we want to carry higher currents with our wire, we need to *pin* the vortex lines, that is, to quench their motion by suitably chosen lattice defects, and achieve a nonequilibrium situation. While the field H_{C2} is an intrinsic property of the metal (or alloy), the critical current measured on a wire is extremely sensitive to the metallurgical state of the sample. This distinction between the factors ruling H_{C2} and I was stressed first by Gorter.

In practice, a favorable defect structure is obtained by the following procedures:

(1) imperfect sintering (e.g., Nb_3Sn)
(2) cold work (e.g., MoRe alloys)
(3) precipitation processes (e.g., lead alloys)

The resulting materials, with high critical currents, are called *hard superconductors*.

The coupling mechanisms between the lines and the defects are only vaguely known at the present time. A rather simple case is met when we have large cavities, due to imperfect sintering, in the superconducting material. A vortex tends to remain pinned to the cavity, since this corresponds to a smaller length of line in the superconducting material, and thus to a smaller line energy. The mechanical stresses realized by cold work impose slight modifications to the condensation energy and to the local density of superconducting electrons n_s. This results in local modifications of λ, ξ, and thus of the line energies $\mathfrak{J}$ and interactions U. These interactions are rather complex, and in the following we only present a phenomenological description of their effects.

Critical State at Zero Temperature

Consider a hard superconductor in an applied magnetic field H (along Oz). In equilibrium the line density would have the value $B(H)/\phi_0$ and be the same at all points. We now consider a metastable situation where the induction B is not equal to B(H) but varies from point to point—say in the x direction. Thus (1) the line density is not constant, (2) there is a macroscopic current $J = (c/4\pi)(\partial B/\partial x)$ flowing in the y direction. The forces acting on the line system can be decomposed in the following way: First, because of the repulsive interactions between lines, the regions of high line density (high B) tend to expand towards the regions of low density. This may be described in terms of the pressure p in our two-dimensional line system.[6] The force (per cm^3) is $-\partial p/\partial x$. This has to be balanced by a pinning force due to the structural defects. This pinning force, however, cannot become arbitrarily large. It must stay below a certain threshold value α_m

$$\left|\frac{\partial p}{\partial x}\right| < \alpha_m \tag{3-70}$$

If at some point $|\partial p/\partial x|$ is larger than α, then the lines start moving and dissipation occurs until condition (3-70) is again satisfied. In practice the line density $(1/\phi_0)B(x)$ will thus adjust itself so that the threshold condition is just realized at all points [equality in (3-70)]. The state thus realized is called the *critical state*, and was first described by Bean. We can get some physical feeling for this critical state by thinking of a sand hill. If the slope of the sand hill exceeds some critical value, the sand starts flowing downwards (avalanche). The analogy is, in fact, rather good, since it has been shown (by careful experiment with pickup coils) that, when the system becomes overcritical, the lines do not move by single units, but rather in the form of avalanches including typically 50 lines or more.

We now proceed to compute explicitly the pressure p of the line system, to be inserted in (3-70). We consider a group of N lines intersecting a surface S in the xy plane. Their energy (per cm along Oz) is

---

[6]We make an isotropic approximation and neglect the tensor properties of p.

$$\mathcal{G} = SG$$

where G is the thermodynamic potential introduced by Eq. (3-64). The pressure is obtained from $\mathcal{G}$ by differentiating with respect to S (that is, to the volume) at constant N.

$$p = \left(-\frac{\partial \mathcal{G}}{\partial S}\right)_N = -G - S\frac{\partial G}{\partial S}$$

For fixed N we have $dS/S = -dB/B$

$$p = -G + B\frac{\partial G}{\partial B} \tag{3-71}$$

Thus when we know the form of the thermodynamic potential G, we know p(B), that is, the pressure-density relation. It is, in fact, possible to write the pressure gradient $\partial p/\partial x$ in a very simple way. Set

$$G(B) = F(B) - \frac{BH}{4\pi}$$

where F(B) describes the line self-energies and interaction. The equilibrium relation B(H) or H(B) is obtained by imposing $\partial G/\partial B = 0$. Thus $H(B) = 4\pi(\partial F/\partial B)$. Now

$$\begin{aligned} \frac{\partial p}{\partial x} &= B\frac{\partial^2 G}{\partial B^2}\frac{\partial B}{\partial x} = \frac{B}{4\pi}\frac{\partial H(B)}{\partial B}\frac{\partial B}{\partial x} \\ \frac{\partial p}{\partial x} &= \frac{B}{4\pi}\frac{\partial H(B)}{\partial x} \end{aligned} \tag{3-72}$$

Thus, if we know H(B), we can immediately compute the pressure gradient as a function of B and $\partial B/\partial x$.

In fact, in materials where $H_{c_1} \ll H_{c_2}$, if $H \gg H_{c_1}$, H(B) is nearly equal to B as is clear from the magnetization curve of Fig. 3-2. In this region we have simply

$$-\frac{\partial p}{\partial x} = -\frac{B}{4\pi}\frac{\partial B}{\partial x} = \frac{BJ_y}{c} \qquad (H \gg H_{c_1}) \tag{3-72'}$$

The critical state is then defined by

$$\left|\frac{B}{4\pi}\frac{\partial B}{\partial x}\right| = \left|\frac{BJ_y}{c}\right| = \alpha_m \qquad (H(B) \gg H_{c_1}) \tag{3-73}$$

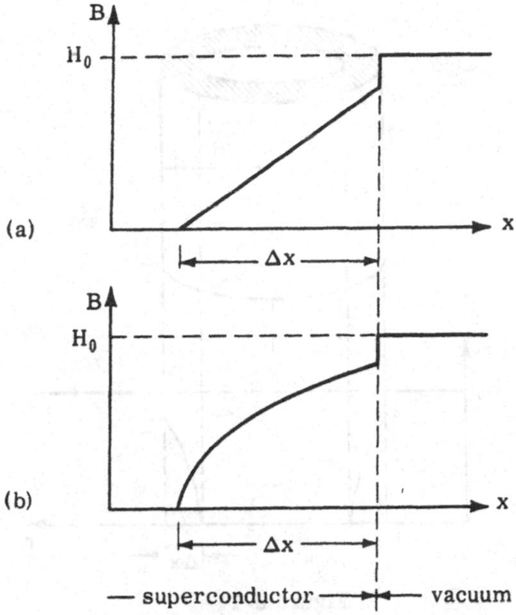

Figure 3-13
Irreversible penetration of flux in a hard superconductor: (a) Bean model; (b) Anderson-Kim model (parabolic profile).

To compute B(x) in the critical state we now need to know what is the dependence of the maximum pinning force α_m on B (that is, on the line density). It was originally assumed by Bean that α_m was linear in B; that is, J_y was a constant in the critical state. (J_y would typically be of order 10^5 A/cm².) A series of experiments by Kim and co-workers on NbZr alloys and on Nb_3Sn indicates that for these systems (in the particular metallurgical condition realized, and for fields in the 10^4 Oe range), a good approximation amounts to taking α_m *independent of B*. This very simple result is not fully explained theoretically at the present time. In particular, we would expect the pinning centers to be most efficient when their size is comparable to the interline distance, and the latter depends on B (like $B^{-1/2}$).

The shape of the profile B(x) in the critical state is represented for the Bean model and for the Kim model (Fig. 3-13). For typical hard superconductors and applied fields of order 10^4 Oe, the total thickness Δx of the zone where flux lines have penetrated is of order 1 mm. To determine B(x) experimentally [or equivalently to determine α_m(B)], the simplest method amounts to measuring the flux ϕ in cylindrical samples of radius R (Fig. 3-14) for increasing values of the applied field H.

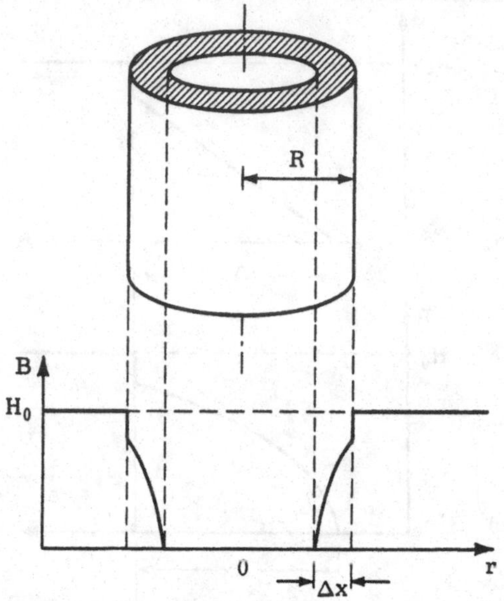

Figure 3-14
Magnetization measurements on a cylinder (external field increasing). The flux lines penetrate only in the hatched area.

(1) If R is much smaller than Δx, the induction is nearly uniform in the sample, $B = B(H)$, $\phi = \pi R^2 B(H)$.[7]

(2) If $R \gg \Delta x$ we have essentially a one-dimensional situation. If x denotes the radial distance, we may write

$$\phi = 2\pi R \int_{R-\Delta x}^{R} dx\, B(x) \tag{3-74}$$

At the edge of the flux ($x = R - \Delta x$), we have $B = 0$, $H(B) = H_{c1}$. At the surface of the cylinder ($x = R$), again assuming no surface barriers, we have the equilibrium value of B corresponding to the external field H. $B = B(H)$. Transforming dx by (3-72) and (3-70), we get

$$\phi = 2\pi R \int_{H_{c1}}^{H} dH\, \frac{B^2(H)}{4\pi\alpha_{mH}}$$

[7]We assume that there is no surface barrier impeding the entrance of vortex lines in the cylinder. Surface barriers do occur sometimes, but their effects can easily be separated.

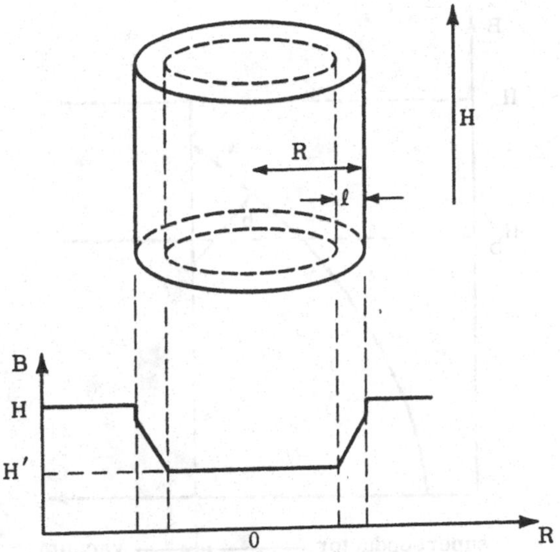

Figure 3-15

Principle of the Kim experiments on hollow cylinders of hard superconductors. An external field H is applied. The field H′ inside the cylinder is measured.

where α_{mH} stands for $\alpha_m[B(H)]$. Of particular interest is the derivative of ϕ

$$\frac{d\phi}{dH} = \frac{R}{2}\frac{B^2(H)}{\alpha_{mH}} \qquad (3\text{-}75)$$

Thus from magnetization measurements in increasing fields, we may derive α_{mH} and α_m (B). Another method, devised by Kim and coworkers, makes use of *hollow* cylinders as shown in Fig. 3-15. A field H is applied on the outside of the cylinder and the field H′ in the cylinder is measured. When H is increased from 0, H′ first stays strictly equal to 0. Then, when the flux front reaches the inner surface of the cylinder, H′ starts to increase (ideally H′ would first jump abruptly to H_{c1}, and then grow steadily). The interest of the method is to give a direct determination of Δx, for that particular value of H where H′ starts to increase.

More complicated situations are met if the field H is alternatively increased and decreased, as shown in Fig. 3-16. Then we meet regions

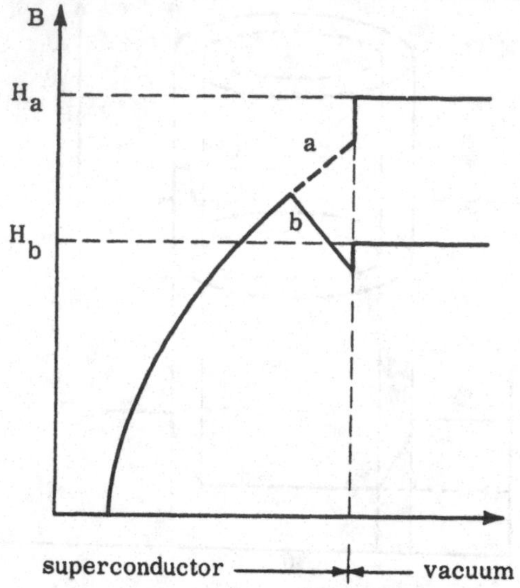

Figure 3-16
Flux distribution in a hard superconductor when the applied field is first raised to H_a (broken curve) and decreases to H_b (full curve).

with $dB/dx > 0$ and regions with $dB/dx < 0$, but the absolute value $|\partial p/\partial x|$ stays equal to α_m. This permits a detailed calculation of all hysteresis cycles when $\alpha_m(B)$ is known.

Flux Creep at Finite Temperatures

At finite temperatures, if $\partial p/\partial x \neq 0$, the vortex lines will tend to move (from the regions of high B towards the regions of low B) by activated jumps across the pinning barriers. We call the average flow velocity of the lines (in the x direction) v_x. Various methods can be used to detect this flow, or "creep":

(1) Magnetic measurements, with thick cylinders or hollow cylinders. In the latter case, for instance, if H has been raised from 0 to some value and then kept constant, we observe that H′ increases slowly in time.

(2) Electrical measurements. If the lines of force are moving, they create electromotive forces that can be measured directly. The most simple situation is represented in Fig. 3-17. A wire (in the y direction) carries a current of density J, and is submitted to an external

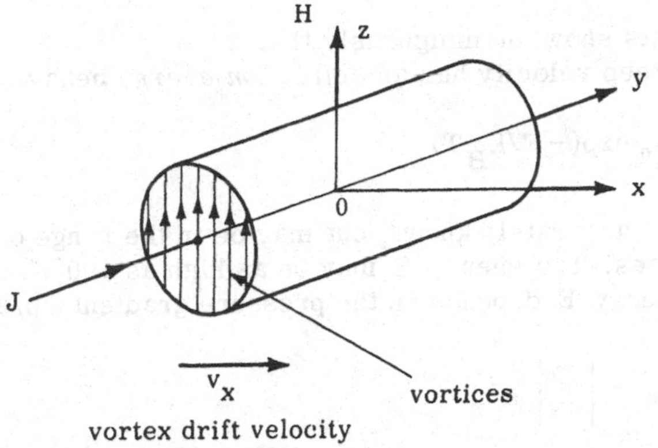

Figure 3-17
Electrical measurements on a hard superconductor wire. There is a current $J = -(c/4\pi)(\partial B/\partial x)$ in the y direction, thus $\partial B/\partial x < 0$. The lines are more closely packed on the left side of the wire: they drift with a velocity v_x towards the positive x axis.

field H in the z direction. Thus we have a nonzero $\partial B/\partial x = 4\pi J/c$ and the lines (pointing along Oz) tend to drift in the x direction. The resulting electric field E_y is along the wire axis (Oy) and in the limit $H \gg H_{c1}$, it is given by

$$E_y = \frac{Bv_x}{c} \tag{3-76}$$

To prove (3-76) we compute the power dissipation per unit volume; this is the work done by the presence of gradient on the lines, that is $(\partial p/\partial x)v_x$. By setting this equal to $E_y J$ and making use of (3-72′), we get (3-76). Thus we need an electric field E_y to maintain the current J. As pointed out by Anderson and Kim, this dissipative effect in the superconducting state explains many features of the resistive behavior of hard superconductors.

The main difference between (1) the magnetic measurements and (2) the electric ones lies in the order of magnitude of the velocities involved. In case (1) the creep typically is measured over intervals of hours or days, and the velocities are of order 1 mm/day or 10^{-6} cm/sec. In case (2) taking $B = 10^4$ $E_y = 1\ \mu V/cm$, we get $v_x \sim 10^{-2}$ cm/sec. The main difficulty of (2) is related to possible inhomogeneities in the wire; experimentally it is found that different portions of the same wire have different E_y's.

The results show unambiguously that

(a) the creep velocity has an *activation energy* behavior

$$v_x = v_0 \exp(-E/k_B T) \tag{3-77}$$

v_0 is not very accurately known, but may be in the range of 10^3 cm/sec in typical cases. The energy E may be as high as 100°K.

(b) the energy E depends on the pressure gradient $\partial p/\partial x$

$$E = E_0 - \left|\frac{\partial p}{\partial x}\right| \rho^4 \tag{3-78}$$

where ρ has the dimensions of a length, and is typically of order 500Å. We can relate E_0 and ρ to the critical pressure gradient α_m, if we notice that for $T \to 0$ the velocity v_x will depart from 0 only when $E = 0$

$$\alpha_m = \frac{E_0}{\rho^4} \tag{3-79}$$

These results (mainly obtained by Anderson, Kim, and co-workers) have various important consequences. First, since v_x varies rapidly with E according to (3-77), it is possible to extend the crictical state concept to finite temperatures. Define a limiting velocity v_{min} below which the line motion cannot be detected. Then, if

$$\frac{E}{k_B T} > \log \frac{v_0}{v_{min}}$$

the line structure is frozen. Thus the critical state at temperature T corresponds to

$$E_0 - \rho^4 \left|\frac{dp}{dx}\right| = k_B T \log \frac{v_0}{v_{min}}$$

$$\left|\frac{dp}{dx}\right| = \alpha_m \left(1 - \frac{k_B T}{E_0} \log \frac{v_0}{v_{min}}\right) \tag{3-80}$$

In general α_m will depend on T (since α_m involves λ, ξ, and the condensation energy, which are all temperature dependent). But if T is much smaller than the transition point T_0 this dependence may be neglected and all the temperature variation in (3-80) comes from the factor $k_B T$. A linear dependence of $|dp/dx|$ on T has indeed been observed experimentally by Kim on various alloys and compounds — the critical currents of hard superconductors are strongly temperature dependent even when $T \ll T_0$.

Another important consequence of (3-77), pointed out by Anderson, is the possibility of severe thermal instabilities. If in a small region of the sample the pinning energy E_0 is slightly smaller than elsewhere, the vortex lines in this region will dissipate a large power, $[(\partial p/\partial x)\, v_x]$ per cm^3, and this will tend to raise the local temperature, if the thermal conductivity of the material is low. This temperature rise will, in turn, increase the line velocity v_x according to (3-77) and may finally result in an instability. These thermal processes have to be taken into account in the design of superconducting coils.

CONCLUDING REMARKS

Our description of pinning and creep has been strictly phenomenological. Of course, we would like to interpret E_0, ρ, and v_0 in terms of microscopic processes. There are two difficulties:

(1) What is the coupling between defects and lines? As pointed out earlier an important term is related to local modifications of the penetration depth λ by strain or by impurity gradients. Another, more obvious, contribution stems from local modifications of the superconducting condensation energy. There may be other contributions.

(2) How can we describe the metastable equilibrium and the irreversible motions of strongly coupled vortex lines in the presence of random perturbations? Coming back to the sand-hill analogy, we need a theory for the equilibrium slope of the sand hill and a theory for the avalanches — both are complicated. For the vortex line system there is an interesting suggestion, due to Frank — creep may take place by a motion of dislocations in the two-dimensional line lattice.

REFERENCES

Vortex line structure:

A. Abrikosov, *Zh. Eksperim. i Teor. Fix.*, **32**, 1442 (1957), translated by *Soviet Phys.—JETP*, **5**, 1174 (1957).

B. B. Goodman, *Rev. Mod. Phys.*, **36**, 12 (1964).

Reversible magnetization curves:

J. D. Livingston, *Phys. Rev.*, **129**, 1943 (1963).

T. Kinsel, E. A. Lynton, and B. Serin, *Phys. Letters*, **3**, 30 (1962).

Neutron diffraction by lines:

D. Cribier, B. Jacrot, L. M. Rao, and B. Farnoux, *Phys. Letters*, **9**, 106 (1964).

Pinning of vortex lines and creep effects:

C. P. Bean, *Phys. Rev. Letters*, **8** (1962), *Rev. Mod. Phys.*, **36**, 31 (1964).

P. W. Anderson, *Phys. Rev. Letters*, **9**, 309 (1962).

Y. B. Kim, C. F. Hempstead, and A. R. Strnad, *Phys. Rev.*, **129**, 528 (1963).

J. Friedel, P. G. de Gennes, and J. Matricon, *Appl. Phys. Letters*, **2**, 199 (1963).

Viscous motion of vortices:

Y. B. Kim, C. F. Hempstead, A. R. Strnad, *Phys. Rev.*, **139**, A1163 (1965).

4

DESCRIPTION OF THE CONDENSED STATE

4–1 INSTABILITY OF THE NORMAL STATE IN THE PRESENCE OF AN ATTRACTIVE INTERACTION

The ground state of a free electron gas corresponds to complete filling of the one-electron energy levels of wavevector $\mathbf{k}$ and energy $\hbar^2 k^2/2m$ up to a certain energy $E_F = \hbar^2 k_F^2/2m$ (the Fermi energy). However, in the presence of an *attractive* interaction, no matter how weak, this state becomes unstable (Cooper, 1957). The instability can be understood by considering two particular electrons of coordinates $\mathbf{r}_1$ and $\mathbf{r}_2$, the other electrons still being treated as a free electron gas. The only effect of this electron gas is to forbid the two electrons to occupy all states $k < k_F$ by the exclusion principle. Let $\psi(\mathbf{r}_1, \mathbf{r}_2)$ be the wavefunction of the two electrons. Consider only states where the center of gravity of the pair $(\mathbf{r}_1, \mathbf{r}_2)$ is at rest; ψ is then only a function of $\mathbf{r}_1 - \mathbf{r}_2$. Expand ψ in plane waves

$$\psi(\mathbf{r}_1 - \mathbf{r}_2) = \sum_{\mathbf{k}} g(\mathbf{k})\, e^{i\mathbf{k}\cdot(\mathbf{r}_1 - \mathbf{r}_2)} \tag{4-1}$$

$g(\mathbf{k})$ is the probability amplitude for finding one electron in the plane-wave state of momentum $\hbar\mathbf{k}$ and the other electron in the state $(-\hbar\mathbf{k})$. Since the states $k < k_F$ are already occupied, the Pauli principle imposes

$$g(\mathbf{k}) = 0 \quad \text{for} \quad k < k_F \tag{4-2}$$

The Schrödinger equation for our two electrons is

$$-\frac{\hbar^2}{2m}(\nabla_1^2 + \nabla_2^2)\,\psi(\mathbf{r}_1, \mathbf{r}_2) + V(\mathbf{r}_1, \mathbf{r}_2)\psi = \left(E + \frac{\hbar^2 k_F^2}{m}\right)\psi \tag{4-3}$$

(E is the energy of the pair relative to the state where the two electrons are at the Fermi level.) On inserting (4-1) into (4-3), we find an equation for $g(\mathbf{k})$,

$$\frac{\hbar^2}{m}k^2 g(\mathbf{k}) + \sum_{k'} g(\mathbf{k}')V_{kk'} = (E + 2E_F)g(\mathbf{k}) \tag{4-4}$$

$$V_{kk'} = \frac{1}{L^3}\int V(\mathbf{r})\, e^{-i(\mathbf{k}-\mathbf{k}')\cdot\mathbf{r}}\, d\mathbf{r}$$

is the matrix element of the interaction between the electronic states $\mathbf{k}$ and $\mathbf{k}'$. L^3 is the volume of the system. Equation (4-4) together with the Pauli condition (4-2) is sometimes called the Bethe-Goldstone equation for the two-electron problem. For $E > 0$ it has a continuous spectrum describing the collisions of the two electrons from the initial state ($\mathbf{k}, -\mathbf{k}$) towards final states ($\mathbf{k}', -\mathbf{k}'$) of the same energy. But if the interaction V is attractive there may also be *bound state solutions*, with $E < 2E_F$. To see this, consider the simplified interaction

$$V_{kk'} = -\frac{V}{L^3} \quad \text{for} \quad \begin{cases} \dfrac{\hbar^2 k^2}{2m} < E_F + \hbar\omega_D \\ \dfrac{\hbar^2 k'^2}{2m} < E_F + \hbar\omega_D \end{cases} \tag{4-5}$$

$$= 0 \quad \text{otherwise}$$

(This interaction is attractive and constant in an energy band $\hbar\omega_D$ above the Fermi level.) Then (4-4) becomes

$$\left(-\frac{\hbar^2 k^2}{m} + E + 2E_F\right)g(\mathbf{k}) = C \tag{4-6}$$

where C is independent of k

$$C = -\frac{V}{L^3}\sum_{k'} g(\mathbf{k}')$$

$$E_F < \frac{\hbar^2 k'^2}{2m} < E_F + \hbar\omega_D \tag{4-7}$$

Comparing (4-6) and (4-7), we find the self-consistency condition

$$1 = \frac{V}{L^3}\sum_{k'} \frac{1}{-E + \dfrac{\hbar^2 k'^2}{m} - 2E_F}$$

$$E_F < \frac{\hbar^2 k'^2}{2m} < E_F + \hbar\omega_D \tag{4-8}$$

If we set

$$\xi' = \frac{\hbar^2 k'^2}{2m} - E_F \tag{4-9}$$

and introduce the density of states per unit energy interval

$$N(\xi') = (2\pi)^{-3} 4\pi k'^2 \frac{dk'}{d\xi'}$$

the condition becomes

$$1 = V \int_0^{\hbar\omega_D} N(\xi') \frac{1}{2\xi' - E} d\xi' \tag{4-10}$$

If we assume $\hbar\omega_D \ll E_F$, $N(\xi')$ can be considered as constant and replaced by its value $N(0)$ at the Fermi level; we can perform the integration

$$1 = \tfrac{1}{2} N(0) V \ln \frac{E - 2\hbar\omega_D}{E} \tag{4-11}$$

and in the limit of weak interactions $N(0)V \ll 1$

$$E = -2\hbar\omega_D \, e^{-2/N(0)V} \tag{4-12}$$

Therefore there exists a two-electron bound state of energy $E < 0$. If we start from a free electron gas, and turn on the interaction V, we predict that the electrons will group themselves in pairs giving up energy to the external world. The normal state is thus unstable.

A few important remarks.

(1) The instability persists even if V is very weak provided that it is attractive. This is a very remarkable difference from the usual two-body problem. If we had only two particles coupled by an attractive interaction of finite range, they would not form bound states unless the attractive interaction exceeded a certain threshold.

For the two-body problem ($k_F = 0$), the density of states $N(\xi)$ varies as $\xi^{1/2}$ as $\xi \to 0$ and the expression

$$f(E) = V \int N(\xi) \frac{d\xi}{-E + 2\xi} \tag{4-13}$$

converges even for $E = 0$. Therefore, if V is small, $f(0) < 1$ and $f(E) < f(0) < 1$ for $E < 0$. The condition $f(E) = 1$ cannot be satisfied in the domain of bound states. On the contrary, for our problem, $N(\xi)$ is nearly constant, $f(0) = \infty$, and there is always a value of $E < 0$ such that $f(E) = 1$.

(2) The binding energy is proportional to $e^{-2/NV}$. It cannot be expanded in powers of V for $V \to 0$. This algebraic difficulty considerably hindered the development of the theory of superconductivity.

(3) In the preceding calculation, we have taken account of the exclusion principle between the electron (1) and the electrons in the Fermi sphere and between (2) and the Fermi sphere. It is also necessary to insure that the wavefunction be antisymmetric in $(\mathbf{r}_1, \mathbf{r}_2)$. With the simplified interaction V, of (4-5), $g(\mathbf{k})$ depends only on ξ, that is, the magnitude of $\mathbf{k}$. Therefore the spatial part of the wave function $\psi(\mathbf{r}_1, \mathbf{r}_2)$ is symmetric from (4-1), and the spin part of the wave function must be antisymmetric, of the form $(1/\sqrt{2})(\alpha_1\beta_2 - \beta_1\alpha_2)$ in Schiff's notation.[1]

This function describes a state where the total spin of the pair is zero, that is, antiparallel coupling. If the interaction $V_{kk'}$, instead of having the simple form (4-5), depended strongly on the angle between k and k′, one could find:

(a) several bound states;

(b) spatially anisotropic solutions and therefore more complicated spin dependences. In practice, in the superconducting metals, it appears that the angular dependence of $V_{kk'}$ is not strong enough to bring in very spectacular effects.

Problem. Calculate the mean square radius ρ of a Cooper pair.

Solution. By definition

$$\rho^2 = \frac{\int |\psi(\mathbf{r}_1 - \mathbf{r}_2)|^2 R^2 \, d\mathbf{R}}{\int |\psi|^2 \, d\mathbf{R}}, \qquad \mathbf{R} = \mathbf{r}_1 - \mathbf{r}_2$$

$$\rho^2 = \frac{\Sigma |\nabla_k g(\mathbf{k})|^2}{k \Sigma_k |g(\mathbf{k})|^2}$$

$$\cong \frac{N(0) \left(\dfrac{\partial \xi}{\partial k}\right)^2_{\xi=0} \displaystyle\int_0^\infty d\xi \left(\frac{\partial g}{\partial \xi}\right)^2}{N(0) \displaystyle\int_0^\infty g^2 \, d\xi}$$

In the approximation of (4-5), $g = C/(E - 2\xi)$. Also, $(\partial k/\partial \xi)_{\xi=0} = 1/\hbar v_F$ where v_F is the Fermi velocity. Performing the integration, one finds

$$\rho = \frac{2}{\sqrt{3}} \frac{\hbar v_F}{E}$$

4–2 ORIGIN OF THE ATTRACTIVE INTERACTION

In a simple electron gas, the only interactions are Coulomb repulsions, and thus not favorable for the Cooper phenomenon. To obtain an attractive matrix element $V_{kk'}$, the electrons must be coupled to

---

[1] L. I. Schiff, *Quantum Mechanics* (New York: McGraw-Hill, 1955), Chapter 9.

another system of particles, or excitations, in the solid. There are many varieties of such excitations we might try for that purpose—phonons, electrons from other bands, spin waves in magnetic media, and so on. Only one of these couplings is really known to be important (at the present time). This is the electron-phonon interaction, first proposed in that connection by Frölich (1950), which we now discuss.

The Quantum Mechanical Picture

We wish to know the matrix element $V_{kk'}$ of the electron-electron interaction between an initial state (I) where two electrons are in the plane wave states **k** and −**k**, and a final state (II) where the electrons are in (**k**′, −**k**′). $V_{kk'}$ will, in general, contain two terms:

(a) There is a direct Coulomb repulsion $U_c(\mathbf{r}_1 - \mathbf{r}_2)$ between the two electrons. The corresponding matrix element is

$$\langle I | \mathcal{H}_c | II \rangle = \int d\mathbf{r}_1 d\mathbf{r}_2 \, e^{-i\mathbf{k}\cdot(\mathbf{r}_1 - \mathbf{r}_2)} U_c(\mathbf{r}_1 - \mathbf{r}_2) \, e^{i\mathbf{k}'\cdot(\mathbf{r}_1 - \mathbf{r}_2)} \tag{4-14}$$

(the wave functions being normalized in a unit volume)

$$\langle I | \mathcal{H}_c | II \rangle = \int U_c(\rho) \, d\rho \, e^{i\mathbf{q}\cdot\rho} = U_q \qquad \mathbf{q} = \mathbf{k}' - \mathbf{k} \tag{4-15}$$

(b) One electron may emit a phonon later reabsorbed by the other electron. This process is displayed in Fig. 4-1. The initial state (I) has energy

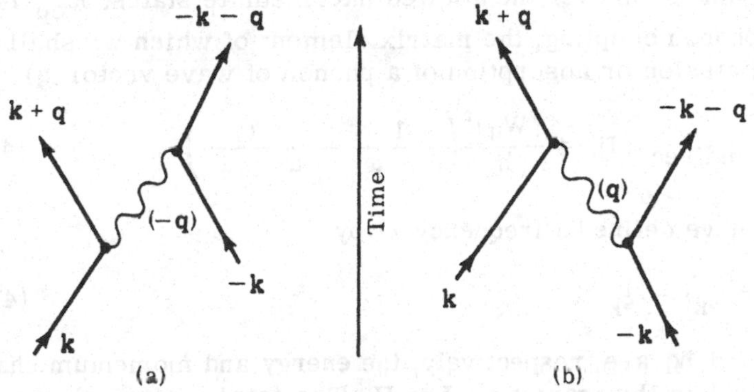

Figure 4-1

Electron-electron interaction via phonons. In process (a) the electron **k** emits a phonon of wave-vector −**q**. The phonon is absorbed later by the second electron. In process (b) the second electron in state (−**k**) emits a phonon **q**, subsequently absorbed by the first electron.

$$E_I = 2\xi_k$$

where ξ_k is defined by Eq. (4-9) (for convenience we always measure the one-electron energies from the Fermi level). The final state (II) has energy

$$E_{II} = 2\xi_{k'}$$

There are two intermediate states allowed by momentum conservation:

(i1): Electron 1 in state $\mathbf{k'} = \mathbf{k} + \mathbf{q}$, electron 2 in $-\mathbf{k}$, one phonon created with wave vector $-\mathbf{q}$, energy $\hbar\omega_q$

$$E_{i1} = \xi_{k'} + \xi_k + \hbar\omega_q$$

(notice that ξ_k and ω_k are even functions of k)

(i2): electron 1 in state $\mathbf{k}$
2 in state $-\mathbf{k'} = -(\mathbf{k} + \mathbf{q})$
one phonon created with wave vector $\mathbf{q}$, energy $\hbar\omega_q$

$$E_{i2} = \xi_{k'} + \xi_k + \hbar\omega_q = E_{i1}$$

The second-order matrix element coupling states (I) and (II) is

$$\langle I|\mathcal{H}_{\text{indirect}}|II\rangle = \sum_i \langle I|\mathcal{H}_{ep}|i\rangle \frac{1}{2}\left(\frac{1}{E_{II} - E_i} + \frac{1}{E_I - E_i}\right)\langle i|\mathcal{H}_{ep}|II\rangle \tag{4-16}$$

Here the sum Σ_i is over the allowed intermediate states. $\mathcal{H}_{ep}$ is the electron phonon coupling, the matrix element of which we shall call W_q (for emission or absorption of a phonon of wave vector $\mathbf{q}$)

$$\langle I|\mathcal{H}_{\text{indirect}}|II\rangle = \frac{|W_q|^2}{\hbar}\left(\frac{1}{\omega - \omega_q} - \frac{1}{\omega + \omega_q}\right) \tag{4-17}$$

where we have defined a frequency ω by

$$\hbar\omega = \xi_{k'} - \xi_k \tag{4-18}$$

Thus $\hbar\omega$ and $\hbar\mathbf{q}$ are, respectively, the energy and momentum change of electron 1 in the transition I → II. The total matrix element is

$$\langle I|\mathcal{H}|II\rangle = U_q + \frac{2|W_q|^2}{\hbar}\frac{\omega_q}{\omega^2 - \omega_q^2} \tag{4-19}$$

When $\omega < \omega_q$ the indirect term is negative (attractive); we have found an attractive interaction, provided that U_q is not too large.

Description in Terms of a Dielectric Constant

It is also possible to present the effective electron-electron interaction in the following way. The bare Coulomb interaction between the two electrons is e^2/r, leading to a repulsive matrix element $4\pi e^2/q^2$. But this interaction is screened—by the other electrons and also by the positive ions in the lattice. If we work near the resonance frequency (ω_q) of the ion motion, the ions give a very large response: They can "overscreen" the negative charge of electron 1; the resultant, positive cloud will be attracted by electron **2**.

In practice, we describe this screening effect by a dielectric constant $\epsilon(\mathbf{q}\omega)$ (which is a function of wave vector and frequency). The effective interaction is the bare interaction divided by ϵ

$$\langle \mathrm{I} | \mathcal{H} | \mathrm{II} \rangle = \frac{4\pi e^2}{q^2 \epsilon(q, \omega)} \tag{4-20}$$

It only remains to calculate ϵ. We do this for a very simple model, and show that Eqs. (4-19) and (4-20) agree.

JELLIUM

Consider a system formed of n electrons per cm^3 (of mass m and charge $-e$) and of ions, of mass M and charge $+Ze$. (The entire system being electrically neutral, the number of ions per cm^3 is n/Z.) We take into account only the electrostatic interactions between particles. Also we treat the ions as fluid, not as an ordered solid. This simple model of a metal has been called "jellium."[2] It neglects mainly

(1) the short-range repulsive interactions between ions due to repulsion of their deep electronic levels;

(2) the fact that the electronic wave functions must be orthogonal to the wave function of the deep-lying levels of each ion. (This gives rise to an effective repulsive electron-ion interaction.)

(3) the possibility of transverse oscillations (transverse phonons) of the ion system, which exist in a real solid, but not in our fluid model.

All these simplifications are rather arbitrary, but the simple model that results is useful for a qualitative understanding of the interactions. The dielectric constant $\epsilon(\mathbf{q}\omega)$ is defined in the following way: Assume that small external charges, of density

$$\delta\rho(\mathbf{r}) = \delta\rho \cos(\mathbf{q}\cdot\mathbf{r})\, e^{i\omega t} + \mathrm{C.C.} \tag{4-21}$$

---

[2]The jellium model was first applied to superconductivity by Pines [*Phys. Rev.*, **109**, 280 (1958)]; the phonon-induced part of that interaction had been earlier derived by Bardeen and Pines [*Phys. Rev.*, **99**, 1140 (1955)]; a description of the electron-ion system in terms of the overall dielectric constant was first worked out by Nozières (unpublished) and generalized slightly by Pines in his 1958 Les Houches lectures.

are added to the system. The latter will build up screening charges, of distribution

$$\rho(\mathbf{r}) = \rho \cos(\mathbf{q}\cdot\mathbf{r})\, e^{i\omega t} + \text{C.C.} \tag{4-22}$$

The dielectric constant is the ratio of the external charge to the total charge

$$\epsilon(\mathbf{q}\omega) = \frac{\delta\rho}{\rho + \delta\rho} \tag{4-23}$$

We must calculate the charge response ρ. In our case

$$\rho = \rho_e + \rho_i$$

is a sum of an electronic component and an ionic component. Our starting equations are

(a) Poisson's equation relating the electrostatic potential V to the charge densities $\delta\rho$, ρ_i, ρ_e:

$$\nabla^2 V = -4\pi(\delta\rho + \rho_i + \rho_e) \tag{4-24}$$

(b) the equation of motion of the ions in the presence of the electric field $\mathbf{E} = -\nabla V$:

$$M\frac{d\mathbf{j}_i}{dt} = nZe^2\,\mathbf{E} \tag{4-25}$$

where $\mathbf{j}_i$ is the current carried by the ions, related to ρ_i by the continuity equation

$$\frac{\partial\rho_i}{\partial t} + \operatorname{div}\mathbf{j}_i = 0 \tag{4-26}$$

In the small motion approximation, the total derivative $d\mathbf{j}_i/dt$ in Eq. (4-25) can be replaced by the partial derivative $\partial\mathbf{j}_i/\partial t$. Eliminating $\mathbf{j}_i$ between (4-25) and (4-26), we obtain

$$\frac{\partial^2\rho_i}{\partial t^2} = \frac{nZe^2}{M}\nabla^2 V \tag{4-27}$$

Writing $\partial/\partial t = i\omega$ and making use of (4-24), we arrive at

$$\omega^2\rho_i(\mathbf{q},\omega) = \omega_i^2[\rho_i(\mathbf{q},\omega) + \rho_e(\mathbf{q},\omega) + \delta\rho(\mathbf{q},\omega)] \tag{4-28}$$

with

$$\omega_i = \sqrt{\frac{4\pi n Z e^2}{M}} \tag{4-29}$$

As we shall see later, ω_i gives essentially the width of the phonon spectrum for our simple model. Typically $\omega_i \sim 10^{13}\ \mathrm{sec}^{-1}$.

(c) We need an equation describing the response of the electron gas (that is, ρ_e) as a function of the electrostatic potential V. There is one simplifying feature in this calculation—the two characteristic frequencies of the electron gas are $E_F/\hbar$ and the plasma frequency

$$\omega_p = \left(\frac{4\pi n e^2}{m}\right)^{1/2} \sim \omega_i \left(\frac{M}{m}\right)^{1/2}$$

Both these frequencies are very high. On the other hand ω is low ($\omega \sim \omega_i$). Thus, in this part of the calculation, we may neglect ω completely and compute ρ_e simply for a *static* perturbation V.

We make a further approximation and derive ρ_e by the method of Thomas and Fermi (this is strictly correct only at long wavelengths $q \ll k_F$). In this approximation the electronic density at $\mathbf{r}$ is proportional to $[E_F + eV(\mathbf{r})]^{3/2}$. Its relative variation for small V is therefore given by

$$\frac{\rho_e}{-ne} = \frac{3}{2}\frac{eV}{E_F} \tag{4-30}$$

For a Fourier component of wave vector $\mathbf{q}$ this may be rewritten, according to Eq. (4-24), in the form

$$\rho_e = \frac{-k_s^2}{q^2}(\rho + \delta\rho) \tag{4-31}$$

where

$$k_s^2 = \frac{6\pi n e^2}{E_F}$$

Collecting (4-28) and (4-31), we have

$$\rho = \rho_i + \rho_e = \left(\frac{\omega_i^2}{\omega^2} - \frac{k_s^2}{q^2}\right)(\rho + \delta\rho)$$

and (4-32)

$$\epsilon(\mathbf{q}\omega) = \frac{\omega^2(k_s^2 + q^2) - \omega_i^2 q^2}{\omega^2 q^2}$$

Let us first determine from this expression the phonon frequency ω_q

for wave vector $\mathbf{q}$. The spontaneous modes of vibration of our system correspond to 0 external charge $\delta\rho = 0$ or $\epsilon(q\omega) = 0$. For each q this defines one frequency ω_q given by

$$\omega_q^2 = \omega_i^2 \frac{q^2}{k_s^2 + q^2} \tag{4-33}$$

For long wavelengths, $\omega_q \sim (\omega_i/k_s)q$ is a linear function of q and the sound velocity ω_i/k_s. This formula gives, in fact, a rather realistic value for the velocity of sound for all the nontransition metals if we choose Z as the valence of the metal. In terms of ω_q we may rewrite Eq. (4-32) in the form

$$\frac{1}{\epsilon(q\omega)} = \frac{q^2}{k_s^2 + q^2}\left[1 + \frac{\omega_q^2}{\omega_i^2 - \omega_q^2}\right] \tag{4-34}$$

and the interaction matrix element is

$$\langle \mathrm{II} \,|\, \mathcal{H} \,|\, \mathrm{I} \rangle = \frac{4\pi e^2}{q^2 \epsilon(q\omega)} = \frac{4\pi e^2}{k_s^2 + q^2} + \frac{4\pi e^2}{k_s^2 + q^2} \times \frac{\omega_q^2}{\omega^2 - \omega_q^2} \tag{4-35}$$

This has exactly the form predicted by Eq. (4-19), but with precise numerical coefficients. The first term is a (screened) Coulomb repulsion. The second term is the phonon-mediated interaction and it is attractive for $\omega < \omega_q$.

For an average phonon: $\omega_q \sim \omega_D$, the Debye frequency of the material. Thus the energy interval in which we find an attraction is of order $\hbar\omega_D = k_B\Theta_D$ where k_B is Boltzmann's constant, and Θ_D (the Debye temperature) is typically of order 300°K ($k_B\Theta_D \sim 0.03$ eV). Note that ω_q (or ω_D) depends on the mass M of the ions. Equations (4-29) and (4-33) show that $\omega_q \sim M^{-1/2}$. Thus two isotopes of the same superconductor have different attractive bandwidths, and their transition temperatures differ slightly—such *isotope effects* were first predicted by Frölich and observed soon after (Maxwell, Reynolds, et al., 1950). They prove that the vibrational motions of the lattice give an important contribution to the electron-electron interactions.

Numerical Values of the Interaction Constants

At the present time we *cannot* compute the matrix elements $V_{kk'}$ in a very reliable way—we know too little about the relevant band structures, electron-phonon couplings, and so on. The calculation is particularly difficult for the following reason: At low frequencies ω, there is a rather strong compensation between the Coulomb repulsion and the indirect interaction. For "jellium," in particular, we notice that $\epsilon(\mathbf{q}\omega) \to \infty$ when $\omega \to 0$. In this model a static perturbation $\delta\rho$ is completely screened. Nothing prevents the positive ions from piling up with an arbitrary density in order to neutralize the charges $\delta\rho$.[3] If the short range repulsion between ions is taken into account, this effect disappears, the screening is not complete, and $\epsilon(\mathbf{q}, 0)$ is then finite. This remark shows that the width of the attractive band and the attractive force are rather sensitive to the details of the physical model. It is important to make exactly the same approximations in the calculations of the direct and indirect interactions to avoid gross errors in the sum.

Problem. Estimate $N(0)V_p$ for simple metals, using the "jellium" approximation.

Solution. We define V_p as being the attractive part of interaction for $\omega = 0$. Then

$$V_p = \frac{4\pi e^2}{k_s^2 + q^2}$$

For q we take the root mean square average $q^2 = \frac{3}{5} q_D^2$ where q_D (the Debye cutoff) is defined by

$$v_0 (2\pi)^{-3} \frac{4\pi}{3} q_D^3 = 1$$

v_0 being the atomic volume. Finally

$$N(0) = \frac{mk_F}{2\pi^2 \hbar^2} \tag{4-36}$$

$$N(0)V_p = \frac{2}{\pi} \frac{me^2}{\hbar^2} \frac{k_F}{q^2 + k_s^2} \tag{4-37}$$

k_s^2 is defined below Eq. (4-31) as a function of the number n of electrons per cm^3

$$n = \frac{k_F^3}{3\pi^2} = \frac{Z}{v_0} \tag{4-38}$$

---

[3]One physical case where this extraordinary screening effect can play an important role seems to be that of protons dissolved in palladium. These protons are very mobile. If a metallic impurity is added to the hydrogenated palladium, the screening charge around the impurity may be made largely of protons and not of electrons.

Finally, we recall that $\hbar^2/me^2 = a_0 = 0.529\text{Å}$ is the Bohr radius. Collecting all these remarks, we find[4]

$$N(0)V_p = \frac{1}{2 + 4.7\, a_0 |v_0 Z|^{-1/3}} \tag{4-39}$$

Problem. Study the sign of the interaction between electrons at low frequencies in a ***magnetic*** medium.

Solution. Suppose that the interaction between an electron situated at the point $\mathbf{r}$, of spin S_i, and the local magnetization $\mathbf{M}(\mathbf{r}')$ of the medium is of the scalar form

$$H_1 = -\Gamma \mathbf{S}_i \cdot \mathbf{M}(\mathbf{r})$$

Since we are only interested in the zero frequency limit, we shall calculate the static response of the magnetic medium to the perturbation H_1. To first order in Γ, this is of the form

$$\delta M_\alpha(\mathbf{r}') = \sum_\beta \chi_{\alpha\beta}(\mathbf{r} - \mathbf{r}')\,\Gamma S_{i\beta}$$

where χ is a generalized susceptibility tensor and $\alpha, \beta = x,y,z$. Consider now the interaction V of δM with a second electron at $\mathbf{r}'$, of spin $\mathbf{S}_j$,

$$V(\mathbf{r} - \mathbf{r}') = -\Gamma \mathbf{S}_j \cdot \delta\mathbf{M}(\mathbf{r}') = -\Gamma^2 \sum_{\alpha\beta} S_{j\alpha}\chi_{\alpha\beta}(\mathbf{r} - \mathbf{r}')\,S_{i\beta}$$

Here again it is useful to consider the Fourier transform

$$V(\mathbf{q}) = -\Gamma^2 \sum_{\alpha\beta} S_{i\alpha}\chi_{\alpha\beta}(\mathbf{q})\,S_{j\beta}$$

The static susceptibility tensor $\chi_{\alpha\beta}(\mathbf{q})$ is symmetric and can be diagonalized by a convenient choice of axes. The three principal susceptibilities $\chi_{\alpha\alpha}(\mathbf{q})$ are necessarily positive. (Otherwise the magnetic material would be unstable with respect to a sinusoidal distortion of the magnetization of wavevector $\mathbf{q}$.)

If the spins $\mathbf{S}_i$ and $\mathbf{S}_j$ are antiparallel (singlet state), the static interaction

---

[4]This result differs from that of Anderson and Morel [*Phys. Rev.*, **125**, 1263 (1962)]. Their equation (14) uses $\omega_q = (\omega_i q)/k_s$ instead of the consistent form

$$\omega_q = \omega_i \frac{q}{(k_s^2 + q^2)^{1/2}}$$

A recent discussion of the effective interaction, with the allowance for periodicity, more accurate screening, and so on, may be found in Pines, *Elementary Excitations in Solids* (New York: Benjamin, 1963), Chapter 5.

will be positive (repulsive). By continuity, this will remain true at finite frequencies small compared to the exchange frequencies of the magnetic medium. (In an ordered system, the change of sign occurs for $\omega = \omega_q$, the lowest spin wave frequency for the wave vector $\mathbf{q}$.) Therefore, in the presence of a magnetic medium, there exists an interaction whose low frequency part is repulsive for Cooper pairs of antiparallel spin. On the contrary, for parallel pairs (triplet states), one can, in principle, find an attractive interaction at low frequencies.

4–3 GROUND STATE AND ELEMENTARY EXCITATIONS

Choice of a Trial Wave Function

The Cooper argument (Section 4-1) suggests that in the presence of attractive interactions the normal ground state of a free electron gas becomes unstable. The argument is of course only suggestive. In particular, it does not allow for the scattering of electrons above the Fermi sea against the electrons inside. We now calculate in more detail the structure of the condensed state, treating all electrons on the same footing.

In Section 4-1 we considered a pair of electrons described by a wave function $\psi(\mathbf{r}_1 - \mathbf{r}_2)$. In order to treat N electrons on the same footing, a natural generalization leads us to look for a wave function of the form

$$\phi_N(\mathbf{r}_1, \mathbf{r}_2, \ldots, \mathbf{r}_N) = \phi(\mathbf{r}_1 - \mathbf{r}_2)\,\phi(\mathbf{r}_3 - \mathbf{r}_4)\cdots\phi(\mathbf{r}_{N-1} - \mathbf{r}_N) \tag{4-40}$$

In the state described by ϕ_N, the electrons are grouped in pairs, each pair having the same wave function ϕ. We can determine the function by minimizing the energy of the state ϕ_N.

Three technical remarks.

(a) The function ϕ_N can only be constructed for an even number N of electrons. For odd N, it would be necessary to place the last electron in a separate state. For an electron gas ($N \sim 10^{23}$), the presence of this electron only gives rise to effects of the order of 1/N on the energy per particle and the other properties of the condensed staté; therefore it will not be important. On the other hand, for superfluids of fermions with small N (atomic nuclei), the even or odd character of N plays an important role especially on the form of the excitation spectrum.

(b) We have not yet included spin indices in ϕ_N. Guided by analogy with the situation in Section 4-1, we select the spin states ↑↓, opposite for the two electrons of a pair. The function ϕ thus chosen will be the most favorable if the interaction $V_{kk'}$ is weakly dependent on the angle between k and k′.

(c) The function ϕ_N must be antisymmetrized, which we represent by an operator A:

$$\phi_N = A\,\phi(\mathbf{r}_1 - \mathbf{r}_2)\cdots\phi(\mathbf{r}_{N-1} - \mathbf{r}_N)(1\uparrow)(2\downarrow)\cdots(N-1\uparrow)(N\downarrow) \tag{4-41}$$

Algebraic Transformations

The form (4-41) of ϕ_N has the merit of being explicit, but on the other hand it would be rather complicated to calculate directly from it. Introduce the Fourier transform of the pair function

$$\phi(r) = \sum_k g_k e^{ik\cdot r} \tag{4-42}$$

$$\phi_N = \sum_{k_1} \cdots \sum_{k_{N/2}} g_{k_1} \cdots g_{k_{N/2}} \cdots A e^{ik_1(r_1 - r_2)} \cdots$$
$$e^{ik_{N/2}\cdot(r_{N-1} - r_N)} (1\uparrow)(2\downarrow)\cdots(N\downarrow) \tag{4-43}$$

The function $A \exp[ik_1(r_1 - r_2)] \cdots \exp[ik_{N/2}\cdot(r_{N-1} - r_N)] \times (1\uparrow) \times (2\downarrow)\cdots(N\downarrow)$ has a simple interpretation. It describes a state where one electron occupies the state $k_1\uparrow$, another the state $(-k_1\downarrow)$, a third $(k_2\uparrow)$, and so on. Because of A, it is conveniently antisymmetrized. This is what is usually called a Slater determinant formed with the states

$$(k_1\uparrow)(-k_1\downarrow)(k_2\uparrow)(-k_2\downarrow)\cdots(k_{N/2}\uparrow)(-k_{N/2}\downarrow)$$

In order to represent such a Slater determinant, we use the more convenient Wigner-Jordan notation

$$a^+_{k_1\uparrow} a^+_{-k_1\downarrow} \cdots a^+_{k_{N/2}\uparrow} a^+_{-k_{N/2}\downarrow} \phi_0 \tag{4-44}$$

The operator $a^+_{k\alpha}$ creates an electron in the state $(k\alpha)$ when operating on the vacuum state designated by ϕ_0. We also use the destruction operator $a_{k\alpha}$, which is the adjoint of $a^+_{k\alpha}$ and is such that $a_{k\alpha}\phi_0 = 0$. Wigner and Jordan have shown that all properties of the Slater determinant are correctly reproduced by the form (4-44) if anticommutation[5] relations are imposed on the operators a and a^+:

$$\begin{aligned} a^+_{k\alpha} a^+_{\ell\beta} + a^+_{\ell\beta} a^+_{k\alpha} &= 0 \\ a^+_{k\alpha} a_{\ell\beta} + a_{\ell\beta} a^+_{k\alpha} &= \delta_{k\ell}\,\delta_{\alpha\beta} \\ a_{k\alpha} a_{\ell\beta} + a_{\ell\beta} a_{k\alpha} &= 0 \end{aligned} \tag{4-45}$$

Our function ϕ_N takes the form

[5]See Landau and Lifshitz, *Nonrelativistic Quantum Mechanics* (New York: Pergamon, 1959), Chap. 9.

$$\phi_N = \sum_{k_1} \cdots \sum_{k_{N/2}} g_{k_1} \cdots g_{k_{N/2}} a^+_{k_1\uparrow} a^+_{-k_1\downarrow} \cdots a^+_{k_{N/2}\uparrow} a^+_{-k_{N/2}\downarrow} \phi_0 \tag{4-46}$$

It is still not very easy to manage. Consider instead the generating function

$$\tilde{\phi} = C\prod_k (1 + g_k a^+_{k\uparrow} a^+_{-k\downarrow}) \phi_0 \tag{4-47}$$

where the product Π_k extends over all plane wave states[5a] and C is a normalization constant. By comparing (4-46) and (4-47), it is easy to see that to within the factor C, ϕ_N is the part of $\tilde{\phi}$ that has N creation operators acting upon ϕ_0, that is, the component of $\tilde{\phi}$ that describes an N particle state. We make a slight transformation on $\tilde{\phi}$ by incorporating the factor C into the product Π_k

$$\tilde{\phi} = \prod_k (u_k + v_k a^+_{k\uparrow} a^+_{-k\downarrow}) \phi_0 \tag{4-48}$$

with

$$\frac{v_k}{u_k} = g_k \qquad u_k^2 + v_k^2 = 1 \tag{4-49}$$

The latter condition assures the normalization of $\tilde{\phi}$. We have assumed g_k, u_k, and v_k to be real. (We see later that this restriction is unimportant.) The function $\tilde{\phi}$ was introduced by Bardeen, Cooper, and Schrieffer in their fundamental paper of 1957. $\tilde{\phi}$ is much simpler than ϕ_N. For large N, all calculations can be performed from $\tilde{\phi}$ rather than ϕ_N. In order to understand this, write the expansion

$$\tilde{\phi} = \sum_N \lambda_N \phi_N \tag{4-50}$$

with $\Sigma_N |\lambda_N|^2 = 1$ to assure normalization. (Note that, with the above choice of phase for u_k and v_k, the coefficients λ_N are, in fact, real.)

For large N and whatever is the detailed form of the g_k's, the form of $|\lambda_N|^2$ as a function of N is that represented in Fig. 4-2. $|\lambda_N|^2$ has a sharp maximum. We can obtain the position of this maximum by calculating the average number of particles in the state

$$N^* = \langle N \rangle = \sum_k 2v_k^2 = \frac{\Omega}{(2\pi)^3} \int dk\, 2v_k^2 \tag{4-51}$$

where Ω is the volume of the sample. In the same way we can calculate the mean square of the distribution $|\lambda_N|^2$. From (4-44), we find

[5a]Note that all factors in the product Π_k commute, as can be seen from (4-45).

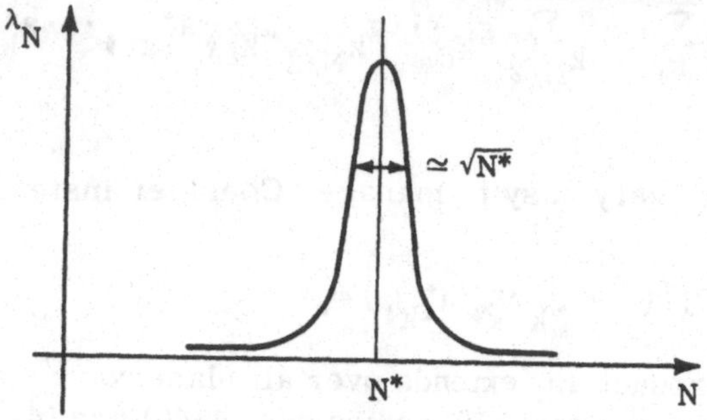

Figure 4-2

The probability amplitude λ_N to find a configuration with N particles in the BCS wave function (4-44). λ_N is strongly peaked around $N = N^*$.

$$\langle N^2 \rangle - \langle N \rangle^2 = \sum_k 4v_k^2 u_k^2 = \frac{\Omega}{(2\pi)^3} \int dk \, 4v_k^2 u_k^2 \tag{4-52}$$

$\langle N^2 \rangle - \langle N \rangle^2$ is proportional to Ω, therefore to N^*—the half-width of the curve is a quantity of the order of $\sqrt{N^*}$. Under these conditions, the relative fluctuations of N are of the order $1/\sqrt{N^*}$—very small. On the contrary, $\sqrt{N^*}$ is still large with respect to 1 and the variations of λ_N are negligible when N varies by one or two

$$\lambda_{N+p} \sim \lambda_N \qquad \text{for } p \ll \sqrt{N^*} \tag{4-53}$$

We now investigate the correspondence between the matrix elements of an arbitrary operator F taken between the ϕ_N or $\tilde{\phi}$.

$$\langle \tilde{\phi} | F | \tilde{\phi} \rangle = \sum_{NN'} \lambda_{N*} \lambda_{N'} \langle \phi_N | F | \phi_{N'} \rangle \tag{4-54}$$

Suppose first that F conserves the number of particles, then

$$\langle \tilde{\phi} | F | \tilde{\phi} \rangle = \sum_N |\lambda_N|^2 \langle \phi_N | F | \phi_N \rangle \tag{4-55}$$

The matrix element $\langle \phi_N | F | \phi_N \rangle$ is a slowly varying function of N and we can replace it by its value at the peak. As $\Sigma_N |\lambda_N|^2 = 1$, we obtain

$$\langle \tilde{\phi} | F | \tilde{\phi} \rangle = \langle \phi_{N*} | F | \phi_{N*} \rangle \tag{4-56}$$

In the same way, if F acting on ϕ_N gives a state of N + p particles,

$$\langle\widetilde{\phi}\,|\,F\,|\,\widetilde{\phi}\rangle = \sum_N \lambda_{N^*+p}\lambda_N \langle\phi_{N+p}\,|\,F\,|\,\phi_N\rangle$$

$$\cong \sum_N |\lambda_N|^2 \langle\phi_{N+p}\,|\,F\,|\,\phi_N\rangle$$

$$= \langle\phi_{N^*+p}\,|\,F\,|\,\phi_{N^*}\rangle \qquad (4\text{-}57)$$

Therefore in practice $\widetilde{\phi}$ gives the same information as ϕ_N.

Calculation of the Energy

Let $\mathcal{H}$ be the Hamiltonian of the interacting electron system. With the wave function ϕ_N we could apply the variational principle directly and look for the minimum of $\langle\phi_N\,|\,\mathcal{H}\,|\,\phi_N\rangle$. With the wave function $\widetilde{\phi}$ there is one slight difference—the number of particles is not fixed, and what we must minimize is

$$(\widetilde{\phi}\,|\,\mathcal{H}\,|\,\widetilde{\phi}) - E_F(\widetilde{\phi}\,|\,N\,|\,\widetilde{\phi}) \qquad (4\text{-}58)$$

where N is the number of particles and E_F is a Lagrange multiplier we call the Fermi level. Now list the contributions to $\mathcal{H} - E_F N$. The kinetic energy term is

$$\mathcal{H}_0 = \sum_{k\alpha} \xi_k a^+_{k\alpha} a_{k\alpha}$$

$$\xi_k = \frac{\hbar^2 k^2}{2m} - E_F \qquad (4\text{-}59)$$

The potential energy term $\mathcal{H}_{int}$ has matrix elements describing the scattering of 2 electrons,

$$(\mathbf{k}\alpha)(\mathbf{k}'\beta) \rightarrow (\mathbf{k} + \mathbf{q}, \alpha)(\mathbf{k}' - \mathbf{q}, \beta)$$

$$\mathcal{H}_{int} = \frac{1}{2} \sum_{\substack{k,k',q\\ \alpha\beta}} V(\mathbf{k} + \mathbf{q}, \mathbf{k}' - \mathbf{q}\,|\,\mathbf{k}, \mathbf{k}')$$

$$\times\, a^+_{k+q,\,\alpha} a^+_{k'-q,\,\beta} a_{k'\beta} a_{k\alpha} \qquad (4\text{-}60)$$

The form (4-60) takes into account conservation of momentum and of total spin.

In the state $\widetilde{\phi}$, the probability of finding a state $\mathbf{k}$ occupied is v_k^2. Therefore the kinetic energy becomes

$$\langle\widetilde{\phi}\,|\,\mathcal{H}_0\,|\,\widetilde{\phi}\rangle = \sum_{k\alpha} v_k^2 \xi_k \qquad (4\text{-}61)$$

The only terms contributing in $\mathcal{H}_{int}$ are

(1) the diagonal terms $V(\mathbf{kk'}\,\mathbf{kk'})$;

(2) the exchange terms $V(\mathbf{kk'}\,\mathbf{k'k})$;

(3) the terms describing the transition of a pair from the state $(\mathbf{k}\uparrow, -\mathbf{k}\downarrow)$ to the state $(\boldsymbol{\ell}\uparrow, -\boldsymbol{\ell}\downarrow)$, $V(\boldsymbol{\ell}, -\boldsymbol{\ell} \mid \mathbf{k}, -\mathbf{k}) = V_{\mathbf{k}\boldsymbol{\ell}}$.

The contributions of (1) and (2) are already present in a normal metal. We see later that they can be simply incorporated into ξ_k. We neglect them for the moment. The interesting contribution comes from (3). It is derived in the following way. We can first analyze the wave function $\tilde{\phi}$ in two parts ϕ_{k1} and ϕ_{k0} in which the pair state $(\mathbf{k}\uparrow - \mathbf{k}\downarrow)$ is, respectively, occupied and unoccupied. From (4-48) we have

$$\tilde{\phi} = v_k\,\phi_{k1} + u_k\,\phi_{k0} \tag{4-62}$$

Similarly we can analyze $\tilde{\phi}$ into four components describing the occupancy of two different pair states $(\mathbf{k}\uparrow - \mathbf{k}\downarrow)$ and $(\boldsymbol{\ell}\uparrow - \boldsymbol{\ell}\downarrow)$

$$\tilde{\phi} = v_k v_\ell\,\phi_{k1\ell1} + v_k u_\ell\,\phi_{k1\ell0} + u_k v_\ell\,\phi_{k0\ell1} + u_k u_\ell\,\phi_{k0\ell0} \tag{4-63}$$

When computing the contribution of $V_{k\ell}$ to the interaction energy $\langle\tilde{\phi}\,|\,\mathcal{H}_{int}\,|\,\tilde{\phi}\rangle$, we must use the components

$$v_k u_\ell \langle\phi_{k1\ell0}\,|\,\mathcal{H}_{int}\,|\,\phi_{k0\ell1}\rangle u_k v_\ell$$

and the remaining matrix element is simply $V_{k\ell}$. Thus we find

$$\langle\tilde{\phi}\,|\,\mathcal{H} - E_F N\,|\,\tilde{\phi}\rangle = 2\sum_k \xi_k v_k^2 + \sum_{k\ell} V_{k\ell}\, u_k v_k u_\ell v_\ell \tag{4-64}$$

In order to minimize $\langle\tilde{\phi}\,|\,\mathcal{H}\,|\,\tilde{\phi}\rangle$ taking account of the relation $u_k^2 + v_k^2 = 1$, it is useful to put

$$u_k = \sin\theta_k \qquad v_k = \cos\theta_k \tag{4-65}$$

$$\langle\tilde{\phi}\,|\,\mathcal{H}\,|\,\tilde{\phi}\rangle = 2\sum_k \xi_k \cos^2\theta_k + \tfrac{1}{4}\sum_{k\ell} \sin 2\theta_k \sin 2\theta_\ell\, V_{k\ell} \tag{4-66}$$

The minimization equations are

$$0 = \frac{\partial}{\partial\theta_k}\langle\tilde{\phi}\,|\,\mathcal{H}\,|\,\tilde{\phi}\rangle = -\,2\xi_k \sin 2\theta_k + \sum_\ell \cos 2\theta_k \sin 2\theta_\ell\, V_{k\ell}$$

or

$$\xi_k \tan 2\theta_k = \tfrac{1}{2}\sum_\ell V_{k\ell} \sin 2\theta_\ell \tag{4-67}$$

We define

$$\Delta_k = -\sum_{\ell} V_{k\ell}\, u_\ell v_\ell \tag{4-68}$$

$$\epsilon_k = \sqrt{\xi_k^2 + \Delta_k^2} \tag{4-69}$$

and we find

$$\tan 2\theta_k = -\frac{\Delta_k}{\xi_k} \tag{4-70}$$

$$2u_k v_k = \sin 2\theta_k = \frac{\Delta_k}{\xi_k} \tag{4-71}$$

$$-u_k^2 + v_k^2 = \cos 2\theta_k = -\frac{\xi_k}{\epsilon_k} \tag{4-72}$$

(The choice of the sign in the last equation is such that for very large positive ξ_k, $u_k = 1$, $v_k = 0$, and the total number of electrons $\Sigma_k v_k^2$ converges.)

With the insertion of the value of $u_\ell v_\ell$ into (4-68), we finally obtain an equation for Δ,

$$\Delta_k = -\sum_{\ell} V_{k\ell} \frac{\Delta_\ell}{2(\xi_\ell^2 + \Delta_\ell^2)^{1/2}} \tag{4-73}$$

We first notice that this equation always has the trivial solution $\Delta_k = 0$, corresponding to

$$v_k = \begin{cases} 1 & \xi_k < 0 \\ 0 & \xi_k > 0 \end{cases} \tag{4-74}$$

The associated function $\widetilde{\phi}$ is simply

$$\widetilde{\phi}_n = \prod_{k<k_F} a_k^+ a_{-k}^+ \phi_0 \tag{4-75}$$

This represents a Slater determinant formed of all states of energy less than $E_F = \hbar^2 k_F^2/2m$; and is the wave function of a noninteracting electron gas. To show explicitly that there can be other solutions we choose the simplified interaction

$$V_{k\ell} = \begin{cases} -V & \text{if } |\xi_k|, |\xi_\ell| \le \hbar\omega_D \\ 0 & \text{otherwise} \end{cases} \tag{4-76}$$

We call this simple form the BCS interaction. V is a positive constant. Then

$$\begin{aligned} \Delta_k &= 0 \qquad \text{for } |\xi_k| > \hbar\omega_D \\ \Delta_k &= \Delta \qquad \text{(independent of k) for } |\xi_k| < \hbar\omega_D \end{aligned} \tag{4-77}$$

We can then rewrite (4-73) by first summing over all orientations of ℓ for a fixed ξ_ℓ. This gives a factor $N(\xi_\ell)$, the density of states per unit energy in the normal state for a given spin direction. Since we are interested in an energy interval of width $\hbar\omega_D \ll E_F$, we can replace $N(\xi_\ell)$ by its value at the Fermi level $N(0)$,

$$\Delta = N(0)V \int_{-\hbar\omega_D}^{\hbar\omega_D} \Delta \frac{d\xi}{2\sqrt{\Delta^2 + \xi^2}} \tag{4-78}$$

$$\frac{1}{N(0)V} = \int_0^{\hbar\omega_D} \frac{d\xi}{\sqrt{\Delta^2 + \xi^2}} = \sinh^{-1}\left(\frac{\hbar\omega_D}{\Delta}\right) \tag{4-79}$$

This equation only has a solution for positive V, that is, an attractive interaction. Then

$$\Delta = \frac{\hbar\omega_D}{\sinh \dfrac{1}{N(0)V}} \tag{4-80}$$

In everything that follows we are only interested in the weak coupling limit $N(0)V \ll 1$, for which

$$\Delta = 2\hbar\omega_D\, e^{-1/N(0)V} \tag{4-81}$$

How is this limit justified? We shall see later that in this simple model the transition temperature is given by $\Delta = 1.75 k_B T_0$. Therefore $e^{-1/N(0)V} = 0.88 T_0/\Theta_D$ where Θ_D is the Debye temperature defined by $\hbar\omega_D = k_B\Theta_D$. For most superconducting metals $\Theta_D \sim 300°K$, $T_0 \sim 10°K$. Therefore we empirically conclude that $N(0)V < 0.3$. (Lead and mercury are two notable exceptions with very low Θ_D's, giving, respectively, $N(0)V = 0.39$ and 0.35.)

Once Δ is known, we can explicitly calculate the kinetic and potential energies. In the weak coupling limit, one finds from (4-64)

$$\langle\tilde{\phi}|\mathcal{H}_0|\tilde{\phi}\rangle = 2\sum_{k<k_F} \xi_k + \frac{\Delta^2}{V} - \frac{N(0)\Delta^2}{2} \tag{4-82}$$

The first term is the kinetic energy of the normal state ($\Delta = 0$). In the same way, we find

$$\langle \widetilde{\phi} | \mathcal{H}_{\text{int}} | \widetilde{\phi} \rangle = - \frac{\Delta^2}{V} \tag{4-83}$$

The energy difference between the state $\widetilde{\phi}$ and the normal state is

$$\langle \widetilde{\phi} | \mathcal{H} | \widetilde{\phi} \rangle - \langle \phi_n | \mathcal{H} | \phi_n \rangle_{(\Delta = 0)} = - \frac{N(0)\Delta^2}{2} \tag{4-84}$$

The condensed state energy is lower than that of the normal state. Thus energy gain is, however, very small, of the order Δ^2/E_F per particle. For $\Delta = 10°K$ and $E_F = 10^{4}°K$, this gives $\Delta^2/E_F = 10^{-2}°K$. We cannot calculate the energy of the normal state with this accuracy. Fortunately the unknown terms are very nearly the same in the normal state and in the condensed state. Each term in (4-82) is incorrect, but the difference is very nearly correct, and it is this difference that is measured experimentally.

Certain simple properties can be determined from the function $\widetilde{\phi}$:

(a) Let us study the probability of finding an electron condensed into the state ($\mathbf{k}\alpha$). It is given by

$$\langle \widetilde{\phi} | a^{+}_{k\alpha} a_{k\alpha} | \widetilde{\phi} \rangle = v_k^2 = \frac{1}{2}\left(1 - \frac{\xi_k}{\sqrt{\xi_k^2 + \Delta_k^2}}\right) \tag{4-85}$$

from (4-70) and (4-55). It is shown as a function of k in Fig. 4-3. v_k^2 gives the momentum distribution in the ground state. Ideally v_k^2 can be measured by the Compton effect, positron annihilation, and so on. In fact, high accuracy would be required to measure v_k^2 directly in the domain of interest. For $k \ll k_F$, $v_k = 1$. For $k \gg k_F$, $v_k = 0$. The distribution function v_k^2 is continuous in contrast to the normal state where it has a discontinuity at k_F. The transition region, where v_k goes from 1 to 0, has an energy width Δ and a momentum width δk.

$$\begin{aligned} &\frac{\hbar^2}{2m}\left[(k_F + \delta k)^2 - k_F^2\right] \cong \Delta \\ &\delta k = \frac{m\Delta}{\hbar^2 k_F} = \frac{\Delta}{\hbar v_F} \end{aligned} \tag{4-86}$$

(b) Let us add two electrons to the states $\mathbf{k}\uparrow$ and $-\mathbf{k}\downarrow$ in the condensed state $\widetilde{\phi}_N$. What is the probability amplitude for obtaining the state ϕ_{N+2}? This is the quantity

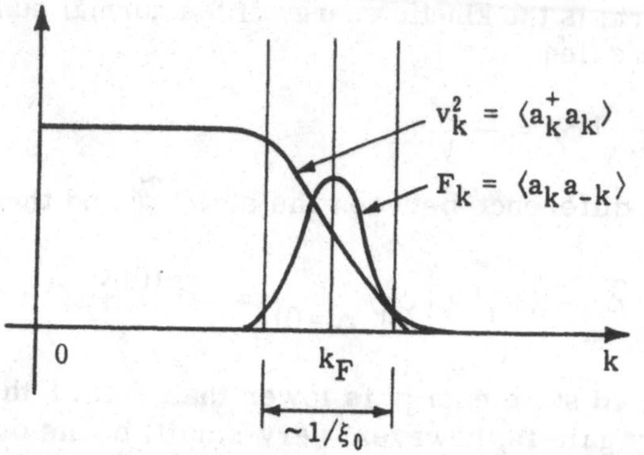

Figure 4-3
The momentum distribution ($|v_k|^2$) of electrons in the BCS ground state. In a normal metal, the distribution drops discontinuously at $k = k_F$. In the superconductor, this drop is smeared out on an interval $\delta k \sim 1/\xi_0 \sim \Delta/\hbar v_F$. Also shown is the condensation amplitude F_k.

$$F_k = \langle \phi_{N+2} | a^+_{k\uparrow} a^+_{-k\downarrow} | \phi_N \rangle = \langle \tilde{\phi} | a^+_{k\uparrow} a^+_{-k\downarrow} | \tilde{\phi} \rangle \tag{4-87}$$

We call F_k the *condensation amplitude* in the state **k**. On using the definition of $\tilde{\phi}$, we find

$$F_k = u_k v_k \tag{4-88}$$

F_k is only nonvanishing in the transition zone and is maximum for $k = k_F$.

First Excited States

We have constructed a ground state function ϕ_N where N particles are coupled in pairs. We now try to add one particle in the plane wave state ($\mathbf{m}\alpha$). We obtain a state

$$\phi_{N+1,\, m\alpha}(\mathbf{r}_1 \cdots \mathbf{r}_{N+1}) = A\phi(\mathbf{r}_1 - \mathbf{r}_2) \cdots \phi(\mathbf{r}_{N-1} - \mathbf{r}_N) \times \exp(i\mathbf{m} \cdot \mathbf{r}_{N+1})(1\uparrow) \cdots (N\downarrow)(N+1, \alpha) \tag{4-89}$$

We introduce the corresponding generating function

$$\widetilde{\phi}_{m\alpha} = \sum_N \lambda_N \phi_{N+1,\, m\alpha} \tag{4-90}$$

By repeating for $\phi_{m\alpha}$ our earlier arguments on ϕ we find

$$\widetilde{\phi}_{m\alpha} = \prod_{k \neq m} (u_k + v_k a^+_{k\uparrow} a^+_{-k\downarrow}) a^+_{m\alpha} \phi_0 \tag{4-91}$$

$\widetilde{\phi}_{m\alpha}$ is orthogonal to $\widetilde{\phi}$. What is the energy of the state $\phi_{m\alpha}$? For the kinetic energy we find

$$\langle \widetilde{\phi}_{m\alpha} | \mathcal{H}_0 | \widetilde{\phi}_{m\alpha} \rangle = \langle \widetilde{\phi} | \mathcal{H}_0 | \widetilde{\phi} \rangle + (1 - 2v_m^2)\, \xi_m \tag{4-92}$$

since $\widetilde{\phi}_{m\alpha}$ has an electron in the orbital state $m\alpha$, while in the state $\widetilde{\phi}$ there is a probability $2v_m^2$ of finding an electron $(m\uparrow)$ or $(m\downarrow)$. For the potential energy, we notice that only the terms corresponding to the transitions $(\mathbf{k}, \alpha)(-\mathbf{k}, -\alpha) \rightarrow (\mathbf{k}'\alpha)(-\mathbf{k}' - \alpha)$ with $\mathbf{k} \neq \mathbf{m}$ and $\mathbf{k}' \neq \mathbf{m}$ contribute since no pair can use the state $\mathbf{m}\alpha$. Therefore

$$\langle \widetilde{\phi}_{m\alpha} | \mathcal{H}_{int} | \widetilde{\phi}_{m\alpha} \rangle = \langle \widetilde{\phi} | \mathcal{H}_{int} | \widetilde{\phi} \rangle - 2 \sum_{\ell} V_{m\ell} u_m v_\ell u_\ell v_m \tag{4-93}$$

The total energy becomes

$$\begin{aligned} \langle \widetilde{\phi}_{m\alpha} | \mathcal{H} | \widetilde{\phi}_{m\alpha} \rangle &= E_0 + (1 - 2v_m^2)\, \xi_m + 2u_m v_m \Delta_m \\ &= E_0 + \frac{\xi_m^2}{\epsilon_m} + \frac{\Delta_m^2}{\epsilon_m} \\ &= E_0 + \epsilon_m \end{aligned} \tag{4-94}$$

Since E_0 is the energy in the ground state $\widetilde{\varphi}$, $\epsilon_m = \sqrt{\Delta_m^2 + \xi_m^2}$ is the energy necessary to place a new particle into the state $\mathbf{m}$. Notice that even when $\xi_m = 0$, ϵ_m is finite and equal to $\Delta_{k_F} = \Delta$.

We can try to construct states with 2, 3, ... n excitations in the same way. For the states with two excitations, for example, we are tempted to take the function

$$\Xi_{m\alpha,\, n\beta} = \prod_{k \neq m,\, n} (u_k + v_k a^+_{k\uparrow} a^+_{-k\downarrow}) a^+_{m\alpha} a^+_{n\beta} \phi_0 \tag{4-95}$$

Unfortunately such functions are not necessarily orthogonal to $\widetilde{\phi}$. For example,

$$\langle \tilde{\phi} | \Xi_{m\uparrow \, -m\downarrow} \rangle = v_m \neq 0 \tag{4-96}$$

Two methods to avoid this difficulty are

(a) Add to Ξ a component $\lambda\tilde{\phi}$ where λ is chosen in such a way as to make the total function orthogonal to $\tilde{\phi}$. This is the original BCS method, which is numerically rather tedious.

(b) Reconsider the wave functions of single excitation $\tilde{\phi}_{m\alpha}$. Let us try to place these functions into the form

$$\tilde{\phi}_{m\alpha} = \gamma^{+}_{m\alpha} \tilde{\phi} \tag{4-97}$$

where $\gamma^{+}_{m\alpha}$ will appear as a creation operator for one elementary excitation. If the operators γ^{+} and their adjoints γ can be found such that

(1) the γ and γ^{+} obey the fermion commutation rules (analogous to (4-45),

(2) $\gamma_{m\alpha} \tilde{\phi} = 0$, that is, $\tilde{\phi}$ is the state of no excitation,

then it is easy to see by manipulating the commutation relations that all the states obtained by applying an arbitrary number of operators γ^{+} to $\tilde{\phi}$ are orthogonal to each other, normalized, and orthogonal to the ground state $\tilde{\phi}$. From the kinematic point of view, these wave functions can therefore conveniently describe states of multiple excitations.

We must now find the γ^{+}. As a first attempt, we notice that

$$\tilde{\phi}_{m\alpha} = \frac{1}{u_m} a^{+}_{m\alpha} \tilde{\phi} \tag{4-98}$$

as is easy to verify from (4-91) by noting that $a^{+}_{m\alpha} a^{+}_{m\alpha} = 0$ and so on. However $(1/u_m) a^{+}_{m\alpha}$ does not obey the above anticommutation rules. As a second attempt, we notice that

$$\tilde{\phi}_{m\downarrow} = \frac{1}{v_m} a_{m\uparrow} \tilde{\phi} \tag{4-99}$$

However, we again encounter the same difficulty with the commutation rules. We then try to define γ^{+} as a linear combination of the preceding forms. If we take

$$\begin{aligned} \gamma^{+}_{m\uparrow} &= u_m a^{+}_{m\uparrow} - v_m a_{-m\downarrow} \\ \gamma^{+}_{m\downarrow} &= u_m a^{+}_{m\downarrow} + v_m a_{-m\uparrow} \end{aligned} \tag{4-100}$$

we find that (4-97) and conditions (1) and (2) are verified. For example, condition (2) becomes

$$\begin{aligned}\gamma_{m\uparrow}\tilde{\phi} &= \prod_{k\neq m}(u_k + v_k a^+_{k\uparrow} a^+_{-k\downarrow})(u_m a_{m\uparrow} - v_m a^+_{-m\downarrow}) \\ &\quad \times (u_m + v_m a^+_{m\uparrow} a^+_{-m\downarrow})\phi_0 \\ &= \prod_{k\neq m}(u_k + v_k a^+_{k\uparrow} a^+_{-k\downarrow}) u_m v_m (a^+_{-m\downarrow} - a^+_{-m\downarrow})\phi_0 \\ &= 0 \end{aligned} \tag{4-101}$$

where we have used the relations $a_{-m\alpha}\phi_0 = 0, a_{m\alpha} a^+_{m\alpha}\phi_0 = \phi_0$, and the anticommutation relations. We thus obtain the operators γ^+, which create excited states that are mutually orthogonal and also orthogonal to the ground state. The use of the γ and γ^+ enormously simplifies all calculations in which the excited states play a role (Bogolubov, Valatin, 1958). We designate the excitations created by the γ^+'s as *quasiparticles*.

Physical remarks.

(a) The fact that $\gamma^+_{m\alpha}$ is a linear combination of $a^+_{m\alpha}$ and $a_{m-\alpha}$ expresses the fact that it is possible to create the state $\phi_{N+1,k\alpha}$ either by adding an electron $(k\alpha)$ to the condensed state ϕ_N or by removing an electron $(-k-\alpha)$ from the state ϕ_{N+2}.

(b) Suppose we start from the ground state ϕ_N. We have seen that if we add a particle in a state **k**, it must at least have an energy equal to $E_F + \Delta$. Such an "injection" experiment is not a purely theoretical invention. It can be accomplished by bringing electrons into a superconductor across a thin oxide layer. This is the previously mentioned tunnel effect (Giaever, 1959), which permits us to determine the gap Δ.

(c) However, except from the tunneling experiments, the excited states are usually studied without changing N (for example, infrared radiation). If N is even, the first excited states of this type correspond to breaking a pair. They correspond to wave functions of the form $\gamma^+_{k\alpha}\gamma^+_{\ell\beta}$ $(k\alpha \neq \ell\beta)$. The excitation energy is easily calculated and becomes $\epsilon_k + \epsilon_\ell$. The minimum of this energy (corresponding to the infrared absorption energy gap) is 2Δ. If N is odd (N = 2K + 1), the state consists of K pairs plus one unpaired electron. There are then 2 kinds of excitations: either a pair is decoupled as previously (gap 2Δ) or the state of the unpaired electron is changed (gap 0). However the absorption intensity of the first process is ~N times greater than that of the second process. In practice, since N ~ 10^{23}, one always measures an absorption threshold at 2Δ.

(d) The excitation energy is at least 2Δ, but the energy per excited particle is Δ. At low temperatures the probability of finding a particle excited into a given state $\mathbf{k}$ (k near k_F) varies as $\exp(-\Delta/k_B T)$.

Case of Two Coupled Superconductors

Consider two superconductors S and S′ separated by an insulating layer. If the layer is thick, the electrons cannot cross it, and S and S′ become completely independent. If the layer is thin (thickness < 30Å), the electrons can traverse it by the tunneling effect. This gives rise to a coupling between the regions S and S′, which we consider now. Algebraically, the transfer of the electrons can be described by adding to the Hamiltonian operator $\mathcal{H}_{SS'}$ of the electrons for S and S′, a small term $\mathcal{H}_T$, which causes an electron to tunnel from S to S′ and vice versa.

$$\begin{aligned} \mathcal{H} &= \mathcal{H}_{SS'} + \mathcal{H}_T \\ \mathcal{H}_T &= \sum_{k\ell} (a^+_{ks} a_{\ell s'} T_{k\ell} + a^+_{\ell s'} a_{ks} T^+_{k\ell}) \end{aligned} \tag{4-102}$$

where a^+_{ks} created an electron in the state $\mathbf{k}$ on the side S and $a_{\ell s'}$ destroys an electron in the state $\boldsymbol{\ell}$ on the side S′. The matrix element $T_{k\ell}$ can be calculated explicitly from a solution of the one-electron Schroedinger equation in the barrier. (Notice that $\mathcal{H}_{SS'}$ is the complete Hamiltonian of S and S′. It contains the electron interactions V.) An eigenstate for $\mathcal{H}_{SS'}$ will be a product of two functions of the type (4-43), one relative to S and one relative to S′. Let

$$\psi_\nu = \phi^{(s')}_{2(N-\nu)} \phi^{(s)}_{2\nu} \qquad \mathcal{H}_{SS'} \psi_\nu = E_\nu \psi_\nu \tag{4-103}$$

ψ_ν describes 2ν electrons coupled in pairs on the side S and $2(N-\nu)$ electrons coupled in S′. The total number of electrons 2N is fixed. On the contrary, ν is not known in advance. How does the energy E_ν vary with respect to ν? If ν is augmented by 1, 2 electrons are created on the side S at the expense of S′. Thermodynamics gives us

$$E_{\nu+1} - E_\nu = 2\left(E_F^{(s)} - E_F^{(s')}\right) \tag{4-104}$$

where $E_F^{(s)}$ is the chemical potential (Fermi level) on the side S. For the moment, let us suppose that there is no voltage applied between S and S′. Then $E_F^{(s)} = E_F^{(s')}$ and the functions ψ_ν are degenerate (E_ν =

E independent of ν). The perturbation $\mathcal{H}_T$ removes this degeneracy. To second order in $\mathcal{H}_T$ there is a matrix element coupling the functions ψ_ν and $\psi_{\nu+1}$

$$J_0 = \sum_{\substack{k,\ell \\ k',\ell'}} \langle \nu + 1 | T_{k\ell} a^+_{ks} a_{\ell s'} | I \rangle \frac{1}{E - E_I} \times \langle I | T_{k'\ell'} a^+_{k's} a_{\ell' s'} | \nu \rangle \tag{4-105}$$

where $|I\rangle$ is an intermediate state with $2\nu + 1$ electrons on the S side and $2(N - \nu) - 1$ electrons on the S′ side. The tunneling Hamiltonian $\mathcal{H}_T$ acting on the state with ν pairs on side S creates a state $|I\rangle$ with one "bachelor electron" of wave vector **k′** on side S. Then $\mathcal{H}_T$ acting on $|I\rangle$ creates another electron of wave vector **k** on side S. The final state onto which this is projected is composed of $(\nu + 1)$ pairs on side S; thus we must save $\mathbf{k} = -\mathbf{k}'$. Similarly $\boldsymbol{\ell} = -\boldsymbol{\ell}'$. Then, using the symmetry relation $T_{-k\,-\ell} = T^+_{k\ell}$ and our results on excited states, we find

$$J_0 = -4 \sum_{k\ell} |T_{k\ell}|^2 \frac{u_k v_k u_\ell v_\ell}{\epsilon_k + \epsilon_\ell} \tag{4-106}$$

To second order in $\mathcal{H}_T$, one can write

$$\mathcal{H}\psi_\nu = E\psi_\nu + J_0(\psi_{\nu+1} + \psi_{\nu-1}) \tag{4-107}$$

The correct wave function will be a linear combination of the ψ_ν. This problem is formally identical to that of the motion of an electron in a linear atomic chain. ψ_ν is then the analog of an atomic orbital localized around the ν th atom. In the tight binding approximation, one attempts to form the wave function as a linear combination of the ψ_ν's. The form of this combination ψ_k is imposed in Bloch's theorem

$$\psi_k = \sum_\nu \psi_\nu e^{ik\nu} \tag{4-108}$$

where k is analogous to a wave vector. The corresponding energy E(k) is

$$E(k) = E + 2J_0 \cos k \tag{4-109}$$

Suppose that we now construct a wave packet of the ψ_k's corresponding to wave vectors in the interval between k and k + Δk. In our "linear chain" analogy, the wave packet has a spatial extent $\Delta\nu \sim 1/\Delta k$. As ν is enormous ($\nu \sim 10^{22}$), we can take $\Delta\nu$ to be very large (for example, $\Delta\nu \sim \sqrt{\nu} \sim 10^{11}$), therefore having a very small Δk and at the same time $\Delta\nu/\nu$ very small ($\sim 10^{-11}$). Therefore it is possible to specify k and ν simultaneously with good accuracy. Such a wave packet moves with group velocity

$$\frac{\hbar d\langle\nu\rangle}{dt} = \frac{\partial E(k)}{\partial k} = -2J_0 \sin k \tag{4-110}$$

For the state described by our wave packet, there is a time variation of ν and therefore *a current passing from S to S'*

$$I = 2e\frac{d\langle\nu\rangle}{dt} = -4e\frac{J_0}{\hbar}\sin k \tag{4-111}$$

In other words, if S and S′ are attached to a generator, a current ($< 4eJ_0/\hbar$) can flow from S to S′ *under zero voltage*. Equation (4-111) shows us in a simple example why the condensed state has superfluid properties. Note that for a normal system ($u_k v_k \equiv 0$) J_0 vanishes (Eq. 4-106). This effect was predicted by Josephson (1961) and observed first by Anderson and Rowell. The present derivation is due to Ferrel (1963). (Typical current values are of order 10^{-2}A/cm^2.)

How are these results modified when a voltage is applied between S and S′? We then have $E_{\nu+1} - E_\nu = 2eV$. This is the analog of a uniform electric field applied on the "chain." The wave packet moves according to the force equation

$$\frac{d}{dt}\langle\hbar k\rangle = 2eV \tag{4-112}$$

Equations (4-111) and (4-112) completely determine the behavior of the junction. For example, if the voltage V is held constant, k varies linearly with time and I is an alternating current of frequency $2eV/\hbar$.

Problem. A Josephson junction SS′ is placed in series with a resistance R and connected to a generator of voltage U. Calculate the currents.

Solution. The voltage across the junction is

$$V = U - RI = U + \frac{4eJ_0R}{\hbar}\sin k$$

From (4-112)

$$\frac{dk}{dt} = \frac{2e}{\hbar}\left(U + \frac{4eJ_0R}{\hbar}\sin k\right) = \omega_0(1 - \lambda\sin k)$$

with

$$\omega_0 = \frac{2eU}{\hbar} \qquad \lambda = -\frac{4eJ_0R}{\hbar U} = \frac{RJ_m}{U}$$

(J_m = maximum supercurrent allowable through the junction).

The force equation can be integrated in the form

$$\omega_0(t - t_0) = \int \frac{dk}{1 - \lambda \sin k}$$

We consider two cases:

(a) If $\lambda > 1$, the denominator has pole for $k = k_0$, where $\sin k_0 = 1/\lambda$. t tends towards ∞ for $k \to k_0$. The current tends toward the limiting value U/R. When this value is reached, the voltage across the junction SS′, U − RI, will fall to zero. There is therefore a regime in which there is a permanent continuous current.

(b) If $\lambda < 1$, the denominator never vanishes. The relation between k and t can be integrated explicitly by setting

$$m = \tan\frac{k}{2} \qquad I = \frac{2J_m m}{1 + m^2}$$

One obtains

$$m = \lambda + \sqrt{1 - \lambda^2} \tan [\tfrac{1}{2}\sqrt{1 - \lambda^2}\, \omega_0 (t - t_0)]$$

m, and consequently k, is a periodic function of time with period $2\pi/\omega_0\sqrt{1 - \lambda^2}$. Notice that the current is not sinusoidal except for $\lambda = 0$.

4–4 CALCULATIONS AT FINITE TEMPERATURE

Construction of the Free Energy

For $T \neq 0$, the states $\gamma_k^+ \tilde{\Phi}$, $\gamma_k^+ \gamma_\ell^+ \tilde{\Phi}$, and so on, are thermally excited. When $k_B T \ll \Delta$, there are only a few of these excitations and they are independent. However, when $k_B T \sim \Delta$, the number of excitations are large. We now study the *average* effect of the interaction between them, and show that it can be simply described by allowing the energy gap Δ to vary with temperature.

We still define the fermion operators γ_k^+ by the transformation (4-100) where the u_k and v_k are not yet determined. We treat the excitations of the system as a gas of independent quasiparticles. At temperature T, there is a probability

$$f_{k\alpha} = \langle \gamma_{k\alpha}^+ \gamma_{k\alpha} \rangle = 1 - \langle \gamma_{k\alpha} \gamma_{k\alpha}^+ \rangle \tag{4-113}$$

of finding a quasiparticle of wave vector **k** and spin α.[6] The symbol

---

[6]In most usual cases $f_{k\alpha}$ will be independent of the index α and written simply f_k.

$\langle\,\rangle$ denotes a thermal average. We shall calculate the free energy F (where the f_k are arbitrary), then minimize F with respect to u_k and v_k and with respect to f_k.

(1) Calculation of the average kinetic energy from (4-92):

$$\langle\mathcal{H}_0\rangle = \sum_{k\alpha} \xi_k [u_k^2 \langle\gamma_{k\alpha}^+ \gamma_{k\alpha}\rangle + v_k^2 \langle\gamma_{k\alpha} \gamma_{k\alpha}^+\rangle]$$

$$= \sum_k 2\xi_k [u_k^2 f_k + v_k^2 (1 - f_k)] \tag{4-114}$$

(2) Potential energy: With our approximations, the only terms that come into play are still those that describe the collisions $(k\uparrow - k\downarrow)$ $(\ell\uparrow - \ell\downarrow)$.

$$\langle\mathcal{H}_{\text{int}}\rangle = \sum_{k\ell} \langle a_{k\uparrow}^+ a_{-k\downarrow}^+ a_{-\ell\downarrow} a_{\ell\uparrow}\rangle V_{k\ell} \tag{4-115}$$

In Eq. (4-115) the operators a^+, a, are linear combinations of the operators γ^+, γ, and the $\gamma\gamma^+$ describe a set of independent fermions. Then we have the simple property

$$\langle 0_1 0_2 0_3 0_4\rangle = \langle 0_1 0_2\rangle\langle 0_3 0_4\rangle + \langle 0_1 0_4\rangle\langle 0_2 0_3\rangle - \langle 0_1 0_3\rangle\langle 0_2 0_4\rangle \tag{4-116}$$

where each of the 0 is an a or an a^+ operator. This property (a particular application of a theorem by Wick) can be checked by writing explicitly the transformation from the a's to the γ's and then taking the averages. It simplifies the algebra considerably. In the present case, we find 3 terms for the interaction. Two of them involve only averages $\langle a_k^+ a_k\rangle$ and lead to a Hartree and an exchange contribution—both are essentially temperature independent and we drop them as before. The remaining term $\langle a^+ a^+\rangle\langle a\, a\rangle$ leads to

$$\langle\mathcal{H}_{\text{int}}\rangle = \sum_{k\ell} V_{k\ell} u_k v_k u_\ell v_\ell (1 - 2f_k)(1 - 2f_\ell) \tag{4-117}$$

(3) Entropy S. The entropy for a system of independent fermions is given by

$$S = -k_B \sum_{k\alpha} [f_k \ln f_k + (1 - f_k) \ln (1 - f_k)] \tag{4-118}$$

(4) The total energy F becomes

$$F = \langle\mathcal{H}_0 + \mathcal{H}_{\text{int}}\rangle - TS \tag{4-119}$$

We first require that F be stationary with respect to u_k and v_k. Again setting $u_k = \sin\theta_k$, $v_k = \cos\theta_k$, we obtain

$$+2\xi_k(1 - 2f_k)\sin 2\theta_k = \sum_\ell V_{k\ell}\cos 2\theta_k \sin 2\theta_\ell (1 - 2f_k)(1 - 2f_\ell) \tag{4-120}$$

If we define

$$\Delta_k = -\tfrac{1}{2}\sum_\ell V_{k\ell}\sin 2\theta_\ell(1 - 2f_\ell) \tag{4-121}$$

we recover the solution (4-69)–(4-72). The only modification arises from the fact that Δ_k (4-121) differs from (4-68) by the excitation occupation number f_ℓ. If we now minimize F with respect to f_k, we obtain

$$2\xi_k(u_k^2 - v_k^2) - 4\sum_\ell V_{k\ell}u_k v_k u_\ell v_\ell(1 - 2f_\ell)$$

$$+ 2k_B T \ln\left(\frac{f_k}{1 - f_k}\right) = 0 \tag{4-122}$$

Then, using (4-69)–(4-72), we find

$$f_k = \frac{1}{1 + \exp(\epsilon_k / k_B T)} \tag{4-123}$$

This is the usual form for the distribution function for fermions of energy ϵ_k. However, the variation of f_k with temperature is not simple because ϵ_k and Δ_k depend on the temperature.

Variation of the Gap with Temperature and Transition Point

The self-consistency equation [analogous to (4-73)] becomes

$$\Delta_k = -\sum_\ell V_{k\ell}\frac{\Delta_\ell}{2\epsilon_\ell}[1 - 2f(\epsilon_\ell)] \tag{4-124}$$

Let us first take for $V_{k\ell}$ the simple BCS interaction (4-76). The condition becomes

$$1 = N(0)V\int_0^{\hbar\omega_D}\frac{d\xi}{\sqrt{\xi^2 + \Delta^2}}\left[1 - 2f(\sqrt{\xi^2 + \Delta^2})\right] \tag{4-125}$$

This is an implicit relation between T and Δ. For $T = 0$, the Fermi function $f(E)$ is zero (E being positive), and one recovers the condition (4-79). The form of $\Delta(T)$ is given in Fig. 4-4. Δ decreases as T increases, and finally vanishes at a certain temperature T_0. Above T_0,

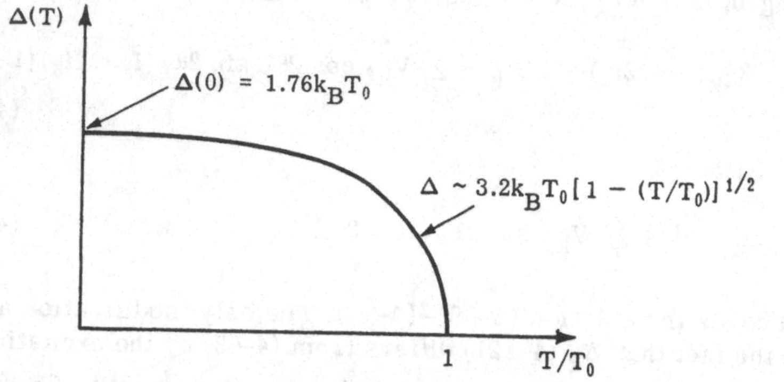

Figure 4-4
Variation of the order parameter Δ with temperature in the BCS approximation.

the normal state ($\Delta_k \equiv 0$) is the only solution of (4-124). T_0 is therefore the transition temperature between the (S) and (N) states (in zero magnetic field). Numerically T_0 is defined by setting $\Delta = 0$ in (4-125)

$$1 = N(0)V \int_0^{\hbar\omega_D} \frac{d\xi}{\xi} \tanh \frac{\xi}{2k_B T_0} \tag{4-126}$$

For $\hbar\omega_D \gg k_B T_0$, it is clear that the integral has the asymptotic form $\ln (\hbar\omega_D/k_B T_0) + C$ since the hyperbolic tangent is equal to unity in the major part of the domain of integration. A detailed calculation gives $C = \ln 1.14$. Therefore

$$1 = N(0)V \ln \frac{1.14\hbar\omega_D}{k_B T_0}$$

or (4-127)

$$k_B T_0 = 1.14\hbar\omega_D \, e^{-1/N(0)V}$$

Taking $\hbar\omega_D = k_B\Theta_D$, where Θ_D is the Debye temperature deduced from specific heat measurements, one can determine the coupling constant $N(0)V$ from T_0. The values for the nontransition metals are given in Table 4-1. In most cases, the coupling constant is rather small and $T_0 \ll \Theta_D$, the only important exceptions being Hg and Pb. (In these two metals, the self-consistent field method is probably inaccurate.) From the exponential form of (4-127), it can be seen that if $N(0)V$ falls below ~0.1, with Θ_D ~200°K, then T_0 falls into the region 10^{-3}°K. For such small values of T_0, superconductivity might be masked by parasitic effects (such as the magnetic field of the earth, nuclear spins, and so on). It is therefore possible that certain metals

Table 4-1

Metal	Θ_D(°K)	T_0(°K)	$[N(0)V]_{exp}$
Zn	235	0.9	0.18
Cd	164	0.56	0.18
Hg	70	4.16	0.35
Al	375	1.2	0.18
In	109	3.4	0.29
Tl	100	2.4	0.27
Sn	195	3.75	0.25
Pb	96	7.22	0.39

not known to be superconducting have, nevertheless, weakly attractive interactions. Finally the qualitative predictions of the "jellium" theory are verified in the sense that N(0)V is large when the valence Z and the atomic volume V_0 are large.[7]

THE ISOTOPE EFFECT

With the BCS interaction (4-76), T_0 is proportional to the cutoff frequency ω_D. As already mentioned, for two isotopes, with the same electronic properties but different ionic masses, ω_D will be different and vary as $M^{-1/2}$. (This is a general result; it comes from the fact that the equation of motion of an ion is of the form $M\, d^2x/dt^2 = F$ where F is a restoring force independent of M.) Equation (4-127) therefore predicts an isotope effect on the transition temperature ($T_0 \propto M^{-1/2}$). This dependence on $M^{-1/2}$ has indeed been observed in a number of nontransition metals (Hg, Pb, Mo, Sn, Tl). However, in many other superconductors (transition metals and compounds), the isotope effect is much smaller, or even nearly absent (Ru, Os). There are various possible explanations for this, notably

(1) The interaction is still mediated by phonons, but the BCS form for $V_{k\ell}$ is oversimplified.

(2) The interaction is mediated by some other low-frequency mode of the solid, involving no lattice motion.

Explanation (2) may well apply in some cases, but it is very hard to prove (or disprove) it experimentally. We discuss only explanation (1).

INCLUSION OF COULOMB REPULSION

In fact, when we discussed the electron-electron interactions earlier in this chapter, we obtained two terms: a repulsive Coulomb term,

---

[7]Numerically, however, the jellium model gives theoretical values for N(0)V which are too large, even when one includes the Coulomb repulsion.

essentially frequency independent, and a phonon term (attractive at frequencies $\omega < \omega_D$, and negligible for $\omega \gg \omega_D$).

As before, we neglect completely the dependence of the matrix $V_{k\ell}$ on the angle between **k** and **ℓ**. We write

$$V_{k\ell} = V(\omega) = V_c - V_p(\omega)$$
$$\omega = \frac{\xi_\ell - \xi_k}{\hbar} \tag{4-128}$$

Equation (4-124) for the transition temperature is now replaced by

$$\Delta(\xi) = -N(0)\int d\xi'\, V\left(\frac{\xi - \xi'}{\hbar}\right)\Delta(\xi')\,\frac{1 - 2f(\xi')}{2\xi'} \tag{4-129}$$

Unfortunately, even with the simple interaction (4-128), this integral equation is hard to solve, and we shall discuss it in a sloppy way. We separate the Coulomb term and call its contribution A

$$A = -N(0)\int d\xi'\, V_c\, \Delta(\xi')\,\frac{1 - 2f(\xi')}{2\xi'} \tag{4-130}$$

A is independent of ξ. The equation for Δ is

$$\Delta(\xi) = N(0)\int d\xi'\, V_p\left(\frac{\xi - \xi'}{\hbar}\right)\Delta(\xi')\,\frac{1 - 2f(\xi')}{2\xi'} + A \tag{4-131}$$

When $|\xi| \gtrsim \hbar\omega_D$ the integral is not large, because the factors $V_p\left(\frac{(\xi - \xi')}{\hbar}\right)$ and $1/\xi'$ cannot be simultaneously large. Thus in this region we may roughly set $\Delta(\xi) = A$. On the other hand, the integral is important when $|\xi|$ is small. Call B the average value of Δ in the region $\hbar\omega_D > |\xi|$. Then we have

$$B \cong N(0)V_p \int_{-\hbar\omega_D}^{\hbar\omega_D} d\xi'\, B\,\frac{1 - 2f(\xi')}{2\xi'} + A$$
$$\cong N(0)V_p \log\frac{\hbar\omega_D}{k_B T_0} + A \tag{4-132}$$

where V_p is some average of $V_p(\omega)$ over the interval $-\omega_D < \omega < \omega_D$. Finally, Eq. (4-130) defining A may be rewritten as

$$A \cong -N(0)V_c\left[B\int_{-\hbar\omega_D}^{\hbar\omega_D} + A\int_{-\hbar\omega_D}^{\hbar\omega_D} + A\int_{-\hbar\omega_D}^{\hbar\omega_c} + A\int_{-\hbar\omega_c}^{\hbar\omega_D}\right]$$

$$\times\ d\xi'\ \frac{1 - 2f(\xi')}{2\xi'}$$

$$\cong -N(0)V_c\left[B \log \frac{\hbar\omega_D}{k_B T_0} + A \log \frac{\omega_c}{\omega_D}\right] \tag{4-133}$$

ω_c is a high-frequency cutoff for the Coulomb interaction (in practice, of the order of $E_F/\hbar$). Writing that (4-133) and (4-130) are compatible, we arrive at the condition

$$1 = \log \frac{\hbar\omega_D}{k_B T_0}\left(K_p - \frac{K_c}{1 + K_c \log \frac{\omega_c}{\omega_D}}\right)$$

$$K_p = N(0)V_p \qquad K_c = N(0)V_c \tag{4-134}$$

An equation of the form (4-134) for the transition temperature was first derived by Bogolubov (1958). It has some remarkable implications:

(1) The Coulomb repulsion described by K_c is not very efficient in counteracting superconductivity, because of the factor

$$\frac{1}{1 + K_c \log \frac{\omega_c}{\omega_D}}$$

In particular, we might have $K_p < K_c$, that is, an over-all interaction constantly repulsive, and still find superconductivity [provided that $K_p(1 + K_c \log(\omega_c/\omega_D)) > K_c$].

(2) The isotope effect is modified: If the ionic mass has a relative variation $\delta M/M$, the Debye frequency is changed according to $\delta\omega_D/\omega_D = -\frac{1}{2}\,\delta M/M$ and the transition temperature shifts according to the law

$$\frac{\delta T_0}{T_0} = \frac{\delta\omega_D}{\omega_D}\left[1 - \frac{K_c^2}{1 + K_c \log \frac{\omega_c}{\omega_D}}\right]$$

The amplitude of the isotope effect is reduced. This reduction should be particularly strong in metals with narrow bands (ω_c small). This may explain why the isotope effect is strongly reduced in transition metals and in related compounds (Garland, 1963).

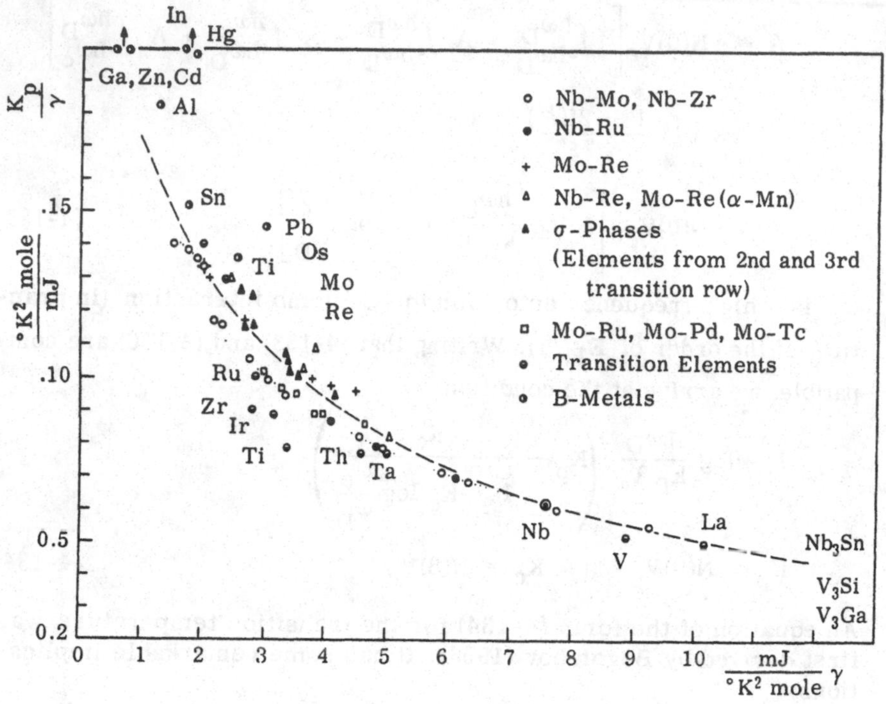

Figure 4-5

An empirical relation between the coupling constant K_p describing the electron-electron interaction via phonons, and the electronic specific heat parameter γ (courtesy J. Muller).

Conversely, from the experimental values of T_0 and of the isotope effect, one can deduce K_p and K_c, assuming that the model is valid. The results for K_p in various metals show a strong correlation between K_p and the electronic specific heat coefficient γ in the normal phase[8] (Muller, 1963) (Fig. 4-5).

Calculation of the Thermodynamic Functions

Combining the expressions for the kinetic and potential energies (4-114) and (4-115) together with the expressions derived for u_k and v_k in the previous section, we obtain for the total energy

[8] γ is proportional to the density of states at the Fermi level. There is, however, one difference between γ and N(0): γ is defined per atom, while N(0) is defined per unit volume.

$$E = \sum_k \frac{1}{2\epsilon_k} [(\epsilon_k + \xi_k)^2 f(\epsilon_k) - (\epsilon_k - \xi_k)^2 (1 - f(\epsilon_k))] \qquad (4\text{-}135)$$

In particular, at absolute zero (f = 0), we recover the energy calculated in Section 4-3. Writing the entropy (4-118) we find

$$TS = 2 \sum_k [\epsilon_k f(\epsilon_k) + k_B T \ln(1 + \exp(\epsilon_k / k_B T))] \qquad (4\text{-}136)$$

The specific heat, for example, is determined by

$$C = T \frac{dS}{dT} \qquad (4\text{-}137)$$

Recalling that ϵ_k depends on T in (4-136), we find

$$C = 2\beta^2 k_B \sum_k f(\epsilon_k)(1 - f(\epsilon_k)) \left[\epsilon_k^2 + \beta \epsilon_k \frac{d\epsilon_k}{d\beta}\right] \qquad (4\text{-}138)$$

In the BCS approximation, $\epsilon_k \, d\epsilon_k/dT = \Delta \, d\Delta/dT$ is independent of k. Then,

$$C = 2\beta^2 k_B N(0) \int_0^\infty d\xi \, f(\epsilon)(1 - f(\epsilon)) \left[\epsilon^2 - T\Delta \frac{d\Delta}{dT}\right]$$

$$\epsilon = \sqrt{\xi^2 + \Delta^2} \qquad (4\text{-}139)$$

The form of this specific heat is represented in Fig. 1-1. At very low temperatures ($\beta\Delta \gg 1$), the term $\Delta \, d\Delta/dT$ is negligible and

$$C \sim 2\beta^2 k_B N(0) \Delta^2 \int_0^\infty d\xi \, e^{-\beta\sqrt{\Delta_0^2 + \xi^2}}$$

$$\sim 2\beta^2 k_B N(0) \Delta^2 e^{-\beta\Delta_0} \int_0^\infty d\xi \, e^{-\beta\xi^2/2\Delta_0} \qquad (4\text{-}140)$$

where $\Delta_0 = \Delta_{T=0}$. The integral is

$$\sqrt{\frac{\pi}{2}} \left(k_D T \Delta_0\right)^{1/2}$$

The dominant factor in C is the term $e^{-\beta\Delta_0}$, which we have already predicted. A measurement of the low-temperature specific heat allows us to determine Δ_0. Also, Eq. (4-139) predicts a discontinuity in the specific heat at the transition temperature because of the $\Delta \, d\Delta/dT$ term

$$C_s - C_n = k_B N(0)\beta_0^2 \left(\frac{d(\Delta^2)}{d\beta}\right)_{T_0}$$

$$\beta_0 = 1/k_B T_0 \tag{4-141}$$

Numerically from (4-125), one obtains $(d\Delta^2/d\beta)_{T_0} = 10.2/\beta_0^3$ and

$$C_s - C_n = 10.2 k_B^2 T_c N(0) \tag{1-142}$$

Finally the Gibbs function is obtained by adding (4-135) and (4-136).

$$G = \sum_k \frac{1}{2\epsilon_k}\left[(\epsilon_k + \xi_k)^2 f + (\epsilon_k - \xi_k)(f-1) - 4f(\epsilon_k^2 + \xi_k^2)\right]$$

$$= \sum_k \left[\left(-\epsilon_k + \frac{\Delta^2}{2\epsilon_k}\right)(2f+1) + \xi_k\right] \tag{4-143}$$

This expression can be transformed by noting that

$$\sum_k \frac{(2f+1)}{2\epsilon_k}\Delta^2 = \sum_k \frac{(2f-1)\Delta^2}{2\epsilon_k} + \sum_k \frac{\Delta^2}{\epsilon_k}$$

The first term on the right is determined from the self-consistency condition (4-124) and is $-\Delta^2/V$. Then

$$G = -2\sum_k \left(\epsilon_k f(\epsilon_k) - \frac{\Delta^2}{V} + \xi_k - \frac{\xi_k^2}{\epsilon_k}\right) \tag{4-144}$$

Knowing G, we can calculate the critical thermodynamic field defined in Chapter 2, from the equation

$$G_n - G_s = \frac{H_c^2}{8\pi} \tag{4-145}$$

In particular, for $T = 0$, we have $H_c = H_{co}$,

$$H_{co}^2 = 4\pi N(0)\,|\,\Delta(0)\,|^2$$

For finite T, from (4-144), we find a curve $H_c(T)$ rather close to the empirical approximation

$$\frac{H_c}{H_{c0}} = 1 - \left(\frac{T}{T_c}\right)^2 \tag{4-146}$$

If we look at finer details, we find that, for superconductors where N(0)V is not too large (weakly coupled superconductors), the detailed theoretical curve of H_c versus T derived from the BCS theory gives even a better fit to the experimental data than does the simplified law (4-146).

Calculation of Transition Probabilities

Suppose that a time-dependent exterior perturbation is applied to a gas of superconducting electrons. Examples:

(1) *Ultrasonics:* A longitudinal acoustic wave modifies the potential energy of each electron by a term $U\theta(rt)$, where θ is the local dilatation of the lattice and U is a constant (the "deformation potential") of the order of several electron volts. The matrix element of this perturbation between the electron plane wave states **k** and **k′** is $U\theta_{k-k'}$ [where θ_q is the Fourier transform of $\theta(\mathbf{r})$]. The perturbation acting on the electron system is then

$$\mathcal{H}_1 = \sum_{k,k',\alpha} U\theta_{k-k'}\, a^+_{k\alpha} a_{k'\alpha} \tag{4-147}$$

(2) *Microwaves:* The effect of an electromagnetic perturbation described by a vector potential A(r, t) is given by replacing $p^2/2m$ by $(1/2m)[\mathbf{p} - (e/c)\mathbf{A}]^2$ in the electronic energy. To first order in **A** the perturbation is $(-e/2mc)(\mathbf{pA} + \mathbf{Ap})$ and as a function of a and a^+ it becomes

$$\mathcal{H}_1 = \sum_{k,k',\alpha} -\frac{e\hbar}{2mc} A_{k-k'} \cdot (\mathbf{k} + \mathbf{k}')\, a^+_{k\alpha} a_{k'\alpha} \tag{4-148}$$

More generally, the perturbations have the form

$$\mathcal{H}_1 = \sum_{\substack{k\alpha \\ k'\alpha'}} B(k\alpha \mid k'\alpha')\, a^+_{k\alpha} a_{k'\alpha'} \tag{4-149}$$

We find two effects in the presence of $\mathcal{H}_1$:

(1) $\mathcal{H}_1$ induces transitions between the different excited states described by the γ^+. To classify these transitions, we write $\mathcal{H}_1$ as a function of the γ, γ^+ operators. Inverting (4-100)

$$a^{+}_{m\alpha} = u_m \gamma^{+}_{m\alpha} + \sum_{\beta} \rho_{\alpha\beta} v_m \gamma_{-m\beta}$$
$$\rho = \begin{pmatrix} 0 & -1 \\ 1 & 0 \end{pmatrix} \tag{4-150}$$

This gives

$$\mathcal{H}_1 = \sum_{\substack{k\alpha \\ k'\alpha'}} B(k\alpha \mid k'\alpha')\{u_k u_{k'} \gamma^{+}_{k\alpha} \gamma_{k'\alpha'} + v_k v_{k'} \sum_{\beta,\beta'} \rho_{\alpha\beta} \rho_{\alpha'\beta'} \gamma_{k\beta} \gamma^{+}_{k'\beta'} + u_k v_{k'} \sum_{\beta} \rho_{\alpha'\beta'} \gamma^{+}_{k\alpha} \gamma^{+}_{-k'\beta'} + u_{k'} v_k \sum_{\beta} \rho_{\alpha\beta} \gamma_{-k\beta} \gamma_{k'\alpha'}\} \tag{4-151}$$

The terms in $\gamma_i^{+}\gamma_j$ and $\gamma_i\gamma_j^{+}$ describe transitions where a quasiparticle in the state i is scattered into the state j (and vice versa). The $\gamma_1^{+}\gamma_2^{+}$ create two quasiparticles and the terms $\gamma_i\gamma_j$ destroy two of them.

(2) $\mathcal{H}_1$ can also modulate the parameters, such as Δ, which describe the structure of the condensed state. and this modulation also leads to an absorption. In many cases this effect is negligible; consider for instance the ultrasonic attenuation problem. The direct perturbation is $U\theta(r)$. If the pair potential is also modulated, it will shift by an amount $\delta\Delta = C\Delta\theta(r)$, where C is a constant of order unity. This gives a modification in the generalized self-consistent field. But U is of order 1-10 eV and $C\Delta$ is of order 10^{-3} eV. Thus the modulation of Δ is unimportant here. In the present paragraph, we neglect it (but we return to it later when we consider the Meissner effect).

We now return to the form (4-151) for $\mathcal{H}_1$ and consider transitions where a quasiparticle passes from the state $(\mathbf{k}'\alpha')$ into the state $(\mathbf{k}\alpha)$. The matrix element $M(k\alpha \mid k'\alpha')$ is the coefficient of $\gamma^{+}_{k\alpha}\gamma_{k'\alpha'}$. The first two terms of (4-151) contribute since $\gamma_1\gamma_2^{+} = -\gamma_2^{+}\gamma_1$ for $1 \neq 2$. Then

$$M(k\alpha \mid k'\alpha') = B(k\alpha \mid k'\alpha') u_k u_{k'} - v_k v_{k'} \sum_{\sigma,\sigma'} \rho_{\sigma'\alpha'} \rho_{\sigma\alpha} B(-k'\sigma' \mid -k\sigma) \tag{4-152}$$

The term $\Sigma_{\sigma\sigma'} B(-k'\sigma' | -k\sigma) \rho_{\sigma'\alpha'} \rho_{\sigma\alpha}$ is essentially the matrix element B where the spins and momenta of the electrons are reversed.

For the interactions $\mathcal{H}_1$, which we consider, it differs at most from $B(k\alpha \mid k'\alpha')$ by a sign

$$\sum_{\sigma\sigma'} B(-k'\sigma' \mid -k\sigma)\rho_{\sigma\alpha}\,\rho_{\sigma'\alpha'} = \eta B(k\alpha \mid k'\alpha') \tag{4-153}$$

$$\text{with } \eta = \begin{cases} +1 & \text{case I} \\ -1 & \text{case II} \end{cases}$$

Then

$$M(k\alpha \mid k'\alpha') = B(k\alpha \mid k'\alpha')[u_k u_{k'} - \eta v_k v_{k'}] \tag{4-154}$$

The factor $[u_k u_{k'} - \eta v_k v_{k'}]$ is called the coherence factor of the transition. The number of transitions $(k'\alpha' \to k\alpha)$ per unit time less the number of inverse transitions is

$$\dot{\nu} = \frac{2\pi}{\hbar} \mid M(k\alpha \mid k'\alpha') \mid^2 \{f(\epsilon_{k'})[1 - f(\epsilon_k)] - f(\epsilon_k)[1 - f(\epsilon_{k'})]\}$$

$$\times\, \delta(\epsilon_k - \epsilon_{k'} - \hbar\omega) \tag{4-155}$$

where we have assumed that $\mathcal{H}_1$ is a sinusoidal perturbation of frequency ω. The power absorbed is

$$W_1 = \sum_{k,\,k'} \dot{\nu}\hbar\omega \tag{4-156}$$

In order to calculate W_1, we first average over the angles of **k** and **k′** and the spin indices of the matrix element B,

$$B^2 = \overline{\mid B(k\alpha \mid k'\alpha') \mid^2}$$

Since $\mid \xi_k \mid$ and $\mid \xi_{k'} \mid$ are small with respect to E_F [this is ensured by the factors f in (4-155)], B can be treated as a constant, and

$$W_1 = 2\pi\omega B^2 \int_\Delta^\infty N_S(\epsilon) N_S(\epsilon')\, d\epsilon\, d\epsilon'\, (uu' - \eta vv')^2$$

$$\times [f(\epsilon') - f(\epsilon)]\, \delta(\epsilon - \epsilon' - \hbar\omega) \tag{4-157}$$

In this formula, $N_S(\epsilon) = N(0) \mid d\xi/d\epsilon \mid$ is the density of states for the Bogolubov excitations

$$N_S(\epsilon) = N(0) \frac{\epsilon}{\sqrt{\epsilon^2 - \Delta^2}} \qquad \epsilon > \Delta \tag{4-158}$$

By using the definitions (4-71) and (4-72) of u and v, we can write

$$[uu' - \eta vv']^2 = \frac{1}{2}\left[1 + \frac{\xi\xi'}{\epsilon\epsilon'} - \eta\,\frac{\Delta^2}{\epsilon\epsilon'}\right] \tag{4-159}$$

Each value of ϵ is obtained for two opposite values of ξ. The term $\xi\xi'/\epsilon\epsilon'$ disappears when summed over these two values. Finally

$$W_1 = 4\pi\omega B^2 N(0)^2 \int_\Delta^\infty d\epsilon \int_\Delta^\infty d\epsilon' \, \frac{\epsilon\epsilon' - \eta\Delta^2}{(\epsilon^2 - \Delta^2)^{1/2}(\epsilon'^2 - \Delta^2)^{1/2}}$$

$$\times [f(\epsilon') - f(\epsilon)]\,\delta(\epsilon - \epsilon' - \hbar\omega) \tag{4-160}$$

An analogous calculation can be made for the power W_2 absorbed by creation and annihilation of two quasiparticles. W_2 is nonzero if $\hbar\omega > 2\Delta$. The final formula for $W = W_1 + W_2$ differs from (4-160) only by the domain of integration,

$$W = 2\pi\omega B^2 N(0)^2 \int_{-\infty}^{\infty}\int_{-\infty}^{\infty} d\epsilon\, d\epsilon' \, \frac{\epsilon\epsilon' - \eta\Delta^2}{(\epsilon^2 - \Delta^2)^{1/2}(\epsilon'^2 - \Delta^2)^{1/2}}$$

$$\times [f(\epsilon') - f(\epsilon)]\,\delta(\epsilon - \epsilon' - \hbar\omega) \tag{4-161}$$

Here ϵ and ϵ' are of arbitrary sign, but $|\epsilon| > \Delta$ and $|\epsilon'| > \Delta$. In practice one always compares W to the absorption W_N, which would be obtained in the normal state [W_N is simply obtained from (4-161) by setting $\Delta = 0$],

$$\frac{W}{W_N} = \frac{1}{\hbar\omega}\int d\epsilon\, d\epsilon' \, \frac{\epsilon\epsilon' - \eta\Delta^2}{(\epsilon^2 - \Delta^2)^{1/2}(\epsilon'^2 - \Delta^2)^{1/2}}$$

$$\times [f(\epsilon') - f(\epsilon)]\,\delta(\epsilon - \epsilon' - \hbar\omega) \tag{4-162}$$

APPLICATIONS

(1) *Absorption of sound:* In this case $\eta = 1$ and the coherence factor is small when ϵ and ϵ' are simultaneously close to Δ or $-\Delta$. If we study the case where $\hbar\omega$ is small compared to Δ or $k_B T$, then (4-162) reduces to

$$\frac{W}{W_N} = -\int_{|\epsilon| > \Delta} d\epsilon \, \frac{\epsilon^2 - \Delta^2}{\epsilon^2 - \Delta^2}\,\frac{\partial f}{\partial \epsilon} = \frac{2}{1 + e^{\beta\Delta}} \tag{4-163}$$

The attenuation is very small at low temperatures and increases rapidly as $T \to T_0$. This therefore gives a way to measure $\Delta(T)$, which has been widely applied, notably by Morse and his co-workers (see Fig. 4-6).

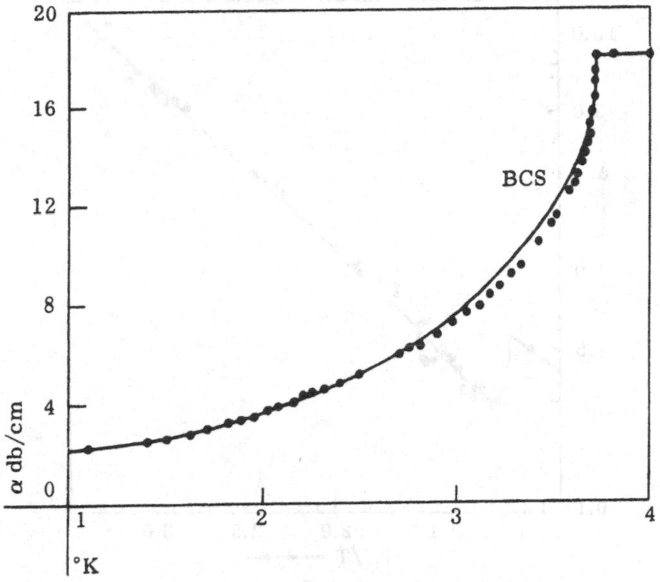

Figure 4-6
Ultrasonic measurements in tin compared with the BCS prediction [after R. W. Morse, *IBM J.*, **6**, 58 (1963)].

(2) *Nuclear Relaxation:* The interaction $\mathcal{H}_1$ between the nuclear spins and the conduction electrons is complicated, but at any rate Case II applies: $\eta = -1$. One can measure the time T_1 for the nuclear spins to equalize their temperature with the electrons in zero field. The ratio T_{1n}/T_1 of the relaxation rates in the superconducting and normal states is still given by (4-162). The frequency ω, which comes into play here, is the precession frequency of the nuclear spins in the local field of the other nuclei. This is small ($\omega \sim 10^4$). Therefore we can let $\omega \to 0$ and obtain

$$\frac{T_{1n}}{T_1} = -\int_{|\epsilon| > \Delta} d\epsilon \, \frac{df}{d\epsilon} \frac{\epsilon^2 + \Delta^2}{\epsilon^2 - \Delta^2} \tag{4-164}$$

Here the coherence factor doesn't vanish at the same time as the denominator (for $|\epsilon| = \Delta$): the integral diverges logarithmically. But in a real metal, Δ_k is anisotropic. The singularity in density of states $N_s(\epsilon)$ is then smeared out somewhat and the integral converges. The results depend in detail on the anistropy of Δ_k. In practice, Δ is rather weakly anisotropic; then the variation of T_{1n}/T_1 with temperature is as shown on Fig. 4-7. The relaxation rate for T slightly less than T_c is *greater than in the normal state*. This remarkable result arises

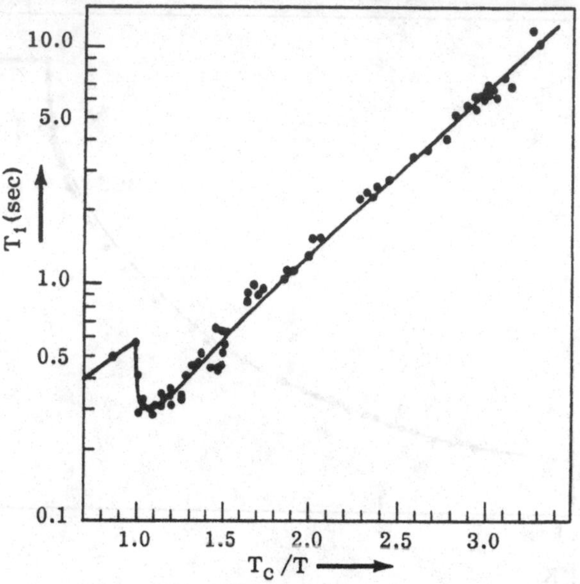

Figure 4-7

Nuclear relaxation rates in aluminum [after A. G. Redfield, *Phys. Rev.*, **125**, 159 (1962)]. Notice the dip of T_1 at temperatures below the transition point. The theoretical curve is calculated on the assumption that the peak in the BCS density of states is smeared out on an energy interval $\Delta/5$.

from the increase in the density of states $N_s(\epsilon)$. At low temperatures $T \ll T_c$, on the contrary, the relaxation becomes very slow because of the population factor $df/d\epsilon$. This type of variation of $1/T_1$ with T was first observed in 1957 by Hebel and Slichter. The prediction of the different coherence factors and of the resulting very different behaviors of the ultrasonic attenuation and nuclear relaxation has been one of the great successes of the BCS theory.

REFERENCES

M. Tinkham, "Superconductivity," in *Low Temperature Physics*, Les Houches lecture notes (New York: Gordon & Breach, 1962).

E. A. Lynton, *Superconductivity* (London: Methuen, 1964, 2nd edition), Chap. 8, 9, 10, 11.

5

THE SELF-CONSISTENT FIELD METHOD

5–1 THE BOGOLUBOV EQUATIONS

In the preceding chapter, we studied a homogeneous electron gas with attractive interactions. We now consider the more general case where the electrons also experience an arbitrary external potential $U_0(\mathbf{r})$ (this will be important to describe the effects of impurities and of the specimen surface) and a magnetic field $\mathbf{H}$ = curl $\mathbf{A}$.

The most natural method, at first sight, to treat $U_0(\mathbf{r})$ would be the following:

(1) Find the wave function w_n of the one-electron Hamiltonians, which are solutions of

$$\xi_n w_n = \left(-\frac{\hbar^2}{2m}\nabla^2 + U_0(\mathbf{r}) - E_F\right) w_n \tag{5-1}$$

(2) Notice that there exist at least two solutions for each energy ξ_n:

$$w_n = w_n(\mathbf{r})\,|\uparrow\rangle \qquad \text{and} \qquad w_{\bar{n}} = w_n^*(\mathbf{r})\,|\downarrow\rangle \tag{5-2}$$

where $|\alpha\rangle$ denotes the spin states.

(3) Construct a trial function of the BCS type describing a pairing of electrons in the degenerate energy states w_n and $w_{\bar{n}}$,

$$\tilde{\phi} = \prod_n (u_n + v_n a_n^+ a_{\bar{n}}^+)\,\phi_0 \tag{5-3}$$

where $a^+_{\bar{n}}$ is the creation operator for an electron in the state $w_{\bar{n}}$. The analysis follows that of Section 3.

In fact, this method has several defects: (a) it is not applicable in the presence of a magnetic field; (b) above all, it does not allow enough freedom in the trial function. In fact, the energy can often be lowered if we pair the electrons in states that are better choices than w_n.

We now describe a more powerful method (Bogolubov, 1959), which is essentially a generalization of the Hartree-Fock equations to the case of superconductivity. We begin by rewriting the Hamiltonian $\mathcal{H}$ of the electron system, not with the operators $a_{k\alpha}$ (which are only

($\alpha = \uparrow$ or $\downarrow$ is again a spin index.) The operators Ψ satisfy the anti-commutation rules

$$\Psi(r\alpha) = \sum_k e^{ik\cdot r}\, a_{k\alpha}$$
$$\Psi^+(r\alpha) = \sum_k e^{-ik\cdot r}\, a^+_{k\alpha} \qquad (5\text{-}4)$$

($\alpha = \uparrow$ or $\downarrow$ is again a spin index.) The operators Ψ satisfy the anticommutation rules

$$\Psi(r\alpha)\Psi(r'\beta) + \Psi(r'\beta)\Psi(r\alpha) = 0$$
$$\Psi^+(r\alpha)\Psi^+(r'\beta) + \Psi^+(r'\beta)\Psi^+(r\alpha) = 0 \qquad (5\text{-}5)$$
$$\Psi^+(r\alpha)\Psi(r'\beta) + \Psi(r'\beta)\Psi^+(r\alpha) = \delta_{\alpha\beta}\,\delta(r - r')$$

The operator associated with the number of particles is

$$N = \sum_{k\alpha} a^+_{k\alpha} a_{k\alpha} = \sum_\alpha \int dr\, \Psi^+(r\alpha)\Psi(r\alpha) \qquad (5\text{-}6)$$

The Hamiltonian $\mathcal{H}$ is also very simply written in terms of Ψ and Ψ^+.[1]

$$\mathcal{H} = \mathcal{H}_0 + \mathcal{H}_1 \qquad (5\text{-}7)$$

$$\mathcal{H}_0 = \int dr \sum_\alpha \Psi^+(r\alpha)\left[\frac{\left(p - \frac{e}{c}A\right)^2}{2m} + U_0(r)\right]\Psi(r) \qquad (5\text{-}8)$$

$$\mathcal{H}_1 = -\tfrac{1}{2}V \int dr \sum_{\alpha\beta} \Psi^+(r\alpha)\Psi^+(r\beta)\Psi(r\beta)\Psi(r\alpha) \qquad (5\text{-}9)$$

---

[1]See Landau and Lifschitz, *Nonrelativistic Quantum Mechanics* (New York: Pergamon, 1959), Chap. 9.

We have taken $U_0(\mathbf{r})$ to be *independent of spin* (it will eventually be necessary to add a spin-dependent term to describe the exchange potential existing in magnetic media). Also, for the electron-electron coupling $\mathcal{H}_1$, we have assumed the simplest form describing an interaction, which is

(a) independent of spin (this is correct for a nonmagnetic material),

(b) pointlike and thus characterized by only one coefficient V (BCS approximation).

Also notice that we have neglected the effect of the magnetic field on the conduction electron spins in $\mathcal{H}_0$. (This is correct for $e\hbar H/mc < \Delta$.) It will be useful to define

$$H_0 - E_F N = \sum_\alpha \int \Psi^+(\mathbf{r}\alpha)\mathcal{H}_e \Psi(\mathbf{r}\alpha)\, d\mathbf{r} \tag{5-10}$$

with

$$\mathcal{H}_e(\mathbf{r}) = \frac{1}{2m}\left(-i\hbar\nabla - \frac{eA}{c}\right)^2 + U_0(\mathbf{r}) - E_F \tag{5-11}$$

Definition of the Effective Potentials

We now replace the interaction $V\Psi^+\Psi^+\Psi\Psi$ by an average potential acting on only one particle at a time (therefore only containing two operators Ψ or Ψ^+). We try an effective hamiltonian of the form

$$\begin{aligned}\mathcal{H}_{\text{eff}} = \int d\mathbf{r} \Big\{ &\sum_\alpha \Psi^+(\mathbf{r}\alpha)\mathcal{H}_e(\mathbf{r})\Psi(\mathbf{r}\alpha) + U(\mathbf{r})\Psi^+(\mathbf{r}\alpha)\Psi(\mathbf{r}\alpha) \\ &+ \Delta(\mathbf{r})\Psi^+(\mathbf{r}\uparrow)\Psi^+(\mathbf{r}\downarrow) + \Delta^*(\mathbf{r})\Psi(\mathbf{r}\downarrow)\Psi(\mathbf{r}\uparrow)\Big\}\end{aligned} \tag{5-12}$$

The term U destroys and creates one electron and therefore conserves the number of particles. On the other hand, the terms in Δ increase or decrease the number of particles by two. This should not disturb us, since $\mathcal{H}_{\text{eff}}$ will operate on wave functions such as $\tilde{\phi}$, which are not eigenfunctions of the number operator. In the simple case of the preceding section, an average such as $\langle\tilde{\phi}\,|\,a_k^+ a_{-k}^+\,|\,\tilde{\phi}\rangle$ is nonzero. Here, in the same way, the mean of the product $\Psi^+\Psi^+$ will be nonvanishing and this term will play an important role. We call Δ the *pair potential*. [Note that terms such as $\Psi(\mathbf{r}\uparrow)\Psi(\mathbf{r}\uparrow)$ vanish identically because of the commutation rules (5-5).]

Energy Levels of the Effective Hamiltonian $\mathcal{H}_{\text{eff}}$

Let us assume, for the moment, that $\mathcal{H}_{\text{eff}}$ is known; we try to determine its eigenstates and corresponding energies. $\mathcal{H}_{\text{eff}}$ is a quadratic form in Ψ and Ψ^+. We can diagonalize it by performing a unitary transformation which is the exact analog of Eq. (4-100)

$$\begin{aligned}\Psi(\mathbf{r}\uparrow) &= \sum_n \left(\gamma_{n\uparrow} u_n(\mathbf{r}) - \gamma^+_{n\downarrow} v^*_n(\mathbf{r})\right)\\ \Psi(\mathbf{r}\downarrow) &= \sum_n \left(\gamma_{n\downarrow} u_n(\mathbf{r}) + \gamma^+_{n\uparrow} v^*_n(\mathbf{r})\right)\end{aligned} \tag{5-13}$$

where the γ and γ^+ are new operators still satisfying the fermion commutation relations

$$\begin{aligned}\gamma^+_{n\alpha}\gamma_{m\beta} + \gamma_{m\beta}\gamma^+_{n\alpha} &= \delta_{mn}\delta_{\alpha\beta}\\ \gamma_{n\alpha}\gamma_{m\beta} + \gamma_{m\beta}\gamma_{n\alpha} &= 0\end{aligned} \tag{5-14}$$

The transformation (5-13) must diagonalize $\mathcal{H}_{eff}$, that is,

$$\mathcal{H}_{eff} = E_g + \sum_{n,\alpha} \epsilon_n \gamma^+_{n\alpha}\gamma_{n\alpha} \tag{5-15}$$

where E_g is the ground state energy of $\mathcal{H}_{eff}$ and ϵ_n is the energy of the excitation n. We can also write this condition by taking the commutator of $\mathcal{H}_{eff}$ with $\gamma_{n\alpha}$ and $\gamma^+_{n\alpha}$

$$\begin{aligned}[\mathcal{H}_{eff}, \gamma_{n\alpha}] &= -\epsilon_n\gamma_{n\alpha}\\ [\mathcal{H}_{eff}, \gamma^+_{n\alpha}] &= \epsilon_n\gamma^+_{n\alpha}\end{aligned} \tag{5-16}$$

These conditions fix the functions u_n and v_n in (5-13). To derive the equations for u and v, we calculate the commutator $[\mathcal{H}_{eff}, \Psi]$, using the definition (5-12) of $\mathcal{H}_{eff}$ and the anticommutation properties of the Ψ, obtaining

$$\begin{aligned}[\Psi(\mathbf{r}\uparrow), \mathcal{H}_{eff}] &= [\mathcal{H}_e + U(\mathbf{r})]\Psi(\mathbf{r}\uparrow) + \Delta(\mathbf{r})\Psi^+(\mathbf{r}\downarrow)\\ [\Psi(\mathbf{r}\downarrow), \mathcal{H}_{eff}] &= [\mathcal{H}_e + U(\mathbf{r})]\Psi(\mathbf{r}\downarrow) - \Delta^*(\mathbf{r})\Psi^+(\mathbf{r}\uparrow)\end{aligned} \tag{5-17}$$

In this equality, we replace the Ψ's by the γ's by means of (5-13). Then we apply the commutation relations (5-16). Comparing the coefficients of γ_n (and γ^+_n) on the two sides of the equation, we obtain the *Bogolubov equations*:

$$\begin{aligned}\epsilon u(\mathbf{r}) &= [\mathcal{H}_e + U(\mathbf{r})]u(\mathbf{r}) + \Delta(\mathbf{r})v(\mathbf{r})\\ \epsilon v(\mathbf{r}) &= -[\mathcal{H}^*_e + U(\mathbf{r})]v(\mathbf{r}) + \Delta^*(\mathbf{r})u(\mathbf{r})\end{aligned} \tag{5-18}$$

The $\begin{pmatrix} u_n \\ v_n \end{pmatrix}$ are eigenfunctions of a linear system with corresponding eigenvalues ϵ_n:

$$\epsilon \begin{pmatrix} u \\ v \end{pmatrix} = \hat{\Omega} \begin{pmatrix} u \\ v \end{pmatrix} \tag{5-19}$$

Remarks. (1) The operator $\mathcal{H}_e^*$ differs from $\mathcal{H}_e$ when a magnetic field is present:

$$\mathcal{H}_e = \frac{1}{2m}\left(-i\hbar\nabla - \frac{eA}{c}\right)^2 + U_0(\mathbf{r}) - E_F$$

$$\mathcal{H}_e^* = \frac{1}{2m}\left(i\hbar\nabla - \frac{eA}{c}\right)^2 + U_0(\mathbf{r}) - E_F \neq \mathcal{H}_e$$

Both $\mathcal{H}_e$ and $\mathcal{H}_e^*$ are Hermitian.

(2) The operator $\hat{\Omega}$ is Hermitian; consequently different eigenfunctions $\begin{pmatrix} u \\ v \end{pmatrix}$ are orthogonal.

(3) If $\begin{pmatrix} u \\ v \end{pmatrix}$ is the solution for the eigenvalue ϵ, $\begin{pmatrix} -v^* \\ u^* \end{pmatrix}$ is the solution for the eigenvalue $-\epsilon$.[2] In agreement with (5-15) we only keep the solutions with positive ϵ.

Choice of the Potentials U and Δ

We now determine $\mathcal{H}_{eff}$ by requiring that the free energy F calculated from the states which diagonalize $\mathcal{H}_{eff}$ be stationary. By definition

$$F = \langle \mathcal{H} \rangle - TS \tag{5-20}$$

where $\mathcal{H}$ is the initial Hamiltonian $\mathcal{H} = \mathcal{H}_0 + \mathcal{H}_1$ (5-7–9) and the average $\langle \mathcal{H} \rangle$ is given by

$$\langle \mathcal{H} \rangle = \frac{\Sigma_\phi \langle \phi | \mathcal{H} | \phi \rangle \exp(-\beta E_\phi)}{\Sigma_\phi \exp(-\beta E_\phi)} \tag{5-21}$$

The matrix elements are taken with respect to the eigenfunctions $|\phi\rangle$ of $\mathcal{H}_{eff}$:

[2] These two properties allow us to verify that the transformation (5-13) is indeed unitary.

$$\mathcal{H}_{\text{eff}} \,|\phi\rangle = E_\phi \,|\phi\rangle \tag{5-22}$$

A general method to calculate $\langle\mathcal{H}\rangle$ is to replace the Ψ's (5-7) by the γ's according to (5-13), then to use the mean value rules

$$\langle \gamma^+_{n\alpha}\gamma_{m\beta}\rangle = \delta_{nm}\delta_{\alpha\beta} f_n$$

$$\langle \gamma_{n\alpha}\gamma_{m\beta}\rangle = 0 \tag{5-23}$$

$$f_n = \frac{1}{\exp(\beta\epsilon_n) + 1}$$

However it will not be necessary to perform this entire calculation. We write $\langle\mathcal{H}\rangle$ in the form

$$\langle\mathcal{H}\rangle = \sum_\alpha \int dr\, \langle \Psi^+(r\alpha)\mathcal{H}_e \Psi(r\alpha)\rangle$$

$$-\sum_{\alpha\beta} \frac{V}{2}\int dr \langle \Psi^+(r\alpha)\Psi^+(r\beta)\Psi(r\beta)\Psi(r\alpha)\rangle \tag{5-24}$$

The product $\langle \Psi^+\Psi^+\Psi\Psi\rangle$ can be simplified thanks to a theorem by Wick, which makes use only of the fact that the Ψ^+ and Ψ are linear functions of the γ^+, γ. The theorem gives

$$\langle\Psi^+(1)\Psi^+(2)\Psi(3)\Psi(4)\rangle = \langle\Psi^+(1)\Psi(4)\rangle\langle\Psi^+(2)\Psi(3)\rangle$$

$$- \langle\Psi^+(1)\Psi(3)\rangle\langle\Psi^+(2)\Psi(4)\rangle$$

$$+ \langle\Psi^+(1)\Psi^+(2)\rangle\langle\Psi(3)\Psi(4)\rangle \tag{5-25}$$

We now vary amplitudes $\binom{u}{v}$ to $\binom{\delta u}{\delta v}$ and the occupation numbers f_n to δf_n. The free energy (5-20) then varies by δF:

$$\delta F = \int dr \Big\{ \sum_\alpha \delta[\langle\Psi^+(r\alpha)\mathcal{H}_e\Psi(r\alpha)\rangle]$$

$$- V\sum_{\alpha\beta}\langle\Psi^+(r\alpha)\Psi(r\alpha)\rangle\; \delta[\langle\Psi^+(r\beta)\Psi(r\beta)\rangle]$$

$$+ V\sum_\alpha \langle\Psi^+(r\alpha)\Psi(r\alpha)\rangle\; \delta[\langle\Psi^+(r\alpha)\Psi(r\alpha)\rangle]$$

$$- V[\langle\Psi^+(r\uparrow)\Psi^+(r\downarrow)\rangle\; \delta(\langle\Psi(r\downarrow)\Psi(r\uparrow)\rangle + \text{C.C.}]\Big\} - T\,\delta S \tag{5-26}$$

where we have assumed $\langle \Psi^+(\mathbf{r}\uparrow)\Psi(\mathbf{r}\downarrow)\rangle = 0$ as is correct for our "non-magnetic" situations. Now notice that the quantity

$$F_1 = \langle \mathcal{H}_{eff} \rangle - TS \tag{5-27}$$

is stationary with respect to δu_n, δv_n, and δf_n since our excitations diagonalise $\mathcal{H}_{eff}$ exactly. Using (5-12), this condition becomes explicitly:

$$\begin{aligned} 0 &= \delta \langle \mathcal{H}_{eff} \rangle - T\ \delta S \\ &= \int d\mathbf{r} \left\{ \sum_\alpha \delta \langle \Psi^+(\mathbf{r}\alpha)(\mathcal{H}_e + U(\mathbf{r}))\Psi(\mathbf{r}\alpha)\rangle \right. \\ &\quad \left. + \left[\Delta(\mathbf{r})\,\delta(\langle \Psi^+(\mathbf{r}\uparrow)\Psi^+(\mathbf{r}\downarrow)\rangle) + \text{C.C.}\right] \right\} - T\ \delta S \end{aligned} \tag{5-28}$$

Comparing (5-26) and (5-28), we see that F will be stationary if we take as effective potentials

$$U(\mathbf{r}) = -V\langle \Psi^+(\mathbf{r}\uparrow)\Psi(\mathbf{r}\uparrow)\rangle = -V\langle \Psi^+(\mathbf{r}\downarrow)\Psi(\mathbf{r}\downarrow)\rangle \tag{5-29}$$

(the standard Hartree-Fock result for a point interaction), and

$$\Delta(\mathbf{r}) = -V\langle \Psi(\mathbf{r}\downarrow)\Psi(\mathbf{r}\uparrow)\rangle = V\langle \Psi(\mathbf{r}\uparrow)\Psi(\mathbf{r}\downarrow)\rangle \tag{5-30}$$

If we replace the Ψ's by the γ's by means of (5-13) and if we use the mean value rules (5-23), we can put these conditions into the explicit form

$$U(\mathbf{r}) = -V \sum_n \left[|u_n(\mathbf{r})|^2 f_n + |v_n(\mathbf{r})|^2 (1 - f_n) \right] \tag{5-31}$$

$$\Delta(\mathbf{r}) = +V \sum_n v_n^*(\mathbf{r})\, u_n(\mathbf{r})\,(1 - 2f_n) \tag{5-32}$$

These conditions assure that the potentials U and Δ are self-consistent.

In practice there is an important distinction between $U(\mathbf{r})$ and $\Delta(\mathbf{r})$. The Hartree-Fock potential $U(\mathbf{r})$ comes from an integral Σ_n involving all states below the Fermi level. Consequently, $U(\mathbf{r})$ is nearly independent of temperature, and U can be approximated by the Hartree-Fock potential calculated *in the normal state*. This represents a considerable simplification. On the contrary, the pair potential $\Delta(\mathbf{r})$ is a sum of terms of the form $u_n v_n^*$. Such terms, as we already have seen for the homogeneous gas, are nonzero only in the neighborhood of the Fermi surface. For this reason $\Delta(\mathbf{r})$ is a strong function of temperature.

For the case of the homogeneous gas considered in the previous chapter, $\Delta(\mathbf{r})$ is spatially constant and the self-consistency equation (5-31) reduces to a determination of the constant Δ. When $\Delta(\mathbf{r})$ varies spatially, it is much more difficult to ensure the self-consistency of Δ. We shall return several times to this problem.

Problem. Discuss the quasiparticle spectrum for a state of uniform current flow, in a pure superconductor. A state of uniform flow is described by a pair potential of the form $\Delta = |\Delta| e^{2i\mathbf{q}\cdot\mathbf{r}}$ where $\mathbf{q}$ is a vector in the direction of flow (the average momentum per electron in this state is $\hbar\mathbf{q}$).

Solution. The solution of (5-18) is then of the form

$$u(\mathbf{r}) = U_k e^{i(\mathbf{k}+\mathbf{q})\cdot\mathbf{r}}$$

$$v(\mathbf{r}) = V_k e^{i(\mathbf{k}-\mathbf{q})\cdot\mathbf{r}}$$

and the eigenvalue equation reads

$$(\epsilon_k - \xi_{k+q})U_k - |\Delta| V_k = 0$$

$$-|\Delta| U_k + (\xi_k + \xi_{k-q}) V_k = 0$$

The positive eigenvalue ϵ_k is given by

$$\epsilon_k = \frac{\xi_{k+q} - \xi_{k-q}}{2} + \left[\left(\frac{\xi_{k+q} + \xi_{k-q}}{2}\right)^2 + |\Delta|^2\right]^{1/2}$$

As we shall see, the region of major interest is $q \sim \Delta/\hbar v_F$, thus $q \ll k_F$. Retaining only terms of first order in q, we can write

$$\frac{\xi_{k+q} - \xi_{k-q}}{2} = \frac{\hbar^2 \mathbf{k}}{m}\cdot\mathbf{q}$$

$$\frac{\xi_{k+q} + \xi_{k-q}}{2} \cong \xi_k$$

$$\epsilon_k = \epsilon_k^0 + \frac{\hbar^2 \mathbf{k}}{m}\cdot\mathbf{q}$$

where ϵ_k^0 is the excitation energy for $q = 0$, $\epsilon_k^0 = \sqrt{|\Delta|^2 + \xi_k^2}$, and $\hbar q/m$ is usually called the superfluid velocity v_s. Notice that the gap in the energy spectrum goes to 0 when $v_s = |\Delta|/\hbar k_F = |\Delta|/p_F$. (In application of these formulas it must be recalled that, in general, the self-consistent value of $|\Delta|$ depends on v_s.)

In principle, this spectrum could be studied in thin films of pure metals; in practice it cannot for the following reasons: The film thickness d must be smaller than the penetration depth λ (to have a uniform current), but d must be large compared with ξ_0 (if not, the diffuse scattering on the film boundaries can bring some important changes in the spectrum). Thus we would have to work with a pure Type II material, with $\lambda \gg \xi_0$. But then comes another difficulty: for $v_s \sim |\Delta|/p_F$, the current density $n_s e v_s$ is high. The field at the surface of the film is of order

$$\frac{1}{c}(n_s e v_s)\, d \sim \frac{\phi_0 d}{\lambda^2 \xi_0} \qquad \text{for } T = 0$$

It is larger than the first penetration field H_{c1} by a factor $\sim d/\xi_0 > 1$. Thus vortex lines come in the film and the situation is not the one desired.

On the other hand, the above calculation can be of use for a discussion of quasiparticles in the Schubnikov phase of bulk, pure, Type II superconductors (M. Cyrot, 1964). Assume that

(a) the Landau-Ginsburg parameter $\kappa = \lambda/\xi$ is much larger than unity,
(b) the field H is much smaller than the upper critical field H_{c2}.

Then the distance d between lines is much larger than ξ, and most of the excitations (with energies $\epsilon \sim \Delta_\infty$, where Δ_∞ is the amplitude of the pair potential far from the line) can be obtained as follows: At point $\mathbf{r}$ compute the superfluid velocity $\mathbf{v}_s(\mathbf{r})$ and write that locally the excitations have a shifted BCS spectrum

$$\epsilon(\mathbf{k}, \mathbf{r}) = \{\Delta_\infty^2 + \xi_k^2\}^{1/2} + \hbar \mathbf{v}_s \cdot \mathbf{k} \qquad \text{where } \xi_k = \frac{\hbar^2(k^2 - k_F^2)}{2m}$$

Since the minimum spatial extension of wave packets made with these excitations is of order ξ, while the fields and the velocities v_s are modulated on a scale $d > \xi$, in general, this procedure is correct.

It fails however for the low-lying excited states ($\epsilon \ll \Delta_\infty$) localized very near one line where v_s and the order parameter Δ vary rapidly. These particular excitations are discussed in the next problem, page 153.

5–2 THEOREMS ON THE PAIR POTENTIAL AND THE EXCITATION SPECTRUM

Gauge Invariance

The one-electron Hamiltonian

$$\mathcal{H}_e(A) = \frac{1}{2m}\left(p - \frac{eA}{c}\right)^2 + U - E_F$$

and its complex conjugate

$$\mathcal{H}_e^*(A) = \frac{1}{2m}\left(p + \frac{eA}{c}\right)^2 + U - E_F$$

depend on the choice of the vector potential **A**. This choice of **A** is not unique. If **A** is replaced by **A′** such as

$$\mathbf{A}' = \mathbf{A} + \nabla\chi(\mathbf{r}) \tag{5-33}$$

where χ is an arbitrary function, then

$$\text{curl } \mathbf{A}' = \text{curl } \mathbf{A} = \mathbf{h}$$

A and **A′** are equally acceptable to describe the field configuration **h(r)**. All physically measurable quantities will have the same values when calculated with **A** or **A′**. We shall prove this property for the quasiparticle excitation energies, that is, for the eigenvalues of Eq. (5-18).

Suppose that the eigenfunctions $\begin{pmatrix} u_n \\ v_n \end{pmatrix}$ of (5-18) with vector potential **A** are known. Now consider Eq. (5-18) with vector potential **A′**. Two things are modified:

(a) the eigenfunctions $\begin{pmatrix} u'_n \\ v'_n \end{pmatrix}$ differ from $\begin{pmatrix} u_n \\ v_n \end{pmatrix}$.

$$\left.\begin{aligned} u'_n(\mathbf{r}) &= u_n(\mathbf{r}) \exp\left[\frac{ie}{\hbar c}\chi(\mathbf{r})\right] \\ v'_n(\mathbf{r}) &= v_n(\mathbf{r}) \exp\left[-\frac{ie}{\hbar c}\chi(\mathbf{r})\right] \end{aligned}\right\} \tag{5-34}$$

(b) the pair potential is modified: if $\Delta(\mathbf{r})$ was associated with **A**, the pair potential $\Delta'(\mathbf{r})$ is given by

$$\Delta'(\mathbf{r}) = \Delta(\mathbf{r}) \exp\left[\frac{2ie}{\hbar c}\chi(\mathbf{r})\right] \tag{5-35}$$

Proof of (5-34): The function u'_n defined by (5-34) satisfies the equation

$$\left(p - \frac{eA'}{c}\right) u_n'(\mathbf{r}) = \left(-i\hbar\nabla - \frac{e}{c}A'\right) \exp\left[\frac{ie}{\hbar c}\chi(\mathbf{r})\right] u_n(\mathbf{r})$$
$$= \exp\left[\frac{ie}{\hbar c}\chi\right]\left(-i\hbar\nabla - \frac{eA'}{c} + \frac{e}{c}\nabla\chi\right) u_n(\mathbf{r}) \qquad (5\text{-}36)$$
$$= \exp\left[\frac{ie}{\hbar c}\chi\right]\left(p - \frac{e}{c}A\right) u_n$$

where we have made use of (5-33). Iterate this property:

$$\left(p - \frac{eA'}{c}\right)^2 u_n' = \exp\left[\frac{ie}{\hbar c}\chi\right]\left(p - \frac{eA}{c}\right)^2 u_n$$
$$\mathcal{H}_e(A')\,u_n' = \exp\left[\frac{ie}{\hbar c}\chi\right]\mathcal{H}_e(A)\,u_n \qquad (5\text{-}37a)$$

Similarly

$$\mathcal{H}_e^*(A')\,v_n' = \exp\left[\frac{-ie}{\hbar c}\chi\right]\mathcal{H}_e^*(A)\,v_n \qquad (5\text{-}37b)$$

Since $\begin{pmatrix} u_n \\ v_n \end{pmatrix}$ is an eigenfunction of (5-18) we have

$$\mathcal{H}_e(A)\,u_n + \Delta v_n = \epsilon_n u_n$$

Multiply by $\exp[ie\chi/\hbar c]$ and make use of (5-37a). The equation becomes

$$\mathcal{H}_e(A')u_n' + \Delta' v_n' = \epsilon_n u_n' \qquad (5\text{-}38a)$$

and similarly

$$-\mathcal{H}_e^*(A')\,v_n' + \Delta' u_n' = \epsilon_n v_n' \qquad (5\text{-}38b)$$

The $\begin{pmatrix} u_n' \\ v_n' \end{pmatrix}$ is the set of eigenfunctions of the vector potential A' and the pair potential Δ'. Furthermore if Δ satisfied the self-consistency relation (5-32) with the set $\begin{pmatrix} u \\ v \end{pmatrix}$, Δ' also satisfies (5-32) with the set $\begin{pmatrix} u' \\ v' \end{pmatrix}$.

Conclusion: The wave functions, and even the pair potential, have changed by the "gauge transformation" $A \to A'$. We say that Δ, for instance, is "gauge covariant." But the eigenvalues ϵ_n are unchanged: We say that they are "gauge invariant." We can prove in the same way that all physically measurable quantities (such as the current density at one point ...) are gauge invariant.

Remarks.

Restrictions on the possible gauge functions χ. The self-consistent field $\Delta(\mathbf{r})$ must be a single-valued function of $\mathbf{r}$, whatever gauge we choose. Take for instance a sample in the form of a hollow cylinder, 0z being the cylinder axis and φ the rotation angle around 0z. Try a gauge function of the form

$$\chi = \frac{\hbar c}{2e} m\varphi \tag{5-39}$$

where m is an arbitrary constant. Then

$$\Delta'(r) = \Delta(\mathbf{r})\, e^{im\varphi} \tag{5-40}$$

Both Δ and Δ' must be single values: This imposes that m be an integer. More generally, for an arbitrary ring shaped specimen, the increment of χ when performing one term around the ring must be $2\pi(c\hbar/2e) = ch/2e$ times an integer.

SPECIAL CHOICES OF GAUGE

Starting from a given **A**, the transformation function χ can be chosen to impose some convenient properties to $\mathbf{A}'$.

Example 1. For a singly connected specimen, we impose

$$\begin{aligned} \operatorname{div} \mathbf{A}' &= 0 \qquad \text{in the sample} \\ \mathbf{A}' \cdot \mathbf{n} &= 0 \qquad \text{at the surface} \end{aligned} \tag{5-41}$$

(where $\mathbf{n}$ is a unit vector normal to the surface). The transformation function χ is derived from the equations

$$\begin{aligned} \nabla^2 \chi &= -\operatorname{div} \mathbf{A} \qquad \text{in the sample} \\ \mathbf{n} \cdot \nabla\chi &= -\mathbf{n} \cdot \mathbf{A} \qquad \text{at the surface} \end{aligned} \tag{5-42}$$

According to general theorems in electrostatics, the solution χ of (5-42) is unique for a singly connected specimen. Thus $\mathbf{A}'$ is well defined. We say that $\mathbf{A}'$ is the vector potential in the *London* gauge.

Example 2. With **A** as vector potential, we have a certain pair potential $\Delta(\mathbf{r})$. Separate the amplitude and phase of Δ:

$$\Delta(\mathbf{r}) = |\Delta(\mathbf{r})|\, e^{i\varphi(\mathbf{r})} \tag{5-43}$$

Then transform to $\mathbf{A}'$ using as gauge function

$$\chi = -\frac{\hbar c}{2e} \varphi \tag{5-44}$$

Since Δ was single valued, this choice χ is always acceptable.[3] The new pair potential is simply

$$\Delta'(\mathbf{r}) = |\Delta(\mathbf{r})|$$

Thus it is always possible to choose a gauge where the pair potential is real. For a singly connected superconductor with no external current leads, this gauge coincides in fact with the London gauge, but it is defined for more general situations.

Flux Quantization

Consider the superconducting ring of Fig. 5-1 (diameter and thickness much larger than the penetration depth). Experiments by Doll and Nabauer, and by Deaver and Fairbanks (1961), have shown that the flux ϕ, which passes through the ring, can only take on certain discrete values:

$$\phi = n\phi_0$$

$$\phi_0 = \frac{ch}{2e} \simeq 2.10^{-7}\ \mathrm{G} \times \mathrm{cm}^2 \tag{5-45}$$

$$n = \text{any integer}$$

(Such measurements of small fluxes are done on metallic films deposited on the surface of a capillary $\sim 10\,\mu$ in diameter. The flux unit ϕ_0 then corresponds to a sizeable field ~ 0.1 G.)

This "flux quantization" is a consequence of the condition that $\Delta(\mathbf{r})$ be a single-valued function.

Proof. Introduce a vector

$$\mathbf{u} = \hbar\nabla\phi - \frac{2e}{c}\mathbf{A} \tag{5-46}$$

where ϕ is the phase of Δ defined as above. Note first that $\mathbf{u}$ is gauge invariant, as can be seen from the transformation laws (5-34), (5-35). Physically, for situations where $|\Delta|$ is constant in space, $\mathbf{u}$ is proportional to the local supercurrent density (take for instance the situation of uniform flow and $\mathbf{A} = 0$ described in the last problem). Now consider an inner point P of the ring (Fig. 5-1). At P, no field penetrates; there is no current and $\mathbf{u} = 0$. Integrate

[3]This property was pointed out to the writer by Dr. P. Marcus.

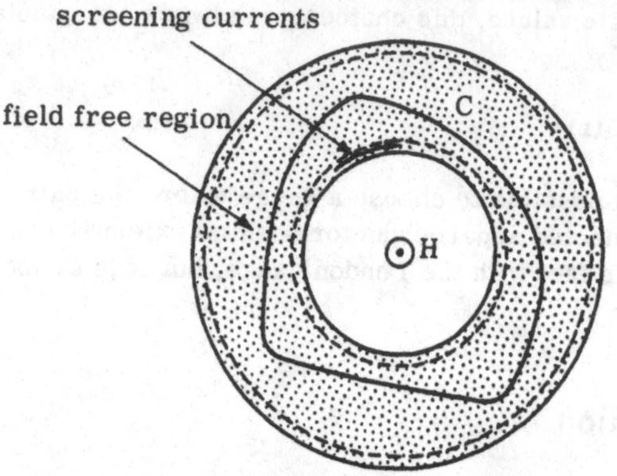

Figure 5-1
Flux trapped in a large superconducting ring. The thickness of the ring is assumed to be very large compared with the penetration depth: then it is possible to find a contour C in the superconductor along which the current and the field are 0.

this relation around the countour C, which comes nowhere close to the surface (**u** = 0 along C)

$$\int_C d\boldsymbol{\ell}\cdot\left(\hbar\nabla\phi - \frac{2e}{c}\mathbf{A}\right) = 0 \tag{5-47}$$

$\int_C \mathbf{A}\cdot d\boldsymbol{\ell} = \phi$ is the flux in the ring. $\int \nabla\phi\cdot d\boldsymbol{\ell} = [\phi]$ is the change of phase of $\Delta(\mathbf{r})$ after one turn around C. Since Δ must be single valued: $[\phi] = 2n\pi$ (n integer). Inserting these results in (5-47) we obtain (5-45).

Flux quantization was first considered by London. He predicted a flux quantization ch/e = twice the experimental value. We now repeat his argument (for a pure metal, at T = 0) and then try and correct it. Let $\phi_0\ (\mathbf{r}_1 \cdots \mathbf{r}_N)$ be the ground state wavefunction of the superconducting electrons i (a state of no current). A state of uniform current flow is obtained if we replace ϕ_0 by

$$\phi_0 \exp i\mathbf{q}\cdot[\mathbf{r}_1 + \mathbf{r}_2 + \ldots + \mathbf{r}_N] \tag{5-48}$$

leading to a current density $ne\hbar\mathbf{q}/m$. A more general state of macroscopic motion would be

$$\phi_0 \exp i[S(\mathbf{r}_1) + \ldots + S(\mathbf{r}_N)] \tag{5-49}$$

where S(**r**) is a slowly varying function in space. If there is also a vector potential **A**, the resulting current has the form

$$j = \frac{ne}{m}\left(\hbar\nabla S - \frac{eA}{c}\right) \tag{5-50}$$

If we assume that the only states that can be realized in the superconductor are of the form (5-49) we immediately find the Meissner effect: Taking the curl of (5-50) leads to the London equation. Furthermore, we find flux quantization: Consider again our superconducting ring. In the interior, far from the surface, $j = 0$, and the vector potential is given by

$$\mathbf{A} = \frac{c\hbar}{e}\nabla S \tag{5-51}$$

If some flux ϕ passes through the annulus, S cannot be single valued. Consider a contour C as shown in Fig. 5-1. If C is nowhere close to the surface, (5-51) is valid along C and the flux enclosed by C is given by

$$\int H d_\sigma = \oint \mathbf{A}\cdot d\boldsymbol{\ell} = \frac{c\hbar}{e}\oint \nabla S \cdot d\boldsymbol{\ell} = \frac{c\hbar}{e}\,\mathcal{S} \tag{5-52}$$

where the integrals $\oint d\boldsymbol{\ell}$ are taken along C, and $\mathcal{S}$ is the increase in S when one makes one complete path around the contour. $\mathcal{S}$ cannot take on arbitrary values. The wave function must be a single-valued function of $r_1 \ldots r_N$. If we allow one of the coordinates r_1 to vary along C, after a complete cycle we must find the same function. Since ϕ_0 is single valued, this imposes the condition that $\mathcal{S} = 2n\pi$ where n is an integer. From (5-52) this gives ch/e as the flux quantum.

This argument has one weak point, however (Brenig, 1961). In reality the states $\phi_0 \exp\{i[S(\mathbf{r}_1) + \ldots + S(\mathbf{r}_N)]\}$ are not the only states of macroscopic motion we can construct in a superconductor. To see this, we have to look into the detailed shape of ϕ. Let us do it on a simple problem: a superconducting electron gas with periodic boundary conditions

$$\phi(x_1 \ldots x_i \ldots x_N) \equiv \phi(x_1 \ldots x_i + L \ldots x_N) \qquad \text{for all } i \tag{5-53}$$

We construct states of translational motion with momentum $\hbar q$ per electron. Pairing electrons $(\mathbf{k} + \mathbf{q}, \uparrow)$ and $(-\mathbf{k} + \mathbf{q}, \downarrow)$, we can build BCS wave functions:

(1) If $qL = \mathcal{S} = 2n\pi$ (n integer), we use one-electron states $\mathbf{k}$ satisfying the usual periodic boundary conditions

$$k_x L = 2n_x\pi \qquad (n_x \text{ integer}) \tag{5-54}$$

Then Eq. (5-53) is obeyed.

(2) If $qL = \mathcal{S} = (2n + 1)\pi$ (n integer), we use one-electron states $\mathbf{k}$ satisfying the opposite boundary condition

$$k_x L = (2n_x + 1)\pi \tag{5-55}$$

Again (5-53) is obeyed.

The functions of class (1) are of the London type: they differ from the BCS function at rest only by the factor $\exp[iq(x_1 + \ldots + x_N)]$. The functions of class (2) are not of the London type. Both classes are acceptable. In all cases the energy is given by

$$E = E_0 + \frac{\hbar^2 q^2}{2m} N \tag{5-56}$$

Equation (5-56) applies to both classes, provided that

$$k_F L >> 1 \quad \text{and} \quad \frac{L}{\xi_0} >> 1 \tag{5-57}$$

When (5-57) is satisfied, all sums Σ_k on the allowed k values can be replaced by integrals which are independent of the choice of conditions (5-54) or (5-55).

Finally, it is easy to transpose these results to the case of an annulus. Again we find $\mathcal{S} = n\pi$ and the resulting flux quantum is ch/2e.

Question. How do the Bogolubov amplitudes $\binom{u}{v}$ change after one turn around the ring?

Answer. This of course depends on the gauge chosen. Choose the gauge where Δ is real. Then, after one turn around the ring, the amplitudes are multiplied by $(-1)^n$, where n is the number of flux quanta in the ring.

Justification: Return again to the simple case of uniform translation with the boundary condition (5-53). In the gauge where $\Delta = |\Delta| e^{2iqx}$, we have

$$\begin{pmatrix} u \\ v \end{pmatrix} = \begin{pmatrix} U_k \; e^{i(k+q)\cdot r} \\ V_k \; e^{i(k-q)\cdot r} \end{pmatrix} \tag{5-58}$$

as discussed in the last problem. For $\mathcal{S} = n\pi$, we have just seen that the one-electron states e^{ikr} must be chosen such that

$$\exp(ik_x L) = (-1)^n$$

Also

$$e^{iqL} = (-1)^n$$

Thus $\binom{u}{v}$ is periodic in this gauge. If we now transform to the gauge where Δ is real, we get the amplitudes

$$\begin{pmatrix} u' \\ v' \end{pmatrix} = \begin{pmatrix} U_k \; e^{ik\cdot r} \\ V_k \; e^{ik\cdot r} \end{pmatrix} \tag{5-59}$$

and $\binom{u'}{v'}$ is multiplied by $(-1)^n$ if we change x into x + L.

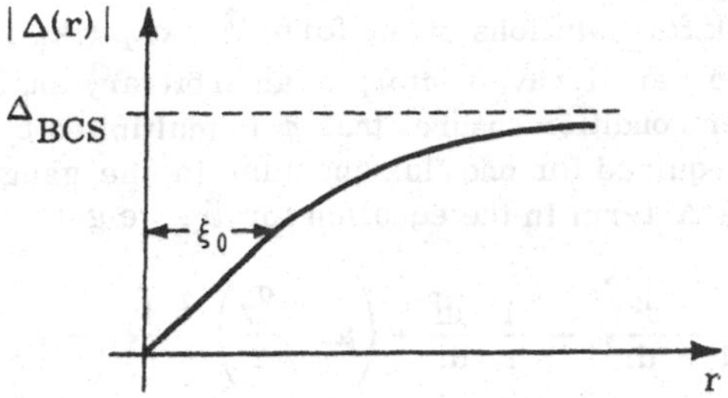

Figure 5-2
The amplitude of the order parameter Δ around a single vortex line in a pure Type II superconductor. Δ vanishes at z = 0 and has a finite slope for $r \to 0$. The region in which Δ is lowered from the BCS value has a radius $\sim \xi_0$.

Problem. Discuss the low energy excitations in the core of a vortex line (C. Caroli, P. G. de Gennes, and J. Matricon, 1964).

Solution. We restrict our attention to *pure* superconductors of Type II. The starting point is Eq. (5-18). We use cylindrical coordinates $(r\,\theta z)$, the line being along the z axis, and a gauge where the pair potential $\Delta(\mathbf{r}) = |\Delta(r)|\,e^{-i\theta}$: We call **A** the vector potential in this particular gauge. $|\Delta(r)|$ vanishes for $r = 0$, then increases (linearly at small r) and finally reaches the BCS value Δ_∞ at distances $r > \xi$. We rewrite (5-18) in a condensed spinor notation $\hat{\phi} = \begin{pmatrix} u \\ v \end{pmatrix}$ and eliminate the phase of Δ by setting

$$\hat{\phi} = \exp(-\tfrac{1}{2}\sigma_z\theta)\,\hat{\psi}$$

where $\sigma_x\,\sigma_y\,\sigma_z$ are Pauli matrices. Equation (5-18) becomes

$$\sigma_z\left\{\frac{1}{2m}\left(p - \sigma_z\frac{eA}{c} - \sigma_z\frac{\hbar}{2}\nabla\theta\right)^2 - E_F\right\}\hat{\psi} + \sigma_x\Delta\hat{\psi} = \epsilon\hat{\psi}$$

Note that $A \sim Hr$ and

$$\frac{\dfrac{eA}{c}}{\hbar\nabla\theta} \sim \left(\frac{H}{\phi_0}\right) r^2$$

(where $\phi_0 = ch/2e$ is the flux quantum). For the excitations on interest, $r \lesssim \xi$ and

$$\frac{eA}{c\hbar\nabla\theta} \sim \frac{H\xi^2}{\phi_0} \sim \frac{H}{H_{c2}} \ll 1$$

Thus we can neglect all magnetic field effects.

We then look for solutions of the form $\hat{\psi} = \exp(ik_F z \cos\alpha)\exp(i\mu\theta)\hat{f}(r)$ where k_F is the Fermi wave vector, α an arbitrary angle, and 2μ an odd integer. The latter condition ensures that ϕ is multiplied by (-1) after one complete turn, as required for one flux quantum in the gauge where Δ is real.

Dropping the Δ term in the equation for $\hat{\psi}$, we get

$$\sigma_z \frac{\hbar^2}{2m}\left\{-\frac{d^2\hat{f}}{dr^2} - \frac{1}{r}\frac{d\hat{f}}{dr} + \left(\mu - \frac{\sigma_z}{2}\right)^2 \frac{\hat{f}}{r^2} - k_F^2 \sin^2\alpha \hat{f}\right\}$$

$$+ \sigma_x \Delta(r)\hat{f} = \epsilon \hat{f}$$

It is possible to solve this equation completely in the region $0 < \mu \ll k_F \xi$, which turns out to be the important one. Consider a radius r_c such that

$$(\mu + 1/2)\, k_F^{-1} \ll r_c \ll \xi$$

For $r < r_c$, the Δ term can be neglected and $\hat{f} = \begin{pmatrix} f_+ \\ f_- \end{pmatrix}$ is given by

$$f_\pm(r) = A_\pm J_{\mu \mp 1/2}\{(k_F \sin\alpha \pm q)\, r\}$$

where J is a Bessel function, A_+, A_- are arbitrary coefficients, and $q = \epsilon/\hbar v_F \sin\alpha$.

For $r > r_c$, we put

$$\hat{f} = \hat{g}(r)\, H_m(k_F r \sin\alpha) + \text{C.C.} \qquad (m = \sqrt{\mu^2 + \tfrac{1}{4}})$$

where H is a Hankel function and $\hat{g}$ a slowly-varying envelope. The equation for $\hat{g}$ may be reduced to

$$-i\sigma_z \hbar v_F \sin\alpha \frac{d\hat{g}}{dr} + \Delta\sigma_x \hat{g} = \left(\epsilon + \frac{\mu\hbar^2}{2mr^2}\right)\hat{g} \qquad v_F = \frac{\hbar k_F}{m}$$

For $\epsilon \ll \Delta_\infty$ and $k_F r \gg \mu$, the right-hand side is a small perturbation. Treating it to first order, we get

$$\hat{g} = \text{const} \times \begin{pmatrix} e^{i\psi/2} \\ -ie^{-i\psi/2} \end{pmatrix} e^{-K}$$

$$K(r) = (\hbar v_F \sin\alpha)^{-1} \int_0^r \Delta(r)\, dr$$

$$\psi(r) = -\int_r^\infty \exp\{2K(r) - 2K(r')\}\left(2q + \frac{\mu}{k_F r'^2 \sin\alpha}\right) dr$$

$$\psi(r_c) \cong -\mu(k_F r_c \sin\alpha)^{-1} + 2qr_c - 2\int_0^\infty dr'\, e^{-2K(r')}$$

$$\times \left(q - \frac{\mu\Delta(r')}{\hbar k_F v_F \sin^2\alpha}\right)$$

Finally we match the solutions at $r = r_c$, making use of the asymptotic forms

$$J_m(z) = const \ z^{-1/2} \sin\left\{z + \frac{m^2}{2z} - \frac{\pi}{2}\left(m - \frac{1}{2}\right)\right\}$$

etc., and obtain the condition (for $\mu \neq 0$)

$$\psi(r_c) = 2qr_c - \mu(k_F r_c \sin \alpha)^{-1}$$

Comparing with our earlier result for $\psi(r_c)$, we see that all the r_c dependent terms cancel out, and find the eigenvalue

$$\epsilon_{\mu\alpha} = \hbar q v_F \sin \alpha = \mu(k_F \sin \alpha)^{-1} \ \frac{\int_0^\infty \frac{\Delta(r)}{r} e^{-2K(r)} dr}{\int_0^\infty e^{-2K(r)} dr}$$

$$= \mu(k_F \sin \alpha)^{-1} \left(\frac{d\Delta}{dr}\right)_{r=0} g(\alpha) \quad (\mu \neq 0, \ \mu \ll k_F \xi)$$

The dimensionless function $g(\alpha)$ thus defined depends on the exact shape assumed for $\Delta(r)$ but is always close to 1. In particular $g(0) = g(\pi) = 1$. Thus the eigenvalues are of order $\mu\Delta_\infty/k_F\xi \sim \mu\Delta^2/E_F$, the lowest one corresponding to $\mu = \frac{1}{2}$. The density of states N_ℓ associated with the levels is (for one spin direction and for a unit length of line)

$$N_\ell(\epsilon) = \frac{1}{2\pi}\left\{\left(\frac{d\Delta}{dr}\right)_{r=0}\right\}^{-1} k_F^2 \int_0^\pi d\alpha \ \frac{\sin^2 \alpha}{g(\alpha)} \quad \left(\frac{\Delta_\infty^2}{E_F} < \epsilon \ll \Delta_\infty\right)$$

Note that $N_\ell(\epsilon) \sim N(0)\xi^2$, that is, each line is equivalent to a normal region of radius $\sim \xi$. The low-lying states occupy only a fraction $\sim (\xi^2/d^2) \sim (B/H_{c2})$ of the volume. They will be of importance mainly at low temperatures T (roughly when $(\xi/d)^2 \exp(\Delta_\infty/T) > 1$). Then the specific heat will be linear in T, the thermal conductivity will be anisotropic (maximum along the lines) and the nuclear relaxation may, in some cases, become limited by the spin diffusion rate.

Excitation Threshold in Inhomogeneous Systems

Consider a *pure* metal in the absence of field or currents. Usually the pair potential Δ is constant throughout the sample. The functions u and v are then plane waves and it is easily verified that the energy gap is Δ. However, it is possible for sufficiently thin samples to impose a spatial variation in $\Delta(\mathbf{r})$ (for example, by depositing metals of different characteristics on the superconductor). We now study the particularly simple case described in Fig. 5-3. The pair potential varies

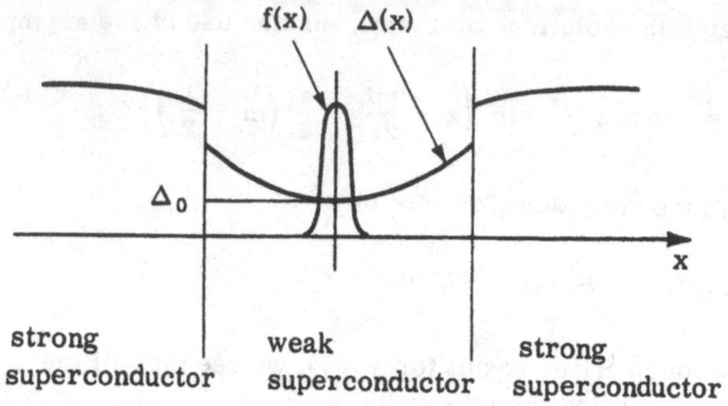

Figure 5-3

The pair potential Δ(x) for a system of three metallic layers, the central layer being the weakest superconductor. Δ has a minimum Δ_0. It is possible to construct a quasiparticle with an amplitude f(x) very localized around Δ_0; such a quasiparticle has an excitation energy essentially equal to Δ_0

only in the x direction and it has a minimum Δ_0 at $x = 0$. For $x \neq 0$, Δ increases; the spatial scale of the variation in Δ is rather large (at least $\sim \xi_0$). We show that the threshold energy ϵ_0 for the quasiparticles, that is, the smallest positive eigenvalue of (5-18), is nearly equal to Δ_0. Writing (5-18) in a more condensed form, we have

$$\epsilon\hat{\psi} = (\mathcal{H}_e \sigma_z + \Delta\sigma_x)\hat{\psi} = \hat{\Omega}\hat{\psi} \tag{5-60}$$

where $\hat{\psi} = \binom{u}{v}$, $\sigma_z = \begin{pmatrix} 1 & 0 \\ 0 & -1 \end{pmatrix}$, and $\sigma_x = \begin{pmatrix} 0 & 1 \\ 1 & 0 \end{pmatrix}$ are the Pauli matrices. We wish to use a variational principle to show that ϵ_0 is close to Δ_0. However $\hat{\Omega}$ is not a positive definite operator [ϵ and $(-\epsilon)$ are simultaneously eigenvalues]. It is therefore necessary to study

$$\hat{\Omega}^2 = \mathcal{H}_e^2 + \Delta^2 + i[\mathcal{H}_e, \Delta]\sigma_y \tag{5-61}$$

[In order to obtain (5-61), we have used $\sigma_x\sigma_z + \sigma_z\sigma_x = 0$ and $\sigma_x\sigma_z - \sigma_z\sigma_x = -2i\sigma_y$.] Take as a trial function

$$\hat{\psi}(r) = \exp(ik_F z)f(x)\hat{\phi}_y \tag{5-62}$$

where z designates an arbitrary direction normal to the x axis and $\hat{\phi}_y$ is a constant spinor such that $\sigma_y\hat{\phi}_y = \pm\hat{\phi}_y$. Since $\mathcal{H}_e$ is the energy measured from the Fermi level, we have simply

$$\mathcal{H}_e \hat{\psi} = \exp(ik_F z)\left(-\frac{\hbar^2}{2m}\right)\frac{d^2 f(x)}{dx^2}\hat{\phi}_y \tag{5-63}$$

(This choice of wave vector leads us to a nodeless function f in the ground state, which is favorable for having a low kinetic energy.) Calculating the average value of $\hat{\Omega}^2$, we obtain

$$\epsilon^2 = \langle\hat{\psi}|\hat{\Omega}^2|\hat{\psi}\rangle = \int dx\, f^*(x)\left[\frac{\hbar^4}{4m^2}\frac{d^4 f}{dx^4} + |\Delta(x)|^2\right]f(x)$$
$$\pm\, i\int dx\, f^*(x)\left[-\frac{\hbar^2}{2m}\frac{d^2}{dx^2}, \Delta(x)\right]f(x) \tag{5-64}$$

If f is taken as real, the last integral vanishes. If f is chosen as a regular function of spatial extent L, then the kinetic energy term gives a contribution $\alpha(\hbar^2/2mL^2)^2$ where α is a numerical coefficient of the order unity. For the pair potential term, we assume that Δ varies parabolically around Δ_0,

$$\Delta(x) = \Delta_0\left(1 + \frac{x^2}{\delta^2}\right) \tag{5-65}$$

with $\delta \gg L$. Then the potential term becomes $\Delta_0^2[1 + \beta(L^2/\delta^2)]$ where β is another numerical coefficient, and finally

$$\epsilon^2 = \Delta_0^2 + \alpha\left(\frac{\hbar^2}{2mL^2}\right)^2 + \beta\Delta_0^2\frac{L^2}{\delta^2} \tag{5-66}$$

Minimizing this expression with respect to L, we obtain

$$\epsilon_0^2 \cong \Delta_0^2(1 + \mu^{2/3}) \tag{5-67}$$

where $\mu \sim \hbar^2/2m\delta^2\Delta_0$. For $\delta \sim \xi \sim \hbar v_F/\Delta_0$, we have $\mu \sim 1/k_F\xi_0 \sim 10^{-2}$ to 10^{-3}. Therefore, in practice, ϵ_0 is very close to Δ_0.

Nonmagnetic Alloys: Anderson's Theorem

We now study the excitation spectrum in a superconducting alloy, in *zero field*. We must solve the system of equations (5-18):

$$\begin{aligned}\epsilon u &= \left(-\frac{\hbar^2}{2m}\nabla^2 + U(\mathbf{r}) - E_F\right)u + \Delta v\\ \epsilon v &= \left(\frac{\hbar^2}{2m}\nabla^2 - U(\mathbf{r}) + E_F\right)v + \Delta^* u\end{aligned} \tag{5-68}$$

U(r) is now the complete Hartree potential for one electron in the normal state. In (5-68), we have assumed, as always, that the electron-electron coupling is nearly a point interaction $[-V\delta(\mathbf{r}_1 - \mathbf{r}_2)]$ and that the potential U(r) does not depend on the spin indices (nonmagnetic alloy). Still U(r) contains the impurity potentials, and (5-68) includes the multiple scattering effects from all of the impurities; it is therefore a very complicated system. The situation is much improved if we assume that $\Delta(\mathbf{r})$ is independent of $\mathbf{r}$ even in the presence of impurities. This is not rigorously correct, but has been shown to be acceptable if the impurities are not too different chemically from the matrix (C. Caroli, 1962). Then a large simplification occurs if we introduce the one-electron wave functions in the normal state $w_n(\mathbf{r})$ defined by:

$$\xi_n w_n(\mathbf{r}) = \left[-\frac{\hbar^2}{2m}\nabla^2 + U(\mathbf{r}) - E_F\right] w_n(\mathbf{r}) \tag{5-69}$$

For pure metal, the $w_n(\mathbf{r})$ are Bloch functions. In an alloy, the $w_n(\mathbf{r})$ are complicated functions describing the successive scattering of an electron by all of the impurities. Fortunately, we shall not need to know these wave functions in detail for the following calculation. When the pair potential $\Delta(\mathbf{r})$ reduces to a constant Δ, the solutions of (5-68) have the same spatial dependence as the $w_n(\mathbf{r})$. If we set

$$\begin{aligned} u_n(\mathbf{r}) &= w_n(\mathbf{r})u_n \\ v_n(\mathbf{r}) &= w_n(\mathbf{r})v_n \end{aligned} \tag{5-70}$$

it is easy to see that (5-68) is satisfied when

$$\begin{aligned} (\epsilon_n - \xi_n)u_n - \Delta v_n &= 0 \\ \Delta^* u_n - (\epsilon_n + \xi_n)v_n &= 0 \end{aligned} \tag{5-71}$$

which implies as usual

$$\epsilon_n^2 = \xi_n^2 + |\Delta|^2 \tag{5-72}$$

The coefficients u_n and v_n depend only on ξ_n. From the normalization condition, they must satisfy

$$|u_n|^2 + |v_n|^2 = 1 \tag{5-73}$$

Using (5-71), one can write explicitly

$$|u_n|^2 = \tfrac{1}{2}\left(1 + \frac{\xi_n}{\epsilon_n}\right)$$
$$|v_n|^2 = \tfrac{1}{2}\left(1 - \frac{\xi_n}{\epsilon_n}\right) \tag{5-74}$$

It is convenient to choose u_n, v_n and $w_n(\mathbf{r})$ to be real. This is possible because the $w_n(\mathbf{r})$ are eigenfunctions of a real operator. This choice of standing waves rather than traveling waves will be useful for the discussion of alloys. The self-consistency equation (5-32) becomes

$$\Delta(\mathbf{r}) = V \sum_n |w_n(\mathbf{r})|^2 \frac{\Delta}{2\sqrt{\Delta^2 + \xi_n^2}} [1 - 2f(\sqrt{\xi_n^2 + \Delta^2})] \tag{5-75}$$

It is useful to introduce the density of states $N(\mathbf{r})$ *at the point* $\mathbf{r}$ in the normal metal, at the Fermi energy

$$N(\mathbf{r}) = \sum_n |w_n(\mathbf{r})|^2 \, \delta(\xi_n) \tag{5-76}$$

The condition (5-75) becomes

$$\Delta(\mathbf{r}) = VN(\mathbf{r}) \int_{-\hbar\omega_D}^{\hbar\omega_D} d\xi \, \frac{\Delta(1 - 2f)}{2\sqrt{\xi^2 + \Delta^2}} \tag{5-77}$$

Thus the self-consistency cannot be exact since Δ was assumed to be constant, but here $\Delta(\mathbf{r})$ is proportional to $N(\mathbf{r})$ and is therefore modified by the impurities. However, if the impurities are chemically similar to the matrix, $N(\mathbf{r})$ is not very different from its average $\bar{N}$ and the self-consistency requirement becomes

$$1 = \bar{N}V \int_{-\hbar\omega_D}^{\hbar\omega_D} d\xi \, \frac{1 - 2f}{2\sqrt{\xi^2 + |\Delta|^2}} \tag{5-78}$$

This is of the same form as for a pure metal. To a first approximation, the excitation spectrum and all of the thermodynamic properties are the same for the alloy as for the pure metal. This property was first pointed out by Anderson (1959). Experimentally, it has been verified that nonmagnetic impurities do not have spectacular effects on the transition temperature in superconducting metals.[4]

---

[4]There exist certain effects that cannot be described by our crude interaction $- V\delta(\mathbf{r}_1 - \mathbf{r}_2)$. However the smallness of these effects shows that this interaction is at least a good starting point.

5–3 THE MEISSNER EFFECT IN METALS AND ALLOYS

The Magnetic Field Treated as a Perturbation

We shall now study the currents $\mathbf{j}$ induced by a weak magnetic field $\mathbf{h} = \text{curl } \mathbf{A}$ in a gas of superconducting electrons. The mathematical machinery necessary for this calculation is contained in the preceding section. We start from the reduced form of the self-consistent field equations with a one-electron energy

$$\mathcal{H}_e = \frac{\left(\mathbf{p} - \frac{e}{c}\mathbf{A}\right)^2}{2m} + U(\mathbf{r}) - E_F \tag{5-79}$$

We do not demand that $U(\mathbf{r})$ be spatially constant, or even periodic. Thus we are able to discuss simultaneously pure metals and alloys. To first order in $\mathbf{h}$ (or $\mathbf{A}$), we set

$$\begin{aligned} u_n(\mathbf{r}) &= u_n^{(0)}(\mathbf{r}) + u_n^{(1)}(\mathbf{r}) \\ v_n(\mathbf{r}) &= v_n^{(0)}(\mathbf{r}) + v_n^{(1)}(\mathbf{r}) \end{aligned} \tag{5-80}$$

where $v_n^{(0)}$ and $u_n^{(0)}$ are the eigenfunctions of the Bogolubov system (5-18) for the case $\mathbf{A} = 0$, and are written more explicitly in Eq. (5-70). Also,

$$\Delta(\mathbf{r}) = \Delta_0 + \Delta_1(\mathbf{r})$$

In general Δ_1 is nonzero, that is, *the self-consistent field Δ is modified when a magnetic field is applied*. This remark, due to Bogolubov, is necessary to preserve the gauge invariance.

A typical case where Δ_1 is nonzero can be found in the following way: Suppose that $\mathbf{A} = \nabla\chi(\mathbf{r})$ where χ is an arbitrary function (then $\mathbf{H} = \text{curl } \mathbf{A} = 0$). One can verify that the system of equations (5-18) and the self-consistency equation (5-32) have a solution

$$\begin{aligned} u_n &= e^{(ie/\hbar c)\chi(\mathbf{r})}\, u_n^0 \\ v_n &= e^{(-ie/\hbar c)\chi(\mathbf{r})}\, v_n^0 \\ \Delta(\mathbf{r}) &= \Delta_0\, e^{(2ie/\hbar c)\chi(\mathbf{r})} \end{aligned} \tag{5-81}$$

and that, as expected, $j = 0$. Expanding to first order in χ (or **A**) we obtain

$$u_n^{(1)} = \frac{ie}{\hbar c} u_n^0 \chi$$

$$v_n^{(1)} = \frac{-ie}{\hbar c} v_n^0 \chi \tag{5-82}$$

$$\Delta_1 = \frac{2ie}{\hbar c} \Delta_0 \chi$$

and thus Δ_1 is nonzero for that particular case. More generally, when Δ_1 is first order in A, the most general form it can have is

$$\Delta_1(\mathbf{r}) = \sum_{\alpha} \int P_\alpha(\mathbf{r},\mathbf{r}') A_\alpha(\mathbf{r}')d\mathbf{r}' \tag{5-83}$$

where the integral is extended over the entire volume of the sample and $\alpha = x,y,z$. If we are considering an isotropic, homogeneous metal, $\mathbf{P}(\mathbf{r},\mathbf{r}')$ must only be a function of $R = |\mathbf{r} - \mathbf{r}'|$ and must be a vector in the direction of **R**. **P(R)** can be completely determined by returning to our special case for which

$$\frac{2ie}{\hbar c} \Delta_0 \chi(\mathbf{r}) = \int \mathbf{P}(\mathbf{r}' - \mathbf{r}) \cdot \nabla\chi(\mathbf{r}')d\mathbf{r}' \tag{5-84}$$

Let us now restrict our attention to a particular class of functions $\chi(\mathbf{r}')$ such that $\chi(\mathbf{r}')$ vanishes at large distances from $\mathbf{r}$, ($|\mathbf{r} - \mathbf{r}'| \to \infty$). Then the integration can be performed by parts to give $-\int \chi(\mathbf{r}') \operatorname{div} \mathbf{P} \times (\mathbf{r}' - \mathbf{r})d\mathbf{r}'$. Therefore we must have $\operatorname{div} \mathbf{P}(\mathbf{r}' - \mathbf{r}) = -(2ie/\hbar c)\Delta_0 \delta(\mathbf{R})$. Since **P** is a radial vector, the unique solution is

$$\mathbf{P}(\mathbf{R}) = -\frac{ie\Delta_0}{2\pi\hbar c}\frac{\mathbf{R}}{R^3} = \frac{ie\Delta_0}{2\pi\hbar c}\nabla\left(\frac{1}{R}\right) \tag{5-85}$$

This form of **P(R)** is also applicable in an alloy if we calculate the average of Δ_1 over all impurity configurations (because on averaging the system becomes homogeneous and isotropic).

Knowing **P(R)**, the following theorem can be proved: If **A** is chosen in the gauge $\operatorname{div} \mathbf{A} = 0$ and if **A** has no component normal to the surface of the sample (London gauge), then Δ_1 is zero. In fact,

$$\begin{aligned}\Delta_1(\mathbf{r}) &= \frac{ie\Delta_0}{2\pi\hbar c}\int \nabla\left(\frac{1}{R}\right)\cdot \mathbf{A}(\mathbf{r}')\, d\mathbf{r}' \\ &= -\frac{ie\Delta_0}{2\pi\hbar c}\left[\int \frac{1}{R}\operatorname{div}\mathbf{A}\, d\mathbf{r}' - \int \frac{1}{R}\mathbf{A}\cdot\mathbf{n}\, d\sigma\right] = 0\end{aligned} \tag{5-86}$$

(**n** is a unit vector normal to the sample surface). It is important to realize that this theorem is valid only when the following conditions occur: (1) isotropic metal, (2) point interaction between electrons. In fact, if the interaction has a finite range, the pair potential $\Delta(\mathbf{r},\mathbf{r}')$ must depend on two arguments. Then the change in Δ takes the form

$$\Delta_1(\mathbf{r},\mathbf{r}') = \int d\mathbf{r}'' K_\alpha(\mathbf{r},\mathbf{r}',\mathbf{r}'') A_\alpha(\mathbf{r}'') \tag{5-87}$$

We no longer have a symmetry argument to allow us to introduce the gradient of a function.[5]

We now explicitly calculate the corrections $u^{(1)}$ and $v^{(1)}$ in the London gauge, where $\Delta_1 = 0$; this choice of gauge greatly simplifies the calculations. We first write the system of equations (5-18) with (5-19) only keeping terms to first order in **A**,

$$\left[\epsilon + \frac{\hbar^2}{2m}\nabla^2 - U(\mathbf{r})\right] u_n^{(1)}(\mathbf{r}) - \Delta v_n^{(1)}(\mathbf{r}) = \frac{ie\hbar}{2mc}(\mathbf{A}\cdot\nabla + \nabla\cdot\mathbf{A})u_n^{(0)}(\mathbf{r})$$
$$\left[\epsilon - \frac{\hbar^2}{2m}\nabla^2 + U(\mathbf{r})\right] v_n^{(1)}(\mathbf{r}) - \Delta^* u_n^{(1)}(\mathbf{r}) = \frac{ie\hbar}{2mc}(\mathbf{A}\cdot\nabla + \nabla\cdot\mathbf{A})v_n^{(0)}(\mathbf{r}) \tag{5-88}$$

We now expand $u_n^{(1)}$ and $v_n^{(1)}$ in terms of the orthonormal set $w_n(\mathbf{r})$ (remember incidentally that we choose the w_n's as real)

$$u_n^{(1)}(\mathbf{r}) = \sum_m a_{nm} w_m(\mathbf{r})$$
$$v_n^{(1)}(\mathbf{r}) = \sum_m b_{nm} w_m(\mathbf{r}) \tag{5-89}$$

replace $u_n^{(0)}$ and $v_n^{(0)}$ by their values (5-70), multiply each of (5-88) by w_m and integrate over **r**. We then obtain

$$a_{nm}(\epsilon_n - \xi_m) - \Delta b_{nm} = iF_{nm} u_n$$
$$-\Delta a_{nm} + b_{nm}(\epsilon_n + \xi_m) = iF_{nm} v_n \tag{5-90}$$

---

[5]Detailed numerical calculations show that the corrections resulting from such a spread-out interaction are small and have little influence on the diamagnetism: See G. Rickayzen, *Phys. Rev.*, **115**, 795 (1959).

where

$$F_{nm} = \frac{e\hbar}{2mc} \int w_m (\mathbf{A}\cdot\nabla + \nabla\cdot\mathbf{A}) w_n \, d\mathbf{r} = -F_{mn} \tag{5-91}$$

The solution of (5-90) is

$$a_{nm} = \frac{iF_{nm}}{\xi_n^2 - \xi_m^2} [(\epsilon_n + \xi_m) u_n + \Delta v_n]$$

$$b_{nm} = \frac{iF_{nm}}{\xi_n^2 - \xi_m^2} [(\epsilon_n - \xi_m) v_n + \Delta u_n] \tag{5-92}$$

Once $u_n^{(1)}$ and $v_n^{(1)}$ are calculated from these formulas, you can return to the self-consistency equation (5-32) and verify directly that in the London gauge Δ_1 is zero, as predicted.

Relation between the Diamagnetic Response and the Conductivity in the Normal State

Knowing the wave functions to first order in **A**, we calculate the current

$$j(r) = \text{Re}\left\{ < \Psi^+ (r) \left(p - \frac{e\mathbf{A}}{c} \right) \Psi(r) > \right\}$$

We limit our discussion to the case where T = 0. To zeroth order in **A**, j = 0. To first order, expressing the Ψ's in terms of the γ's, taking the average and expanding $u = u^0 + u'$, and so on, we find

$$j(r) = -\frac{e\hbar i}{2m} \sum_n [v_n^{*(0)} \nabla v_n^{(1)} + v_n^{*(1)} \nabla v_n^{(0)} - \text{C.C.}]$$

$$= -\frac{e\hbar i}{2m} \sum_{n,m} v_n [w_n \nabla w_n b_{nm} + w_m \nabla w_n b_{nm}^* - \text{C.C.}] \tag{5-93}$$

$$= -\frac{e\hbar i}{2m} \sum_{n,m} v_n (b_{nm} - b_{nm}^*)(w_n \nabla w_m - w_m \nabla w_n) \tag{5-94}$$

From (5-92), we determine the following properties

$$b_{nm}^* = -b_{nm}$$

$$v_n b_{nm} = iF_{nm} R_{nm}$$

where

$$R_{nm} = \frac{1}{2\epsilon_n(\xi_n^2 - \xi_m^2)}[(\epsilon_n - \xi_n)(\epsilon_n - \xi_m) + \Delta^2]$$

The symmetric part of R_{nm} is

$$\tfrac{1}{2}L(\xi_n, \xi_m) = \frac{R_{nm} + R_{mn}}{2} = \frac{\Delta^2 + \xi_n \xi_m - \epsilon_n \epsilon_m}{\epsilon_n \epsilon_m(\epsilon_n + \epsilon_m)}$$

The relation between the current and vector potential takes the form

$$j_\mu(\mathbf{r}) = \sum_\nu \int d\mathbf{r}' \; A_\nu(\mathbf{r}') \; S_{\mu\nu}(\mathbf{r},\mathbf{r}') \tag{5-95}$$

where $\mu, \nu = x,y,z$ and

$$S_{\mu\nu}(\mathbf{r},\mathbf{r}') = \left(\frac{e\hbar}{2m}\right)^2 \frac{1}{c} \sum_{n,m} L(\xi_n, \xi_m) p_{\mu nm}(\mathbf{r}) p_{\nu nm}(\mathbf{r}')$$

$$- \frac{ne^2}{mc} \delta(\mathbf{r} - \mathbf{r}') \, \delta_{\mu\nu} \tag{5-96}$$

and we have defined

$$p_{\mu nm}(\mathbf{r}) = \left[w_n(\mathbf{r}) \frac{\partial}{\partial r_\mu} w_m(\mathbf{r}) - w_m(\mathbf{r}) \frac{\partial}{\partial r_\mu} w_n(\mathbf{r}) \right]$$

For a pure infinite metal, the w_n are plane waves and the p_{nm} can be calculated immediately. However, in practice, we must take account of the presence of impurities and the surface of the sample. It is then impossible to construct the w_n explicitly.

However, $S_{\mu\nu}$ can be obtained without much work by noticing that the electrical conductivity $\sigma_{\mu\nu}$ in the *normal* state involves the same matrix elements. This observation also explains the success of Pippard's phenomenological model, which led to a (j,A) relation for a superconductor analogous to the (j,E) relation in a normal metal.

More precisely, the conductivity $\sigma_{\mu\nu}(\mathbf{r},\mathbf{r}', \Omega)$ at a frequency Ω is defined in the following way: At each point of the sample $\mathbf{r}'$, an electric field $\mathbf{E}(\mathbf{r}')\, e^{-i\Omega t}$ + C.C. is applied. The current induced at the point $\mathbf{r}$, $\mathbf{j}(\mathbf{r})e^{-i\Omega t}$ + C.C., is measured. The current is a linear function of the fields

$$j_\mu(\mathbf{r}) = \int d_3r' \, \sigma_{\mu\nu}(\mathbf{r}\mathbf{r}', \Omega) E_\nu(\mathbf{r}')$$

In practice we know the explicit form of $\sigma_{\mu\nu}$. The field current relation is given by the Chambers equation

$$j_\mu(\mathbf{r}) = \frac{e^2 v_F}{2\pi} N(0) \sum_\nu \int d\mathbf{r}' E_\nu(\mathbf{r}') \frac{R_\mu R_\nu}{R^4}$$

$$\times \exp[(i\Omega/v_F - 1/\ell)R] \qquad R = |\mathbf{r}' - \mathbf{r}| \tag{5-97}$$

where $\mathbf{R} = \mathbf{r}' - \mathbf{r}$. In (5-97), ℓ represents the transport mean free path. The integral extends *only over the volume of the sample*. Three assumptions are necessary to establish (5-97):

(1) The reflection of the electrons from the surface must be diffuse, which is realized in most cases.

(2) The fields $\mathbf{E}$ and the currents $\mathbf{j}$ are transverse (div $\mathbf{E} = 0$, div $\mathbf{j} = 0$). $\mathbf{E}$ and $\mathbf{j}$ have no components normal to the surface. This restriction doesn't disturb us since in the superconducting state our perturbation (the vector potential $\mathbf{A}$) will also be transverse (London gauge).

(3) The frequency Ω is less than $k_B T/\hbar$. From (5-97) we see that σ has the form

$$\sigma_{\mu\nu}(\mathbf{r}\mathbf{r}') = \frac{e^2 v_f}{2\pi} N(0) \frac{R_\mu R_\nu}{R^4} \exp(i\Omega R/v_F)\, e^{-R/\ell}$$

(r,r' inside the sample)

$$= 0 \qquad \text{(if } \mathbf{r} \text{ or } \mathbf{r}' \text{ is outside the sample)} \tag{5-98}$$

In fact, we only need the real part of $\sigma_{\mu\nu}$, which gives the power dissipated W

$$W = \sum_{\mu\nu} \int d\mathbf{r}\, d\mathbf{r}' \,[E^*_\mu(\mathbf{r})E_\nu(\mathbf{r}') + \text{C.C.}]\, \text{Re}\{\sigma_{\mu\nu}(\mathbf{r},\mathbf{r}',\Omega)\} \tag{5-99}$$

(Re denotes real part of)

We now calculate $\sigma_{\mu\nu}$ from the one-electron wave functions in the normal metal $w_n(\mathbf{r})$ and show that Re $\sigma_{\mu\nu}$ can be expressed by a formula analogous to (5-96) for $S_{\mu\nu}$. In the normal metal the one-electron energy has the form

$$\mathcal{H}_e = \frac{1}{2m}\left(\mathbf{p} - \frac{e}{c}\mathbf{A}\right)^2 + U_0(\mathbf{r}) \tag{5-100}$$

where $U_0(\mathbf{r})$ is the lattice potential energy including impurities and surfaces. On the other hand, the alternating electric field is not in U_0, but derived from the vector potential by

$$\mathbf{E} = -\frac{1}{c}\frac{\partial \mathbf{A}}{\partial t} = \frac{i\Omega}{c}\mathbf{A} \tag{5-101}$$

Because the field $\mathbf{E}$ is a weak perturbation,

$$\mathcal{H}_e = \mathcal{H}_0 + \mathcal{H}_1$$

$$\mathcal{H}_0 = \frac{p^2}{2m} + U_0 \tag{5-102}$$

$$\mathcal{H}_1 = -\frac{e}{2mc}(\mathbf{p}\cdot\mathbf{A} + \mathbf{A}\cdot\mathbf{p})$$

The eigenfunctions of $\mathcal{H}_0$ are the $w_n(\mathbf{r})$. The oscillating perturbation $\mathcal{H}_1$ induces transitions, $n \to m$, such that $\xi_m - \xi_n = \pm\hbar\Omega$. The transition probability per unit time is given by

$$g_{nm} = \frac{2\pi}{\hbar} f(\xi_n)[1 - f(\xi_m)]\,|\langle n|\mathcal{H}_1|m\rangle|^2$$

$$\times\,[\delta(\xi_n - \xi_m + \hbar\Omega) + \delta(\xi_n - \xi_m - \hbar\Omega)] \tag{5-103}$$

where

$$\langle n|\mathcal{H}_1|m\rangle = \frac{ei\hbar}{2mc}\int w_n(\nabla\cdot\mathbf{A} + \mathbf{A}\cdot\nabla)w_m\,d\mathbf{r}$$

$$= \frac{ei\hbar}{2mc}\sum_\nu \int A_\nu(\mathbf{r})p_{\nu nm}(\mathbf{r})\,d\mathbf{r} \tag{5-104}$$

The power dissipated is

$$W = \sum_{n,m} g_{nm}(\xi_m - \xi_n) \tag{5-105}$$

Comparing this equation with (5-99), we find

$$\operatorname{Re}\{\sigma_{\mu\nu}(r,r',\Omega)\} = \frac{2\pi}{\hbar}\left(\frac{e\hbar}{m}\right)^2 \sum_{n,m} \frac{f(\xi_n) - f(\xi_m)}{\hbar\Omega}$$

$$\times\, p_{\mu nm}(r)p_{\nu nm}(r')\,\delta(\xi_n - \xi_m - \hbar\Omega) \tag{5-106}$$

In order to compare this with Chamber's formula we must consider the region $\hbar\Omega \ll k_B T$. In this region,

$$\frac{f(\xi_n) - f(\xi_m)}{\hbar\Omega} = \frac{f(\xi_n) - f(\xi_m)}{\xi_n - \xi_n} \cong \frac{\delta f}{\delta \xi}$$

which is nearly $-\delta(\xi)$. Then (5-106) reduces to

$$\operatorname{Re}\{\sigma_{\mu\nu}\} = \frac{2\pi}{\hbar}\left(\frac{e\hbar}{m}\right)^2 N(0) \sum_m \overline{p_{\mu nm}(r)p_{\nu nm}(r')}$$

$$\times\, \delta(\xi_n - \xi_m - \hbar\Omega) \tag{5-107}$$

where the symbol $\overline{(\ )}$ denotes an average over one-electron states at the Fermi surface, in the normal state.[6]

Calculation of the Diamagnetic Currents

Equations (5-107) and (5-96) show that the conductivity $\sigma_{\mu\nu}$ in the normal state involves the same matrix elements as the diamagnetic response $S_{\mu\nu}$ in the superconducting state. Replacing Σ_n by $N(0)\int d\xi_n$ in (5-96), we write the fundamental relation

$$S_{\mu\nu}(\mathbf{r},\mathbf{r}') = \frac{\hbar}{2\pi c}\int d\xi d\xi' L(\xi,\xi') \operatorname{Re}\left\{\sigma_{\mu\nu}\left(\mathbf{r},\mathbf{r}', \frac{\xi - \xi'}{\hbar}\right)\right\}$$

$$-\, \frac{ne^2}{mc}\,\delta(\mathbf{r})\,\delta_{\mu\nu} \tag{5-108}$$

We now insert Chambers' formula for $\sigma_{\mu\nu}$ to obtain

---

[6] Note that the average depends strongly on $\hbar\Omega$ but only slightly on ξ_m when $|\xi_n| \ll E_F$.

$$S_{\mu\nu}(\mathbf{r},\mathbf{r}') = \frac{3ne^2}{4\pi mc\,\xi_0}\,\frac{R_\mu R_\nu}{R^4}\,e^{-R/\ell}\,J(R) - \frac{ne^2}{mc}\,\delta(R)\,\delta_{\mu\nu} \qquad (5\text{-}109)$$

where we have set $\xi_0 = \hbar v_F/\pi\Delta(0)$, $\mathbf{R} = \mathbf{r}' - \mathbf{r}$, and

$$J(R) = \frac{1}{\pi^2\Delta(0)}\int_{-\infty}^{\infty} d\xi\, d\xi'\, L(\xi,\xi')\cos\left[\frac{(\xi-\xi')R}{\hbar v_F}\right] \qquad (5\text{-}110)$$

From the form of $L(\xi,\xi')$ we predict that the important contribution to the integral occurs when $\xi - \xi' \sim \Delta$. This leads to a range for I(R) of the order of $hv_F/\Delta(0) \sim \xi_0$ (the choice of the numerical coefficient in the definition of ξ_0 will be useful later). For an explicit calculation of I(R), we change variables to: $\xi = \Delta \sinh\theta$, $\xi' = \Delta\sinh\theta'$, then set $\alpha = (\theta - \theta')/2$, $\beta = (\theta + \theta')/2$ to obtain

$$J(R) = \frac{1}{\pi^2}\int d\alpha\, d\beta\,\frac{\sinh^2\alpha}{\cosh\alpha\,\cosh\beta}\cos\left[\frac{2\Delta(0)R}{hv_F}\sinh\alpha\cosh\beta\right]$$

$$= \frac{1}{\pi^2}\int_{-\infty}^{\infty}\frac{d\beta}{\cosh\beta}\int_{-\infty}^{\infty} du\left(1 - \frac{1}{1+u^2}\right)\cos\left(\frac{2\Delta(0)R}{\cosh\beta}\right)$$

$$(u = \sinh\alpha) \qquad (5\text{-}111)$$

and

$$J(R) = \frac{1}{\pi}\frac{hv_F}{\Delta(0)}\,\delta(R) - \frac{1}{\pi}\int_{-\infty}^{\infty}\frac{d\beta}{\cosh\beta}\exp(-2\Delta(0)R\cosh\beta/hv_F)$$

$$(5\text{-}112)$$

The first term represents a contribution localized at the origin. On performing the $\mathbf{r}'$ integration in (5-95), we find that this term exactly cancels $(-ne^2/mc)\,\delta(R)\,\delta_{\mu\nu}$. Then, finally, we have

$$S_{\mu\nu}(\mathbf{r},\mathbf{r}') = \frac{-3ne^2}{4\pi mc\,\xi_0}\,\frac{R_\mu R_\nu}{R^4}\,e^{-R/\ell}\,I(R) \qquad (5\text{-}113)$$

where

$$I(R) = \frac{2}{\pi}\int_{-\infty}^{\infty}\frac{d\beta}{\cosh\beta}\exp[-(2/\pi)(R/\xi_0)\cosh\beta] \qquad (5\text{-}114)$$

Formulas (5-95), (5-113), and (5-114) give a complete solution to the problem of the currents induced by a static, weak, magnetic field in a superconductor.

Conclusions.

(1) For $A \neq 0$ (in the London gauge), permanent currents exist in a superconductor. Unfortunately this crucial result has been obtained only after a long calculation and the physical content is obscure; it will become clearer in Chapter 6.

(2) In a pure metal ($\ell = \infty$), the range of the kernel $S_{\mu\nu}(r, r')$ is governed by $I(R)$ and is of the order of ξ_0. Pippard's formula assumes $I(R) = \exp(-R/\xi_0)$. The exact result (5-114) is not an exponential but only differs from it slightly. Note in particular that

$$\left.\begin{aligned} I(0) &= 1 \\ \int_0^\infty I(R)\, dR &= \xi_0 \end{aligned}\right\} \quad \text{for } T = 0$$

which are common to the exact result and Pippard's formula.

(3) For a pure infinite metal, if $\mathbf{A}(\mathbf{r}')$ varies slowly relative to ξ_0, we can take it outside of the integral (5-95), and integrate to obtain the London relation

$$\begin{aligned} \mathbf{j}(\mathbf{r}) &= \frac{-ne^2}{mc}\mathbf{A}(\mathbf{r}) \\ \text{curl } \mathbf{j} &= \frac{-ne^2}{cm}\mathbf{H} \end{aligned} \tag{5-115}$$

(4) In the presence of impurities, the kernel $S(\mathbf{r}, \mathbf{r}')$ is reduced by the factor $e^{-R/\ell}$, where ℓ is the transport mean free path of the electrons in the normal state. This result was also predicted by Pippard before the development of the microscopic theory.

(5) In an impure metal, if $\mathbf{A}$ varies slowly with respect to $(\xi_0^{-1} + \ell^{-1})^{-1}$, $\mathbf{A}$ can still be taken out from under the integral sign and we then obtain a new London relation, which is usually written in the form

$$\begin{aligned} -c\Lambda_0 \mathbf{j}(\mathbf{r}) &= \mathbf{A}(\mathbf{r}) \\ -c\Lambda_0 \text{ curl } \mathbf{j}(\mathbf{r}) &= \mathbf{H} \end{aligned} \tag{5-116}$$

where

$$\Lambda_0^{-1} = \frac{ne^2}{m}\int_0^\infty \frac{dR}{\xi_0} I(R)\, e^{-R/\ell} < \frac{ne^2}{m} \qquad (T = 0) \tag{5-117}$$

In particular for $\ell \ll \xi_0$, $I(R)$ can be taken to be unity and we find

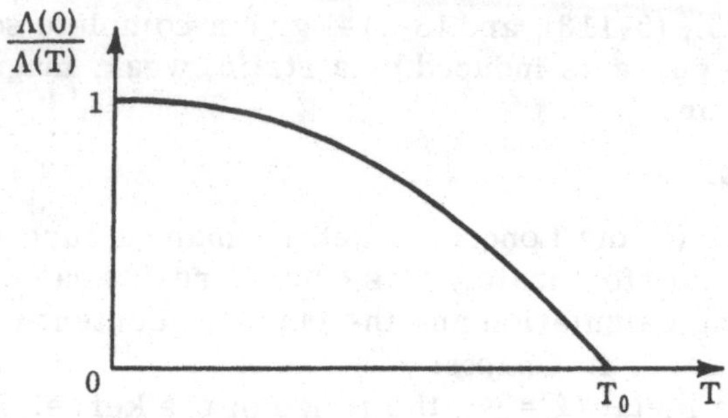

Figure 5-4
Strength of the diamagnetic response of the superconductor as a function of temperature.

$$\Lambda_0^{-1} = \frac{ne^2}{m}\left(\frac{\ell}{\xi_0}\right) \qquad (\ell \ll \xi_0) \tag{5-118}$$

(6) We have limited this calculation to $T = 0$. It can be performed in the same way for $T \neq 0$. The equation (5-95) still applies, but the definitions of $L(\xi, \xi')$ must be generalized for $T \neq 0$. The integration must be performed numerically with the following results: The spatial form of the kernel $S_{\mu\nu}(\mathbf{r}, \mathbf{r}')$ is *nearly independent of temperature*. Its range of pure metal remains ξ_0 even for $T \to T_0$. On the other hand, its normalization changes: The supercurrents become weaker as T increases, and finally vanish at $T = T_0$. This normalization is usually defined in terms of the coefficient Λ_T given by

$$\int S_{\mu\nu}(\mathbf{r}, \mathbf{r}', T)\, d\mathbf{r}' = \frac{1}{c\Lambda_T} \tag{5-119}$$

where Λ_T depends only on temperature. The temperature dependence of Λ_0/Λ_T is given in Fig. 5-4. We shall come back later to a discussion of parameters related to Λ_T near $T = T_0$ in terms of the Landau-Ginsburg equations.

REFERENCES

The self-consistent field method and the Anderson theorem:

P. W. Anderson, *Proceedings of the 8th Conference on Low Temperature Physics* (Toronto: University of Toronto Press, 1961).

Flux quantization and long range order in superconductors:

C. N. Yang, *Rev. Mod. Phys.*, **34**, 694 (1962).

Meissner effect: a simple approach relating the Meissner effect to the energy gap in BCS superconductors:

M. Tinkham, "Superconductivity" in *Low Temperature Physics*, Les Houches Lecture Notes 1961. (New York: Gordon and Breach, 1965.)

6

PHENOMENOLOGICAL LANDAU-GINSBURG EQUATIONS

6–1 INTRODUCTION

Let us first recapitulate our knowledge of the microscopic behavior of a superconductor in the presence of a static magnetic field **H**. Our first method of determining the currents **j** induced by the field was to use the London equation

$$\text{curl } \mathbf{j} = \frac{-c}{4\pi\lambda^2(T)} \mathbf{H} \tag{6-1}$$

However, Eq. (6-1) is applicable only when the following conditions are satisfied:

(1) $\lambda(T)$ must be much larger than $(1/\xi_0 + 1/\ell)^{-1}$, where $\xi_0 = 0.18\,\hbar v_F/k_B T_c$ and ℓ is the mean free path;

(2) **H** is small and can be treated as a perturbation;

(3) the superconducting electron density $n_s(\mathbf{r})$ is nearly constant in space.

Finally we were able to relax the first condition by replacing (6-1) by the nonlocal Pippard equation. However conditions (2) and (3) remain and we are not able to attack such problems as that of a wall separating a normal region ($H = H_c$, $n_s = 0$) from a superconducting region ($H = 0$, n_s finite) in a Type I superconductor.

In principle, we can return to the general method:

(a) Write the equations of motion for the excitations

$$\epsilon u(\mathbf{r}) = \frac{1}{2m}\left[\left(\mathbf{p} - \frac{e}{c}\mathbf{A}\right)^2 - E_F\right] u(\mathbf{r}) + \Delta(\mathbf{r})\, v(\mathbf{r})$$

$$\epsilon v(\mathbf{r}) = -\frac{1}{2m}\left[\left(\mathbf{p} + \frac{e}{c}\mathbf{A}\right)^2 - E_F\right] v(\mathbf{r}) + \Delta^*(\mathbf{r})\, u(\mathbf{r})$$

(b) Assume a spatial form of the pair potential $\Delta(\mathbf{r})$. (c) Calculate u and v, which give us current densities and other physical properties. (d) Verify that the form chosen for $\Delta(\mathbf{r})$ is correct by means of the self-consistency equation

$$\Delta(\mathbf{r}) = \sum_n u_n v_n^* [1 - 2f(\epsilon_n)]$$

This program is, however, extremely tedious and difficult in practice.

There is considerable simplification if we restrict our attention to the neighborhood of the transition temperature T_0. For $T \to T_0$: (1) the penetration depth $\lambda(T)$ becomes large [in the London limit $\lambda(T)$ is proportional to $n_s^{-1/2}$ and n_s tends to zero as $T \to T_0$]; (2) the thickness of the walls separating N and S regions also becomes large. Empirically, in a first-kind material, this depth varies roughly as $\xi_0 [T_0/(T_0 - T)]^{1/2}$ (for a pure metal).

More generally, in many physical situations the current $\mathbf{j}(\mathbf{r})$ and the pair potential $\Delta(\mathbf{r})$ have only *slow spatial variations* if one is sufficiently close to the transition temperature. Under these conditions, the physical properties can be determined from more simple equations, which we now describe.

6-2 CONSTRUCTION OF THE FREE ENERGY

We consider a homogeneous superconducting metal or alloy. We first form the free energy for a configuration where the pair potential $\Delta(\mathbf{r})$ and the vector potential $\mathbf{A}(\mathbf{r})$ have arbitrary, but slowly varying, spatial forms. Then, by minimizing the energy, we obtain explicit equations for $\Delta(\mathbf{r})$ and for the current $\mathbf{j}(\mathbf{r})$.

(a) First return to the trivial case when (1) Δ is the same at each point, and (2) there is no magnetic field. If we fix an arbitrary value for Δ, we can determine u and v and calculate, by the methods of Chapter 4, the free energy F (per cm^3) as a function of Δ. In particular, for small Δ, which is the case in the region near T_0, we find

$$F = F_n + A(T)\,|\Delta|^2 + \frac{B(T)}{2}\,|\Delta|^4 + \cdots \tag{6-2}$$

for Δ constant and zero field. Only the even terms in Δ appear in the expansion. F_n is the free energy in the normal phase; A and B are coefficients which in the BCS approximation are given by

$$A(T) = N(0)\frac{T - T_0}{T_0} \qquad T \to T_0 \qquad (6\text{-}3)$$
$$B(T) = 0.1066\frac{N(0)}{(k_B T_0)^2}$$

More generally, we assume that the exact free energy can be represented by an expansion of the type (4-2) with the coefficients A and B having the following properties: A is negative for $T < T_0$ and vanishes at $T = T_0$, the slope $(dA/dT)T_0$ being finite. B is positive and finite at $T = T_0$ and can, in practice, be replaced by its value at T_0.

For $T < T_0$, the minimum of F corresponds to a nonvanishing value of Δ,

$$\Delta = \Delta_0(T) \qquad (6\text{-}4)$$
$$\Delta_0^2 = -\frac{A}{B} = 9.38 k_B^2 T_0 (T_0 - T)$$

This is the value that will occur in equilibrium in the absence of a magnetic field. The corresponding free energy is

$$F = F_s = F_n - \frac{A^2}{2B} \qquad (6\text{-}5)$$

Because of our assumptions about A and B, the difference $F_n - F_s = H_c^2/8\pi$ is proportional to $(T_0 - T)^2$. This is characteristic of all second-order transitions.

(b) We now allow Δ to have a slow variation from point to point, however, still keeping $\mathbf{H} = \mathbf{A} = 0$. How will the free energy (6-2) be modified? Since in the absence of a magnetic field equilibrium corresponds to Δ = const, the first significant terms will be of the form $(\partial\Delta/\partial x)^2$ or $(\partial\Delta/\partial x)(\partial\Delta/\partial y)$, and so on. For simplification, from now on we only consider *cubic crystals*. Then,

$$F = F_n + A|\Delta|^2 + \frac{B}{2}|\Delta|^4 + C\left[\left|\frac{\partial\Delta}{\partial x}\right|^2 + \left|\frac{\partial\Delta}{\partial y}\right|^2 + \left|\frac{\partial\Delta}{\partial z}\right|^2\right] \qquad (6\text{-}6)$$

We take C to be nonzero and positive for $T = T_0$. In fact, a microscopic calculation will show later that for a pure metal $C \sim N(0)\xi_0^2$, that is, the "torsion energy" represented by the last term in (6-6) is of the order $N(0)\Delta^2 \xi_0^2[(1/\Delta)(\partial\Delta/\partial x)]^2$.

The relation (6-6) between the free energy F and the order parameter Δ in the neighborhood of the transition temperature has been assumed as a starting point by Landau for a general theory of second-order phase transitions. (In a ferromagnet, for instance, the order parameter is the magnetization.[1])

Problem. Calculate the fluctuation in the pair potential Δ in the normal phase ($T > T_0$), in zero field.

Solution. The probability of finding a configuration where the pair potential has a determined spatial form $\Delta(\mathbf{r})$ is proportional to $\exp(-\mathfrak{F}/k_B T)$, where $\mathfrak{F}$ is the free energy of the sample calculated from (6-6). For $T > T_0$, it suffices to keep only the terms of order of Δ^2 in (6-6). In the gauge where $A = 0$ and Δ is real, we have

$$\mathfrak{F} = \int d\mathbf{r}\, [A|\Delta(\mathbf{r})|^2 + C|\nabla\Delta(\mathbf{r})|^2]$$
$$= L^3 \sum_k |\Delta_k|^2 (A + Ck^2)$$

where

$$\Delta_k = L^{-3} \int d\mathbf{r}\, e^{ikr}\, \Delta(\mathbf{r})$$

and L^3 is the volume of the sample. (As we shall see later, this equation is valid, in a pure superconductor, when $\xi_0 k \ll 1$.) One then finds the thermal averages

$$\langle \Delta_k \rangle = 0$$

$$\langle |\Delta_k|^2 \rangle = \frac{\int \exp(-\mathfrak{F}/k_B T)\, |\Delta_k|^2\, d\Delta_k}{\int \exp(-\mathfrak{F}/k_B T)\, d\Delta_k} = \frac{1}{2L^3} \frac{k_B T}{A + Ck^2}$$

On the other hand $\langle|\Delta_k|^2\rangle$ is related to the spatial correlation function of Δ:

$$\langle |\Delta_k|^2 \rangle = L^{-6} \int d\mathbf{r}\, d\mathbf{r}' \,\langle \Delta(\mathbf{r})\, \Delta(\mathbf{r}') \rangle\, e^{ik\cdot(\mathbf{r}-\mathbf{r}')}$$
$$= L^{-3} \int d\mathbf{R}\, \langle \Delta(0)\, \Delta(\mathbf{R}) \rangle\, e^{ik\cdot R}$$

[$\langle \Delta(\mathbf{r})\, \Delta(\mathbf{r}') \rangle$ only depends on $\mathbf{r} - \mathbf{r}'$.] On taking the inverse of this expression we find the correlation function

---

[1] For magnetic materials, recent experiments and theoretical calculations indicate that the Landau hypothesis on the analytic form of F is, in fact, too restrictive; the real situation is more complicated. However, in superconductors, (6-6) is excellent. The complications encountered in the magnetic case stem from short-range order effects. These effects are large in magnetic materials, but are very small in superconductors (see problem above).

$$\langle \Delta(0)\, \Delta(R) \rangle = \frac{1}{(2\pi)^3} \int dk \frac{k_B T}{2(A + Ck^2)} e^{-ik\cdot R} = \frac{k_B T}{8\pi CR} e^{-qR}$$

where $q^2 = A/C$, $q > 0$. For a pure metal, ($C \sim N(0)\,\xi_0^2$):

$$\frac{\langle \Delta(0)\, \Delta(R) \rangle}{(k_B T_0)^2} \sim \frac{k_B T}{E_F} \frac{1}{k_F R} e^{-qR}$$

which is very small. Thus short-range order effects are indeed negligible in most superconductors.

(c) Eq. (6-6) must be completed by adding a vector potential **A**, which gives rise to the field **h** = curl **A**. The free energy must be independent of the gauge chosen for **A**. (If we set $\mathbf{A}' = \mathbf{A} + \nabla\chi$, the magnetic field derived from **A**′ is the same as that derived from **A**. If H is unchanged, the free energy F must also be unchanged.) As we have seen in Chapter 5, the pair potential does change according to the law

$$\Delta'(\mathbf{r}) = \Delta(\mathbf{r})\, e^{2ie\chi(\mathbf{r})/\hbar c}$$

Thus in order to insure the invariance of F, we must replace (6-6) by

$$F = F_n + A|\Delta|^2 + \frac{B}{2}|\Delta|^4 + C\left|\left(-i\nabla - \frac{2e\mathbf{A}}{\hbar c}\right)\Delta\right|^2 + \frac{h^2}{8\pi} \qquad (6\text{-}7)$$

(The last term $h^2/8\pi$ represents the magnetic field energy in the vacuum.) There is a formal analogy between the terms of order $|\Delta|^2$ in (6-7) and the equation giving the energy density of a particle of charge 2e described by a wave function $\psi(\mathbf{r})$. More precisely, if we set

$$\psi(\mathbf{r}) = \frac{(2mC)^{1/2}}{\hbar}\Delta(\mathbf{r}), \qquad \alpha = \frac{\hbar^2}{2m}\frac{A}{C}, \qquad \beta = \left(\frac{\hbar^2}{2m}\right)^2 \frac{B}{C^2}$$

$$\frac{\alpha^2}{2\beta} = \frac{H_c^2}{8\pi} \qquad (6\text{-}8)$$

Then

$$F = F_n + \alpha|\psi|^2 + \frac{\beta}{2}|\psi|^4 + \frac{1}{2m}\left|\left(-i\hbar\nabla - \frac{2e\mathbf{A}}{c}\right)\psi\right|^2 + \frac{h^2}{8\pi} \qquad (6\text{-}9)$$

This is the free energy form proposed by Landau and Ginsburg in 1951 well before the development of the microscopic theory. At that time the physical significance of $\psi(\mathbf{r})$ was far from being clear; furthermore, the value e^* of the charge that appears in $-i\hbar\nabla - e^*\mathbf{A}/c$

was not known (initially Landau and Ginsburg took $e^* = e$). This construction of (6-9) independent of any detailed theory of the superconducting state represented a tour de force of physical intuition.

Notice that the choice of the free electron mass m as a factor in (6-8) and (6-9) is entirely arbitrary. We could just as well have chosen the mass of the sun.[2] On the contrary, the presence of the charge 2e in the factor $2e\mathbf{A}/c$ is not a convention, but expresses a fundamental property of the pair potential: $\Delta = V\langle\psi\uparrow(\mathbf{r})\psi\downarrow(\mathbf{r})\rangle$ is the average of the product of *two* annihilation operators.

6–3 EQUILIBRIUM EQUATIONS

We must now minimize the free energy with respect to: (1) the order parameter $\Delta(\mathbf{r})$; and (2) the magnetic field distribution, therefore $\mathbf{A}(\mathbf{r})$. We set $\mathcal{F} = \int F\, d\mathbf{r}$ where the integral is extended over the volume of the specimen. If we vary $\psi(\mathbf{r})$ by $\delta\psi(\mathbf{r})$ and $\mathbf{A}(\mathbf{r})$ by $\delta\mathbf{A}(\mathbf{r})$, we obtain the variation in the free energy which after an integration by parts becomes

$$\delta\mathcal{F} = \int d\mathbf{r}\left\{\delta\psi^*\left[\alpha\psi + \beta|\psi|^2\psi + \frac{1}{2m}\left(-i\hbar\nabla - \frac{2e\mathbf{A}}{c}\right)^2\psi\right] + \text{C.C.}\right\}$$
$$+ \int d\mathbf{r}\,\delta\mathbf{A}\cdot\left\{\frac{1}{4\pi}\,\text{curl}\,\mathbf{h} - \frac{e}{mc}\left[\psi^*\left(-i\hbar\nabla - \frac{2e}{c}\mathbf{A}\right)\psi + \text{C.C.}\right]\right\} \tag{6-10}$$

In the second term, notice that $(1/4\pi)\,\text{curl}\,\mathbf{h}$ is the current density $\mathbf{j}/c$ from Maxwell's equations. By setting $\delta\mathcal{F} = 0$, we obtain the conditions

$$\alpha\psi + \beta|\psi|^2\psi + \frac{1}{2m}\left(-i\hbar\nabla - \frac{2e\mathbf{A}}{c}\right)^2\psi = 0 \tag{6-11}$$

$$\mathbf{j} = \frac{e\hbar}{im}(\psi^*\nabla\psi - \psi\nabla\psi^*) - \frac{4e^2}{mc}\psi^*\psi\mathbf{A} \tag{6-12}$$

Equations (6-11) and (6-12) are the fundamental Landau-Ginsburg equations. The first gives the order parameter and the second gives the currents, that is, the diamagnetic response of a superconductor.

Remarks.

(1) The relation (6-12) giving the current is identical to that which would occur for particles of charge 2e, mass m, and wavefunction $\psi(\mathbf{r})$.

(2) Boundary condition: In deriving (6-10) we neglected the surface integral

---

[2]Some authors do in fact use 2m instead of m. Our normalization follows the original paper of Landau and Ginsburg.

$$\int \delta\psi^* \left(-i\hbar\nabla - \frac{2eA}{c}\right)\psi\, \frac{i\hbar}{2m} \cdot d\sigma + \text{C.C.}$$

If we keep this term and require that it vanishes, we obtain the boundary condition

$$\left(-i\hbar\nabla - \frac{2eA}{c}\right)_n \psi = 0 \tag{6-13}$$

where the subscript n designates the component normal to the surface. This procedure was followed by Landau and Ginsburg in their original paper. It implicitly assumes that the form (6-9) for the free energy remains valid even in the neighborhood of a surface. We shall see, in fact, from the microscope theory that for a superconductor-insulator interface (6-13) is correct, but that for a superconductor-normal metal junction the boundary condition is severely modified.

On substituting (6-13) into the equation (6-12) for the current, we see that $j_n = 0$; that is, there is no current entering or leaving the superconductor. If we do the inverse, and impose $j_n = 0$, we do not obtain (6-13), but only

$$\left(-i\hbar\nabla - \frac{2eA}{c}\right)_n \psi = i\lambda\psi$$

where λ is an arbitrary real constant. In practice, this is the type of boundary condition encountered for a superconductor-normal metal junction.

6-4 THE TWO CHARACTERISTIC LENGTHS

The Landau-Ginsburg equations (6-11) and (6-12) introduce two characteristic lengths, which we now discuss.

(a) First consider a situation where there are neither currents nor magnetic fields. Choose the gauge in which ψ is real. Then in one dimension, for example, (6-11) becomes simply

$$\frac{-\hbar^2}{2m}\frac{d^2\psi}{dx^2} + \alpha\psi + \beta\psi^3 = 0 \tag{6-14}$$

There are two obvious solutions: (1) $\psi = 0$ which describes the normal state, (2) $\psi = \psi_0$ with

$$\psi_0^2 = -\frac{\alpha}{\beta} > 0 \tag{6-15}$$

which describes the usual superconducting state. This second solution exists and is lower in energy when $\alpha < 0$, that is, $T < T_0$. However we would like to consider more general solutions. For example, if, by means of exterior constraints, we impose that $\psi(x)$ have a different value than ψ_0 at one point, how does the order parameter deform about this point?

In order to fix the length scale, it is useful to write (6-14) in terms of reduced variables

$$\psi = \psi_0 f \tag{6-16}$$

$$\frac{\hbar^2}{2m|\alpha|} = \xi^2(T) \tag{6-17}$$

where $\xi(T)$ has the dimensions of a length. Equation (6-14) becomes

$$-\xi^2(T)\frac{d^2 f}{dx^2} - f + f^3 = 0 \tag{6-18}$$

The natural unit of length for the variation of f is $\xi(T)$, which we call the *coherence length at the temperature T*. What is its order of magnitude? First notice that $\hbar^2/2m|\alpha| = -C/A$ from (6-8). For a *pure material* we have already mentioned that $A \simeq [(T - T_0)/T_0] N(0)$ and $C \sim N(0)\,\xi_0^2$; thus

$$\xi(T) = 0.74\,\xi_0\left(\frac{T_0}{T_0 - T}\right)^{1/2} \tag{6-19}$$

(where we have inserted the exact numerical coefficient).

Conclusion: The variations in ψ, which take place within the length $\xi(T)$, are slow with respect to ξ_0 if T is close to T_0.

Problem. From (6-18) calculate the energy of a NS wall in a Type I superconductor where the penetration depth is negligible. The field distribution and $|\psi|$ are represented in Fig. 6-1.

Solution. By hypothesis, the field does not penetrate into the region where ψ is nonzero. We can therefore use (6-18) with the boundary conditions

$$f = 0, \qquad x = 0$$

$$f \to 1, \qquad x \to \infty$$

By multiplying (6-18) by df/dx and integrating, we obtain the first integral

$$-\xi^2(T)\left(\frac{df}{dx}\right)^2 - f^2 + \tfrac{1}{2}f^4 = \text{const}$$

In order to satisfy the boundary condition as $x \to \infty$, we must take the constant to be equal to $-\frac{1}{2}$. This gives

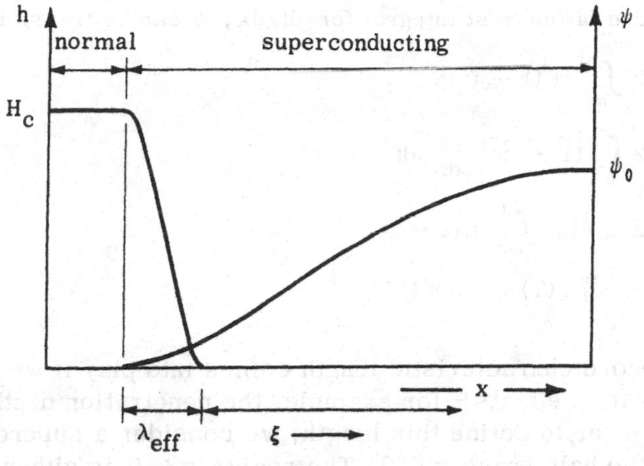

Figure 6-1
Spatial variation of the field and order parameter near a N-S wall for $\xi(T) \gg \lambda(T)$. The effective penetration depth λ_{eff} is of order $\sqrt{\lambda\xi}$ (see problem, page 178).

$$\xi^2(T)\left(\frac{df}{dx}\right)^2 = \tfrac{1}{2}(1 - f^2)^2$$

The solution satisfying our boundary conditions is

$$f = \tanh\left[\frac{x}{\sqrt{2}\,\xi(T)}\right]$$

Energy per unit area of the wall:

$$F_p = \int_0^\infty dx \;\; \frac{\hbar^2}{2m}\;|\nabla\psi|^2 + \alpha|\psi|^2 + \frac{\beta}{2}|\psi|^4$$

The surface energy $(H_c^2/8\pi)\,\delta$ is the difference between F_p and the condensation energy, which we would find if the medium were uniformly superconducting from the point $x = 0$:

$$\frac{H_c^2}{8\pi}\,\delta = F_p - \int_0^\infty\left(-\frac{H_c^2}{8\pi}\right)dx$$

Inserting the reduced parameter f and the relations (6-8), we obtain

$$\delta = \int_0^\infty dx\left[2\xi^2(T)\left(\frac{df}{dx}\right)^2 + (1 - f^2)^2\right]$$

By making use of the first integral for df/dx, δ can be transformed into

$$\delta = 2\int_0^{\infty} dx\,(1 - f^2)^2$$

$$= 2\int_0^1 (1 - f^2)^2 \frac{dx}{df}\, df$$

$$= 2\sqrt{2}\,\xi(T)\int_0^1 df\,(1 - f^2)$$

$$\delta = \frac{4}{3}\sqrt{2}\,\xi(T) = 1.89\,\xi(T)$$

(b) A second characteristic length comes into play if we introduce electromagnetic effects—for example, the penetration depth in weak fields. In order to define this length, we consider a superconductor occupying the half-space $z > 0$. The region $z < 0$ is either empty or filled with an insulator where the boundary condition (6-13) applies.

In weak fields, and to first order in h, $|\psi|^2$ in the region $z > 0$ can be replaced by its equilibrium value in the absence of a field $|\psi_0|^2$ defined by (6-15). Because ψ_0 is independent of position, when we take the curl of both sides of (6-12), we obtain a London-type equation:

$$\text{curl } \mathbf{j} = -\frac{4e^2}{mc}\psi_0^2 \mathbf{h} \tag{6-20}$$

On adding Maxwell's equations, we find, as in Chapter 2, nonvanishing solutions only if $\mathbf{h}$ is in the xy plane; for example, taking $\mathbf{h}$ along the x axis,

$$h_x = h_x(0)\, e^{-z/\lambda(T)} \tag{6-21}$$

$$\lambda(T)^{-2} = \frac{16\pi e^2 \psi_0^2}{mc^2} \tag{6-22a}$$

$$= \frac{32\pi e^2}{\hbar^2 c^2} C\Delta_0^2 \tag{6-22b}$$

where Δ_0 is the equilibrium value of the pair potential. Equation (6-22b) allows C to be related to directly measurable quantities.[3] Note that $\lambda(T)$ is proportional to ψ_0^{-1} and therefore to $[T_0/(T_0 - T)]^{1/2}$. For a pure metal in the free electron approximation, the microscopic calculation in the BCS approximation gives

$$\lambda(T) = \frac{1}{\sqrt{2}}\lambda_L(0)\left(\frac{T_0}{T_0 - T}\right)^{1/2} \qquad T \to T_0 \tag{6-23}$$

[3]Note that for a noncubic crystal it is necessary to take into account the tensorial nature of C.

where $\lambda_L(0)$ is the London penetration depth at absolute 0, given in terms of the number of electrons per cm^3 n by the equation $\lambda_L^{-2}(0) = 4\pi ne^2/mc^2$. Except for the numerical coefficient, (6-23) can be predicted from (6-22) if we guess (as stated earlier) that $C \sim N(0)\,\xi_0^2$ and that

$$\Delta_0^2 \sim k_B^2 T_0 (T_0 - T)$$

Conclusions. The Landau-Ginsburg hypothesis, in particular the form of the free energy (6-7) in the presence of a field, has led us to a local relation (6-12) between the current and vector potential. We know from the microscopic analysis of Chapter 5 that in the special case where $|\Delta|$ is constant and $\mathbf{h}$ small, the exact relation is nonlocal; that is, the current density $\mathbf{j}(\mathbf{r})$ depends on $\mathbf{A}(\mathbf{r}')$ for $|\mathbf{r} - \mathbf{r}'| \lesssim \xi_0$ in a pure metal. In order for the local approximation to be valid, it is necessary that **A** or the current have a slow variation on the scale of ξ_0, therefore, $\lambda(T) >> \xi_0$,

$$\lambda_L(0)\left(\frac{T_0}{T_0 - T}\right)^{1/2} >> \xi_0 \tag{6-24}$$

For a pure metal, this condition will be satisfied if T is sufficiently close to T_0. [However, for certain nontransition metals, notably aluminum, $\lambda_L(0)$ is so much smaller than ξ_0 that the temperature interval allowed by (6-24) is very small.]

(c) We have defined two characteristic lengths $\xi(T)$ and $\lambda(T)$, which determine the behavior of a superconductor near the transition point. They both diverge as $(T_0 - T)^{-1/2}$ as $T \to T_0$. It is therefore particularly interesting to form their ratio

$$\kappa = \frac{\lambda(T)}{\xi(T)} \tag{6-25}$$

By using the definitions (6-17) and (6-22a) of $\xi(T)$ and $\lambda(T)$, one obtains

$$\kappa = \frac{mc}{2e\hbar}\left(\frac{\beta}{2\pi}\right)^{1/2} = \frac{\hbar c}{4Ce}\left(\frac{B}{2\pi}\right)^{1/2} \tag{6-26}$$

κ is called *the Landau-Ginsburg parameter* of the substance. When $\kappa \gtrsim 1$ $(\lambda < \xi)$, the material is of the first kind, when $\kappa \lesssim 1$ $(\lambda > \xi)$, material is of the second kind. We see later that the exact separation between the two types of behavior occurs for $\kappa = 1/\sqrt{2}$. For a pure substance we show later that

$$\kappa = 0.96\,\frac{\lambda_L(0)}{\xi_0} \tag{6-27}$$

It is interesting to notice that κ can be defined directly from experimentally obtained numbers on the penetration depth and thermodynamic field H_c measured in the region of validity of the Landau-Ginsburg equations. By making use of (6-22a), (6-15), (6-9), and (6-8), one can put (6-26) in the form

$$\kappa = 2\sqrt{2}\,\frac{e}{\hbar c} H_c(T)\lambda^2(T) \tag{6-28}$$

Another way of writing (6-28), often useful for numerical calculations, is the following

$$H_c(T) = \frac{\phi_0}{2\pi\sqrt{2}\,\xi(T)\lambda(T)} \tag{6-29}$$

where $\phi_0 = ch/2e$ is the flux quantum.

In this chapter we also study other methods that allow the determination of κ.

6–5 SITUATIONS WHERE $|\psi|$ IS CONSTANT

We now apply the Landau-Ginsburg equations to some concrete examples. First, consider the particularly simple cases where the amplitude $|\psi|$ of the order parameter is the same at all points. Such a situation has already been encountered (the penetration depth of a weak magnetic field in a bulk sample). The examples we now consider are rather different. They concern thin samples (films, wires, and so on) where ψ cannot vary through the depth without catastrophically increasing the term $|\nabla\psi|^2$ in the free energy. But allow the field $\mathbf{h}$ (or currents $\mathbf{j}$) to be strong and $|\psi|$, though constant, will not necessarily be equal to its unperturbed value ψ_0.

Critical Current in a Thin Film

The situation envisaged is that represented in Fig. 6-2a. The film of thickness d carries a current density $\mathbf{j}$ along the x axis. We assume that

$d \ll \xi(T)$ in order to have $|\psi|$ constant[4]

$d \ll \lambda(T)$ in order to have $\mathbf{j}$ constant

Then our equations become particularly simple. If we set $\psi = |\psi|\, e^{i\phi(\mathbf{r})}$ where $|\psi|$ is independent of $\mathbf{r}$, the current becomes

[4]It is easy to verify that $|\psi|$ = constant is compatible with the boundary condition (6-13).

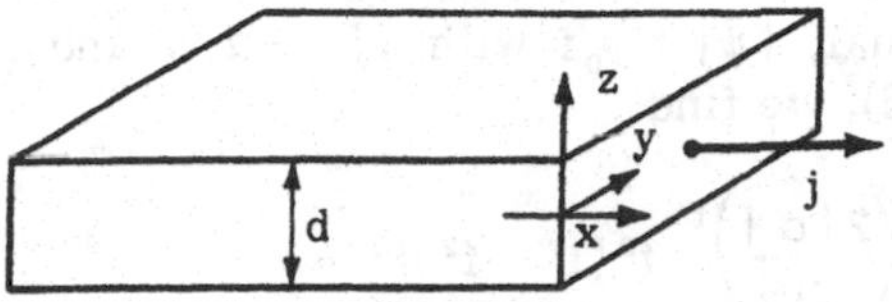

(a)

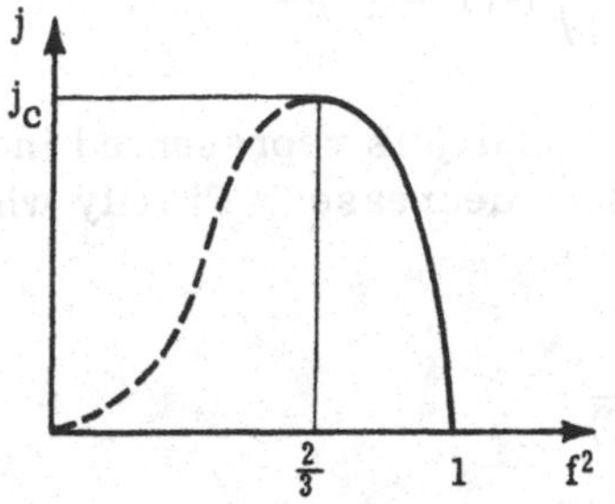

(b)

Figure 6-2
(a) Geometry to measure the critical current of a thin film (thickness d). The current flows along the x axis. (b) Current density (normalized to the critical current density) as a function of the order parameter f (normalized to the order parameter in zero current). For $j > j_c$, there is no solution to (6-35) and f falls abruptly to zero.

$$j = \frac{2e}{m} |\psi|^2 \left(\hbar \frac{\partial \phi}{\partial x} - \frac{2e}{c} A_x\right) = 2e |\psi|^2 v \qquad (6\text{-}30)$$

$$v = \frac{1}{m}\left(\hbar \frac{\partial \phi}{\partial x} - \frac{2e}{c} A_x\right) \qquad (6\text{-}31)$$

where v is the velocity of the "particles" described by the wave function ψ. The free energy (6-9) is simply

$$F = F_n + |\psi|^2 \left[\alpha + \frac{\beta}{2} |\psi|^2 + \frac{1}{2} mv^2\right] + \frac{h^2}{8\pi} \qquad (6\text{-}32)$$

and contains a term with the form of a kinetic energy. The minimum of F with respect to $|\psi|$ is obtained for

$$\alpha + \beta |\psi|^2 + \tfrac{1}{2} mv^2 = 0 \qquad (6\text{-}33)$$

If we set, as usual, $|\psi| = \psi_0 f$ with $\psi_0^2 = -\alpha/\beta$ and eliminate v between (6-30) and (6-33), we find

$$j = 2e\psi_0^2 \left(\frac{2|\alpha|}{m}\right)^{1/2} f^2(1 - f^2)^{1/2} \tag{6-34}$$

$$= 2e\psi_0^2 \left(\frac{\hbar}{m\xi(T)}\right) f^2(1 - f^2)^{1/2} \tag{6-35}$$

The relation between f and j is represented in Fig. 6-2b. For $j = 0$, $f = 1$. As j increases, f decreases. Finally when j becomes larger than j_c given by

$$j_c = \frac{4e\psi_0^2}{3\sqrt{3}} \frac{\hbar}{m\xi(T)} \tag{6-36}$$

there do not exist any solutions with $f \neq 0$. Thus for $j > j_c$ the film becomes normal and f changes abruptly from 0.8 to 0. The current j_c is the critical current of the film. It is given in (6-36) as a function of ψ_0^2 and $\xi(T)$, that is, as a function of the Landau-Ginsburg parameters α and β. In order to determine its order of magnitude, it is interesting to rewrite j_c in terms of the velocity v_c. When $j = j_c$, the velocity is $v_c \sim h/m\xi(T)$. For a pure metal

$$\xi(T) \sim \xi_0 \left(\frac{T_0}{T_0 - T}\right)^{1/2} \sim \frac{\hbar v_F}{k_B T_0} \left(\frac{T_0}{T_0 - T}\right)^{1/2} \sim \frac{\hbar v_F}{\Delta_0(T)} \tag{6-37}$$

and consequently

$$v_c \sim \frac{\Delta_0(T)}{m v_F} \tag{6-38}$$

It is also interesting to compare (6-38) to the maximum velocities associated with the currents giving rise to the Meissner effect in bulk superconductors. If H is an applied field, then the current near the surface is $\sim cH/\lambda$, where λ is the penetration depth. If one is sufficiently close to T_0, λ is correctly given by (6-22a); the velocity v near the surface becomes

$$v \sim \frac{j}{2e\psi_0^2} \sim \frac{cH}{2\lambda e\psi_0^2} \tag{6-39}$$

In a first kind superconductor, the maximum field is $H_c = (4\pi/\beta)^{1/2}|\alpha|$ which, by using the definition (6-25, 6-26) of κ, becomes

$$v \sim \frac{\kappa\hbar}{m\lambda} \sim \frac{\hbar}{m\xi(T)}$$

Thus the order of magnitude is the same as (6-38). In a type 2 superconductor, the maximum field for which there is a complete Meissner effect is $H_{C1} \sim H_C/\kappa$ and $v \sim \hbar/m\lambda(T) \ll \hbar/m\xi(T)$. The supercurrents are less stable because of the possible formation of vortex lines.

The critical velocity is proportional to $(T_0 - T)^{1/2}$ and the critical current varies as $v_C\psi_0^2$ and thus as $(T_0 - T)^{3/2}$.

Experimentally, the measurement of j_C is very delicate. (a) It is necessary to control the magnetic field distribution by shielding the film with other superconducting foils. (b) In practice, the transition appears at a "hot point" and then the remainder of the film becomes normal by a purely thermal effect. The situation is improved by using supports that are good thermal conductors and operating in a pulsed regime. (c) The transition as determined by a current-voltage diagram is not abrupt, but continuous.

At any rate, certain careful experiments (Newhouse and Bremer) on thin tin films [$\xi_0 \sim 2300$Å, $d \sim 3000$Å, $T_0 - T \sim 3 \times 10^{-2}$K, $\lambda_L(T) \sim d$, $\xi(T) \gg d$] seem to be in agreement with theory. Typically $j_C \sim 10^4$A/cm^2 in this temperature region. The variation of $|\psi|$ or $|\Delta|$ with current could, in principle, be followed by a tunneling measurement of the gap. (Detailed calculations show that, for the present geometry, the gap in the excitation spectrum is nearly equal to $|\Delta|$.)

There are two precautions that must be taken in the discussion of critical current experiments:

(1) When $\lambda(T) \sim d$, it is necessary to take account of the variation of the current density through the thickness of the film. This calculation is feasible because in the region of validity of the Landau-Ginsburg equations when $|\psi|$ is constant in space, the current $\mathbf{j}$ obeys a simple London equation.

(2) When $d < \xi(T = 0)$ in addition to $d < \xi(T)$, the bulk Landau-Ginsburg equations no longer apply (see problem, page 225).

The Experiment of Little and Parks

Consider now a superconducting film deposited on a cylindrical insulating support of radius R as is shown in Fig. 6-3. The film has a thickness $d \ll R$ ($d \sim 300$-1000Å). A uniform magnetic field $\mathbf{H}$ (typically several tens of gauss) is applied along the axis of the cylinder. How does the transition temperature vary with H?

We shall calculate it by a method due to Tinkham (1962) based on the Landau-Ginsburg equations. We still assume that $d \ll \xi(T)$ and $d \ll \lambda(T)$ and we take $|\psi|$ to be constant in the film. As usual we set $\psi = |\psi|\, e^{i\theta(\mathbf{r})}$ and the free energy is still given by (6-32).

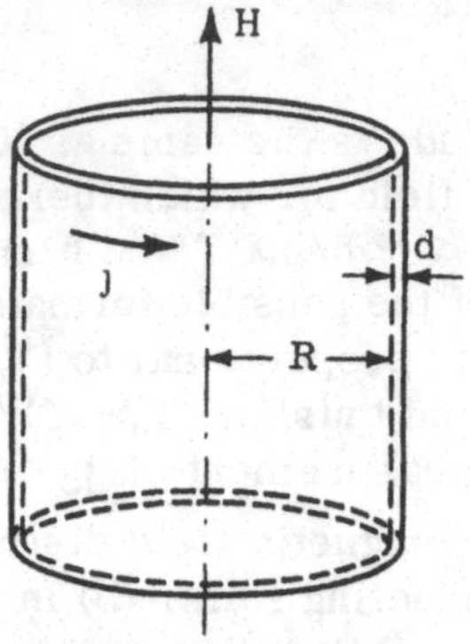

Figure 6-3
Experimental geometry for the Little-Parks experiment. One measures the critical temperature of a thin film (thickness d) deposited on a cylinder (radius $R \gg d$) in an axial field H.

If the cylinder does not carry a current along its axis, **v** will be a tangential vector of constant length. How is **v** determined? In the preceding case, the current at the boundaries of the film was fixed and **v** was determined from the current density by $\mathbf{v} = j/2e\,|\psi|^2$. Here we fix the exterior magnetic field H. In order to relate it to v we study the velocity circulation $\oint \mathbf{v}\cdot d\boldsymbol{\ell}$ around the cylinder. The contour of integration will be a circle of radius R and the circulation becomes

$$\oint \mathbf{v}\cdot d\boldsymbol{\ell} = 2\pi R v = \frac{\hbar}{m}[\theta] - \frac{2e}{mc}\oint \mathbf{A}\cdot d\boldsymbol{\ell} \tag{6-40}$$

$[\theta]$ represents the change in phase after a complete revolution about the cylinder. $\psi(\mathbf{r})$, proportional to the self-consistent field $\Delta(\mathbf{r})$, must be a single-valued function as explained in Chapter 5. Therefore $[\theta] = 2\pi n$, where n is an arbitrary integer. The second term in (6-40) is proportional to

$$\oint \mathbf{A}\cdot d\boldsymbol{\ell} = \int \mathbf{H}\cdot d\boldsymbol{\sigma} = \phi = \pi R^2 H \tag{6-41}$$

which represents the flux contained in the interior of the cylinder. Finally,

$$v = \frac{\hbar}{mR}\left[n - \frac{\phi}{\phi_0}\right] \tag{6-42}$$

where $\phi_0 = hc/2e$ is the quantum of flux.

For H fixed, ϕ is fixed, but v can take on an infinity of discrete values from (6-42). However, we see from (6-32) that the free energy will be minimized by choosing the integer n in (6-42) such that $|v|$ is minimum.

$$v = \min\left\{\frac{\hbar}{mR}\left|n - \frac{\phi}{\phi_0}\right|\right\} \tag{6-43}$$

Thus v is a periodic function of H with period $\phi_0/\pi R^2$. For $R = 7 \times 10^3$Å, $\phi_0/\pi R^2 \cong 14$G. Since v is known, we can minimize the free energy with respect to $|\psi|$ and find

$$|\psi|^2 = \left(-\alpha - \frac{mv^2}{2}\right)\beta^{-1} \tag{6-44}$$

This solution exists when $-\alpha > \frac{1}{2}mv^2$. The transition point T_H therefore corresponds to $-\alpha(T_H) = \frac{1}{2}mv^2$ and T_H is also a periodic function of H. Since $\alpha(T) \sim (T - T_0)$, we see finally that the theoretical curve T_H is a series of arcs of parabolas. The maximum displacement of the transition point corresponds to

$$v = \frac{\hbar}{2mR}$$

$$-\alpha(T_H) = \frac{\hbar^2}{2m\xi^2(T_H)} = \frac{1}{2}m\left(\frac{\hbar}{2mR}\right)^2 \tag{6-45}$$

$$\xi(T_H) = 2R$$

For a pure substance film (for $d > \xi_0$), we already have the relation $\xi(T) = 0.74\,\xi_0[T_0/(T_0 - T)]^{1/2}$, which gives

$$(T_0 - T_H)_{max} = 0.55\,T_0\left(\frac{\xi_0}{2R}\right)^2 \tag{6-46}$$

Remarks:

(1) Equation (6-45) shows that the conditions $d \ll \xi(T)$ and $d \ll \lambda(T)$ will be satisfied in practice if

$$d \ll 2R \qquad d \ll 2R\kappa$$

(2) One often uses films for which $d < \xi_0$. In this case, (6-46) is modified to $(T_0 - T_H) \sim T_0[\xi_0 d/(2R)^2]$. We see the justification for this formula later (problem, page 225).

The effects we are describing were discovered by Little and Parks (1961)

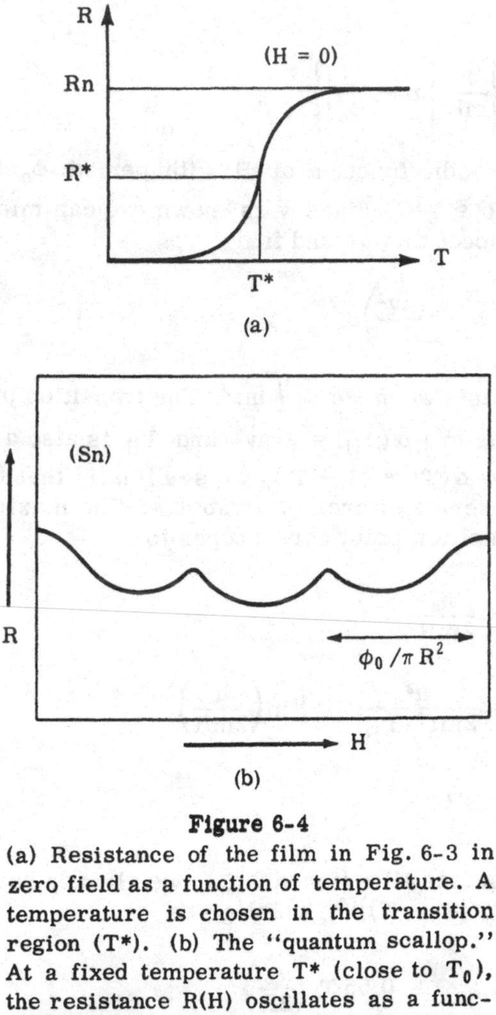

Figure 6-4

(a) Resistance of the film in Fig. 6-3 in zero field as a function of temperature. A temperature is chosen in the transition region (T*). (b) The "quantum scallop." At a fixed temperature T* (close to T_0), the resistance R(H) oscillates as a function of field. One period corresponds to one quantum of flux in the cylinder $\Delta H = \phi_0/\pi R^2$. [After Little and Parks, *Phys. Rev.*, **133A**, 97 (1964).] The parabolic "background" is probably due to a slight misorientation of the field with respect to the axis of the cylinder.

on tin films. In order to have a measurable displacement $T_0 - T_H$, it is evidently necessary to have a small R. As we have already mentioned, R is typically of the order of 0.7μ (the support is a thin plastic fiber) and $d \sim 350Å$. The resistance is measured along the axis of the cylinder with as small a current as possible.

(3) In zero field, the relation $R(H = 0, T)$ between resistance and temperature has the form represented in Fig. 6-4a. The transition is spread over $0.05°K$.

(4) A temperature T^* is chosen in the transition region and the resistance $R(H, T^*)$ is measured as a function of field. The result is the curve in Fig. 6-4b, which displays the oscillation with a period of 14 G for $R = 0.7\mu$. The experiment therefore gives a direct determination of the quantum of flux ϕ_0. On the contrary, the amplitude of the oscillations in R(H) is not well understood. The dissipation mechanisms that take place in the transition zone are not fully elucidated. [In particular, in the presence of a field H, the curve R(T) is not only shifted along the temperature axis, but also deformed.]

In Fig. 6-4b, there is a continuous increase proportional to H^2 superimposed on the oscillations, which is related to the diamagnetism of the sample. In the preceding calculation we have assumed that **H** is parallel to the axis of the cylinder and $d \ll \lambda(T)$. Under these conditions, h does not vary through the thickness of the film, and the lines of the force are nearly undisplaced when the sample undergoes the superconducting transition—the diamagnetism is negligible. This is not the case if H is not perfectly aligned along the axis of the cylinder. Then, in the superconducting state, the lines of force are more severely distorted and the energy is increased by a term proportional to H^2. This term is responsible for the continuous increase. (It is experimentally verified that the H^2 contribution is very sensitive to the orientation of the cylinder.)

Plane Film in a Parallel Field

We consider a film of thickness $d\ [< \xi(T)]$ in an external field H applied parallel to the plane of the film and having the same value at both faces. In this case, the film does not carry a macroscopic current. This is the difference with the previously discussed experiments. In order to understand this, we consider the Maxwell equation curl $\mathbf{h} = (4\pi/c)\mathbf{j}$ and take **h** along the y axis, **j** along the x axis, the film being defined by the planes $z = \pm d/2$. Then

$$\frac{\partial h(z)}{\partial z} = -\frac{4\pi}{c} j$$

$$0 = h\left(\frac{d}{2}\right) - h\left(-\frac{d}{2}\right) = -\frac{4\pi}{c}\int_{-d/2}^{d/2} j\, dz \qquad (6\text{-}47)$$

There are currents that cancel on the average and we wish to examine their distribution in detail. We begin with the London equation which, as we have seen previously, is a consequence of the Landau-Ginsburg equation in the case where $|\psi|$ is spatially constant. It is written as

$$\frac{\partial j}{\partial z} = -\frac{4e^2}{mc}|\psi|^2 h \tag{6-48}$$

Therefore

$$\frac{\partial^2 h}{\partial z^2} = \frac{f^2 h}{\lambda^2(T)} \tag{6-49}$$

(as usual we set $f = |\psi|/\psi_0$). Application of the boundary condition $h(\pm d/2) = H$ gives

$$h = H\frac{\cosh(zf/\lambda(T))}{\cosh(\epsilon f/2)} \tag{6-50}$$

where $\epsilon = d/\lambda(T)$. From h, we can calculate j from (6-47) or (6-48). As usual, we introduce the drift velocity

$$v = \frac{j}{2e|\psi|^2} = -\frac{2eH\lambda(T)}{mcf}\frac{\sinh(zf/\lambda(T))}{\cosh(\epsilon f/2)} \tag{6-51}$$

A knowledge of v allows us to calculate the free energy from (6-32). We must determine the average value of the kinetic energy through the thickness of the film

$$\overline{v^2} = \frac{1}{d}\int_{-d/2}^{d/2} v^2\,dz = \frac{1}{2}\left[\frac{2eH\lambda(T)}{mcf\cosh(\epsilon f/2)}\right]^2\left(\frac{\sinh \epsilon f}{\epsilon f} - 1\right) \tag{6-52}$$

On minimizing (6-32) with respect to f, we obtain

$$f^2 = 1 + \frac{m\overline{v^2}}{2\alpha} \tag{6-53}$$

Using our basic equations (6-8), (6-15), and (6-22), we can place this result in the form

$$\left(\frac{H}{H_c}\right)^2 = 4f^2(1 - f^2)\frac{\cosh^2\frac{\epsilon f}{2}}{\frac{\sinh \epsilon f}{\epsilon f} - 1} \tag{6-54}$$

Equation (6-54) gives the decrease in f as H increases. The behavior for $\epsilon f \ll 1$ is

$$\left(\frac{H}{H_c}\right)^2 = \frac{24}{\epsilon^2}(1 - f^2), \qquad d < \frac{\lambda(T)}{f} \tag{6-55}$$

On the other hand, for $\epsilon f \gg 1$, we later see that f is close to 1 in the important region and then (6-54) reduces to

$$\left(\frac{H}{H_c}\right)^2 = 2\epsilon(1 - f^2), \qquad d > \lambda(T) \tag{6-56}$$

What is the critical field of the film, that is, the maximum value allowed for H? If the transition to the normal state is first order, we must calculate and compare the Gibbs potential G (at fixed H and T) in the superconducting phase [with f given by (6-54)] and in the normal phase.

We first calculate the free energy F_s (per cm³) in the superconducting state from (6-32) and (6-52). Upon using the equilibrium condition (6-53) and the equations (6-8), (6-15), we can put it in the form

$$F_s = F_n - \frac{H_c^2}{8\pi} f^4 + \frac{\overline{h^2}}{8\pi} \tag{6-57}$$

where

$$\overline{h^2} = \frac{1}{d}\int_{-d/2}^{d/2} h^2\, dz = H^2 \frac{\epsilon f + \sinh \epsilon f}{\epsilon f(1 + \cosh \epsilon f)} \tag{6-58}$$

We also need the induction B (which is the average of h)

$$B = H \frac{2}{\epsilon f} \tanh \frac{\epsilon f}{2} \tag{6-59a}$$

The reader can verify that $dF/dB = H/4\pi$ and therefore that H is the thermodynamic field as could be predicted from the analysis of Chapter 2. We finally form the Gibbs potential

$$G_s = F_s - \frac{BH}{4\pi} = F_n + \frac{H^2}{8\pi}\left[\frac{\sinh \epsilon f + \epsilon f}{\epsilon f(1 + \cosh \epsilon f)} - \frac{4}{\epsilon f}\tanh\left(\frac{\epsilon f}{2}\right)\right]$$

$$- \frac{H_c^2}{8\pi} f^4 \tag{6-59b}$$

On the other hand, $G_n = F_n - H^2/8\pi$ and the critical field H is obtained by setting $G_n = G_s$ or

$$\left(\frac{H_\ell}{H_c}\right)^2 \left[1 + \frac{\epsilon f_\ell + \sinh \epsilon f_\ell}{\epsilon f_\ell(1 + \cosh \epsilon f_\ell)} - \frac{4}{\epsilon f_\ell}\tanh\left(\frac{\epsilon f_\ell}{2}\right)\right] = f_\ell^4 \tag{6-60}$$

Equations (6-54) and (6-60) define H_ℓ and the corresponding order parameter $\psi_\ell = \psi_0 f_\ell$. One can eliminate H_ℓ to obtain

$$1 + \frac{1}{6}\frac{f_\ell^2}{1 - f_\ell^2} = \frac{1}{3}\frac{\epsilon f_\ell [\cosh \epsilon f_\ell - 1]}{\sinh(\epsilon f_\ell) - \epsilon f_\ell} \tag{6-61}$$

In Fig. 6-5a we plot both sides of (6-61) as a function of f_ℓ on the interval $(0 \rightarrow 1)$. For $f \rightarrow 0$, the left side behaves as $1 + \frac{1}{6}f_\ell^2$ and the right-hand side as $1 + (\epsilon f_\ell)^2/30$. If $\epsilon > \sqrt{5}$, the right-hand side increases faster than the left. Since the left-hand side diverges for $f = 1$, there must be an intersection between $f_\ell = 0$ and $f_\ell = 1$. On the contrary, there will be no intersection for $\epsilon < \sqrt{5}$ (see Fig. 6-5b).

Therefore, if $d > \sqrt{5}\,\lambda(T)$, we have a first-order transition with a critical field H_ℓ deduced from (6-61) and (6-60). If $d < \sqrt{5}\,\lambda(T)$, G_s always remains less than G_n when the solution exists; ψ decreases as H increases, according to (6-54) and finally vanishes for

$$H = H_\ell' = \frac{\sqrt{24}}{\epsilon} H_c \tag{6-62}$$

The transition that occurs at H_ℓ' is of the second order. These two types of behavior, depending on the ratio $d/\lambda(T)$, were predicted by Ginsburg in 1952. They have been qualitatively verified in a series of tunneling experiments by Douglass (1961) (metal used: aluminum, $\xi_0 = 16{,}000$Å). He showed that

(1) The energy gap ϵ_0 in the excitation spectrum was a decreasing function of the field.

(2) For $T \cong 0.75T_0$ in films with thicknesses $d > 3{,}500$Å, the gap dropped abruptly from a finite value to 0 when the critical field was reached.

(3) At the same temperature, in films with $d < 3{,}500$Å, ϵ_0 dropped smoothly to 0 when the field was increased up to the critical value.

These results are represented on Fig. 6-5c. The distinction between first- and second-order transitions is very apparent. Unfortunately it is not possible to go beyond this qualitative statement for the following reasons:

Both $\lambda(T)$ and d are much smaller than ξ_0 in all the films. The currents j_s have a rapid variation on the scale ξ_0 (note that j_s reverses its sign when one goes from one side of the film to the other). In such a situation the Landau-Ginsburg equations cannot be applied, even if one allows for a change of κ with thickness.[5] Note: the energy

[5]The space variations of j_s depend on two parameters d and $\lambda(T)$. Thus, in general, an "effective κ" for the present geometry would depend on $\lambda(T)$ and lose its intrinsic significance. Furthermore, even in the limit $\lambda(T) >> d$, this effective κ would be numerically different from the one required by a critical current measurement on the same film.

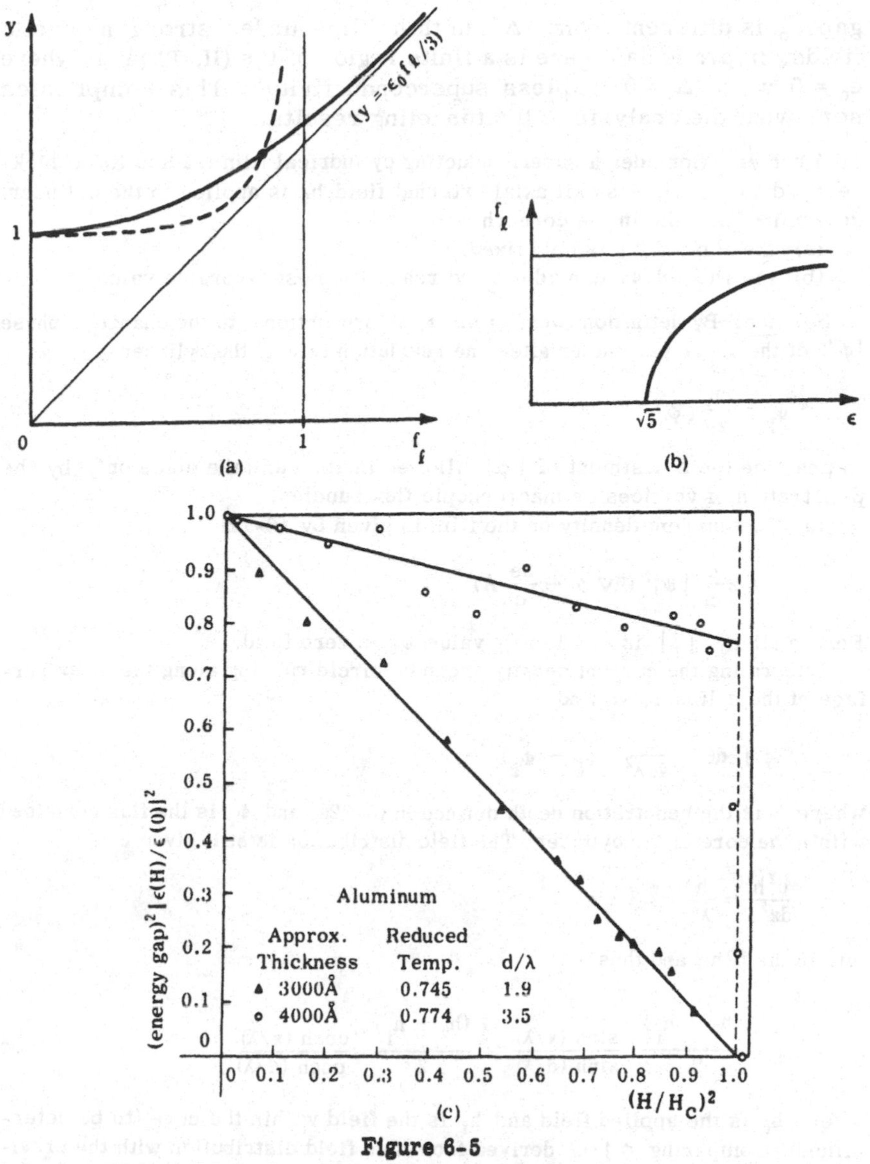

Figure 6-5

(a) A plot of the left- and right-hand sides of Eq. (6-61) as a function of the normalized order parameter f_ℓ. Left side: broken curve. Right side: continuous curve. The two curves have an intersection if $d > \sqrt{5}\lambda(T)$. (b) The dependence of the order parameter at the first-order transition as a function of $\epsilon = d/\lambda(T)$. Note that for $\epsilon < \sqrt{5}$ there is a second-order transition and the critical value f_ℓ is zero. (c) Tunneling measurement of the energy gap versus magnetic field in two aluminum samples of different thicknesses. For the film with $\epsilon > \sqrt{5}$, note the first-order transition; for the thinner film we see a much less abrupt behavior. [After D. Douglas, Jr., *IBM J. Res. Develop.*, **6**, 47 (1962).]

gap ϵ_0 is different from $|\Delta|$ in thin films under strong magnetic fields. In particular there is a finite region of the (H, T) plane where $\epsilon_0 = 0$ while $\Delta \neq 0$ (gapless superconductivity). This complicates somewhat the analysis of the tunneling results.

Problem. Consider a superconducting cylindrical film (radius R) of thickness 2d (d << R). A small axial external field h_0 is applied to the cylinder; determine the field in the core when

(a) the fluxoid Φ_f is held fixed,

(b) the fluxoid Φ_f can adjust and reach its most favorable value.

Solution. By definition, the fluxoid Φ_f is proportional to the change in phase $[\phi]$ of the order parameter after one revolution around the cylinder

$$\Phi_f = \frac{c\hbar}{2e}[\phi]$$

In practice the adjustment of $[\phi]$ allowed in (b) can take place only by the penetration of vortices or macroscopic flux bundles.

(a) The current density on the film is given by (6-12)

$$\mathbf{j} = 2\frac{e}{m}|\psi|^2\left(\hbar\nabla\phi - \frac{2e}{c}\mathbf{A}\right)$$

For small h_0, $|\psi|$ is equal to its value ψ_0 in zero field.

Integrating the current density around a circle running along the *inner* surface of the cylinder, we find

$$\oint \mathbf{j}\cdot d\boldsymbol{\ell} = \frac{c}{4\pi\lambda^2}(\Phi_f - \Phi_i)$$

where λ is the penetration depth defined in (6-22a) and Φ_i is the flux contained within the core of the cylinder. The field distribution is still given by

$$\frac{d^2h}{dx^2} = \frac{h}{\lambda^2}$$

within the film, and thus

$$h = \frac{(h_0 - h_i)}{2}\frac{\sinh(x/\lambda)}{\sinh(d/\lambda)} + \frac{(h_0 + h_i)}{2}\frac{\cosh(x/\lambda)}{\cosh(d/\lambda)}$$

where h_0 is the applied field and h_i is the field within the core (to be determined). Comparing $\oint \mathbf{j}\cdot d\boldsymbol{\ell}$ derived from this field distribution with the previous equation involving the fluxoid, we find

$$h_i = \frac{\dfrac{\Phi_f}{\pi\lambda R} + \dfrac{2h_0}{\sinh(2d/\lambda)}}{2\coth(2d/\lambda) + R/\lambda}$$

$$\simeq \frac{\Phi_f}{\pi R^2} + \frac{2h_0\lambda}{R\sinh(2d/\lambda)} \qquad \left(R \gg \frac{\lambda^2}{d}\right)$$

If the fluxoid is held fixed, for example, $\Phi_f = 0$, we find

$$h_i \simeq \frac{2h_0 \lambda}{R \sinh (2d/\lambda)}$$

h_i is, in general, quite small compared to the applied field h_0, except of course for extremely thin films ($d \leq \lambda^2/R$).

(b) Let us now assume that the fluxoid is allowed to vary. For a fixed applied field h_0, we must minimize the Gibbs potential $\mathcal{G}$ with respect to the fluxoid. $\mathcal{G}$ may conveniently be written as

$$\mathcal{G} = \frac{\lambda^2}{8\hbar} \int \left(\frac{dh}{dx}\right)^2 dr + \frac{1}{8\pi} \int (h - h_0)^2 \, dr + \text{const}$$

Using the explicit form derived above for the field, one finds that the minimum corresponds to $h_i = h_0$.

Thus in thermodynamic equilibrium the field within the core is equal to the applied field. However it often occurs in A.C. experiments that the fluxoid cannot adjust itself to its equilibrium value during the time of observation and we then find a situation where the core field is small.

6–6 SITUATIONS WHERE $|\psi|$ VARIES SPATIALLY

Nucleation in the Bulk

We consider a superconducting metal in a high magnetic field. For sufficiently high fields, superconductivity is destroyed and the field is uniform in the sample. If we continuously decrease the field, at a certain field $H = H_{C2}$, superconducting regions begin to nucleate spontaneously. We see that H_{C2} does not coincide with the thermodynamic field H_C; depending on the case it can be either larger or smaller.

In the regions where the nucleation occurs, superconductivity is just beginning to appear and therefore $|\psi|$ is small. The Landau-Ginsburg equation can be linearized to give

$$\frac{1}{2m}\left(-i\hbar\nabla - \frac{2eA}{c}\right)^2 \psi = -\alpha\psi \qquad (6\text{-}63)$$

Also, in (6-63) we can set curl **A** = **H** where **H** is only the applied field. This is valid because the supercurrents are of order $|\psi|^2$ and the resulting corrections to the field are therefore negligible in the linear approximation. Equation (6-63) is then formally identical to the Schrödinger equation for a particle of charge 2e and mass m in a uniform magnetic field.

First consider the case of an *infinite medium*. The particle has a constant velocity v_z along the field and moves in a circle in the xy plane with the frequency

$$\omega_c = 2eH/mc \tag{6-63'}$$

The energies corresponding to the bound states of (6-63) have the form $\frac{1}{2}mv_z^2 + (n + \frac{1}{2})\hbar\omega_c$ where n is a positive integer. In particular, the lowest level corresponds to $v_z = 0$, $n = 0$, and gives $-\alpha = e\hbar H/mc$.

The field H thus obtained is the field H_{c2}. (The other levels with n or v_z nonzero will give, for the same α, lower values of H.) By using the relation $\alpha^2/2\beta = H_c^2/8\pi$ and the definition (6-26) of the Landau-Ginsburg parameter, one finally obtains

$$H_{c2} = \kappa\sqrt{2}\, H_c$$

$$= \frac{\phi_0}{2\pi\,\xi(T)^2} \tag{6-64}$$

Discussion of this formula.

(1) When $\kappa > 1/\sqrt{2}$, ($H_c < H_{c2}$), a condensed phase ($\psi \neq 0$) will appear in the bulk of the sample for fields $H < H_{c2}$. This phase cannot correspond to a complete exclusion of the magnetic flux, since a complete Meissner effect is energetically unfavorable at fields $H > H_c$ and we can choose $H_c < H < H_{c2}$. In fact, what we obtain here below H_{c2} is the Schubnikov phase discussed in Chapter 3. As predicted there, we see that the upper critical field is of order $\phi_0/\xi(T)^2$. It corresponds to a close packing of vortex lines, each of them carrying the flux unit ϕ_0 and having a core radius $\xi(T)$.

(2) When $\kappa < 1/\sqrt{2}$ ($H_{c2} < H_c$), if we decrease the field we first meet the value H_c, at which a complete Meissner effect takes place—we are dealing here with a first type superconductor.

$$\begin{aligned} \kappa &< \frac{1}{\sqrt{2}} \quad \text{first type} \\ \kappa &> \frac{1}{\sqrt{2}} \quad \text{second type} \end{aligned} \tag{6-65}$$

Nucleation on the Sample Surface

The calculation that led to (6-64) is only valid in an infinite medium. It neglects boundary effects. In fact, for an ideal substance, nucleation always takes place first on the surface (D. Saint-James and P. G. de Gennes, 1963). We now analyze in greater detail the effects of the sample surface in a simple case. The surface is assumed to be a plane [radius of curvature much larger than $\xi(T)$] and the boundary separates the superconductor from either an insulator or a vacuum such that the boundary conditions (6-13) apply. We consider two cases:

(1) The field (along z) is normal to the surface (xy plane). Then the function ψ corresponding to the lowest level of (6-63) is also an eigenfunction of $\Pi_z = -i\hbar\, \partial/\partial z - 2eA_z/c$ with eigenvalue 0. Therefore (6-13) is automatically satisfied and the surface does not modify the nucleation field.

(2) The field is in the surface of the sample; **H** is directed along the z axis, the surface is in the yz plane and the superconductor occupies the region $x > 0$. We select the gauge $A_z = A_x = 0$, $A_y = Hx$ and look for solutions of the form

$$\psi = e^{iky} f(x) \tag{6-66}$$

Equation (6-63) becomes

$$-\frac{\hbar^2}{2m}\frac{d^2 f}{dx^2} + \frac{1}{2m}\left(\hbar k - \frac{2eHx}{c}\right)^2 f = -\alpha f \tag{6-67}$$

and the boundary condition is

$$\left(\frac{df}{dx}\right)_{x=0} = 0 \tag{6-68}$$

Equation (6-67) is analogous to the Schrödinger equation of a harmonic oscillator of frequency ω given by (6-63'), the equilibrium position being the point

$$x_0 = \frac{\hbar kc}{2eH} \tag{6-69}$$

However (6-68) introduces a complication. If $x_0 \gg \xi(T)$, the wavefunction will be localized around x_0 and nearly zero at the surface; therefore (6-68) will automatically be satisfied. In this case, we have

$$f \cong \exp\left[-\frac{1}{2}\left(\frac{x - x_0}{\xi(T)}\right)^2\right] \tag{6-70}$$

the eigenvalue still being given by (6-63). If x_0 is zero, on the other hand, the function (6-70) still satisfies the boundary condition (6-68), and (6-63) is still obtained. We now show that between the two extreme cases, namely for $x_0 \sim \xi(T)$, the lowest eigenvalue is lower than (6-63). In order to see this, we can replace (6-67), valid for $x > 0$, by an equation valid over the interval $-\infty < x < \infty$

$$-\frac{\hbar^2}{2m}\frac{d^2 f}{dx^2} + V(x) f = -\alpha f \tag{6-71}$$

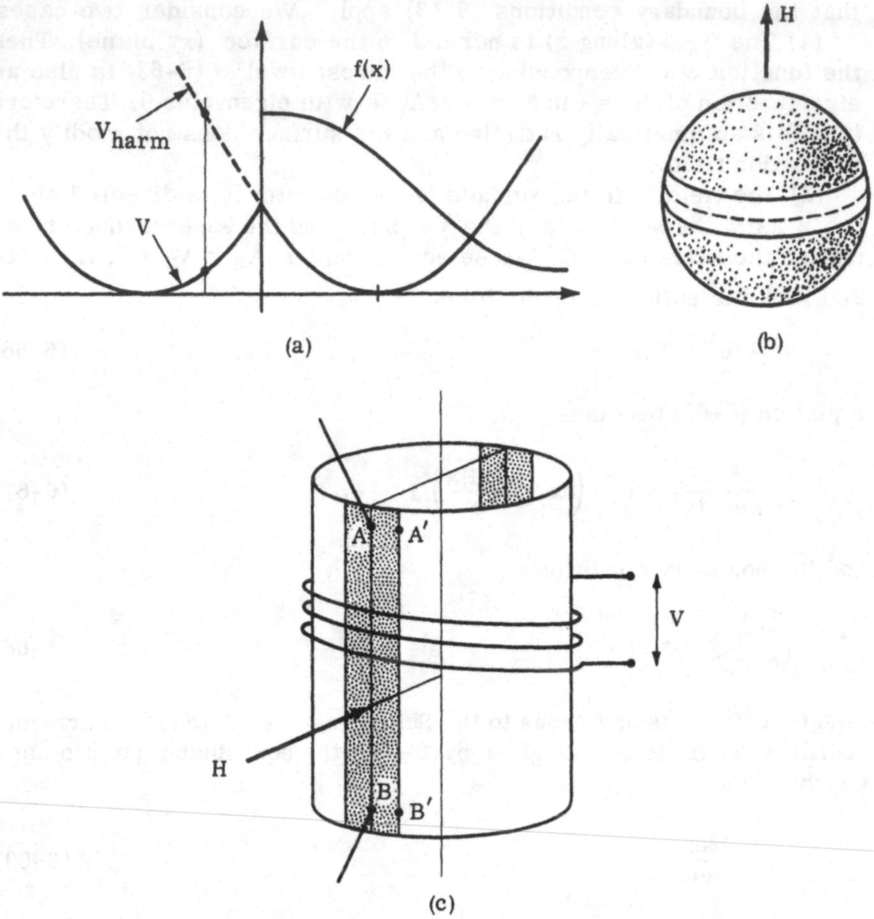

Figure 6-6

(a) The harmonic potential $V_{harm} = \text{const}\,(x - x_0)^2$ and the symmetrical potential V. Since $V < V_{harm}$, the first level in V is lower than the first level in V_{harm}. Also shown is the shape of the eigenfunction f(x). (b) Surface superconductivity in a sphere. The band around the equator represents the superconducting sheath when H is just below H_{c3}. As H decreases, this band broadens and reaches the poles at H_{c2}. (c) Principle of the determination of H_{c2} and H_{c3} on hollow cylinders [after J. P. Burger, G. Deutscher, E. Guyon, and A. Martinet, *Solid State Commun.*, **2**, 101 (1964)]. A field $H > H_{c2}$ is applied normal to the cylinder axis. Two normal strip appears on the cylinder (shaded areas). Their onset is detected with a coil wound around the cylinder: the self-inductance of the coil increases abruptly when the strips appear. The width of the strips is measured with the resistive probes AB. The current flows from A to A′, then in the superconducting regions from A′ to B′, and finally from B′ to B. The resistance is proportional to the length AA′.

The potential V(x) coincides with that of (6-67) for $x > 0$. For $x < 0$, we continue it symmetrically (Fig. 6-6a):

$$V(x) = \begin{cases} \dfrac{2e^2H^2}{mc^2}(x - x_0)^2 & x > 0 \\ V(-x) & x < 0 \end{cases} \tag{6-72}$$

The eigenfunction corresponding to the lowest eigenvalue of (6-71) has no nodes, is an even function of x, and automatically satisfies (6-68). For $x < 0$, we see from the figure that the potential V(x) is always less than $(2e^2H^2/mc^2)(x - x_0)^2$ and therefore the lowest eigenvalue of V(x) is lower than (6-63). Hence, *nucleation is favored by the presence of a surface.*

A detailed calculation shows that the optimum value of x_0 is $0.59\xi(T)$ and that the corresponding eigenvalue becomes

$$-\alpha = 0.59\,\frac{e\hbar H}{mc} \tag{6-73}$$

The eigenfunction is a rather complicated Weber function. But, as first noticed by C. Kittel, a good approximation to the exact eigenvalue (6-73) can be obtained more simply from the variational principle applied to the Schrödinger equation (6-67). Taking as a trial function a gaussian of adjustable width

$$f(x) = e^{-rx^2}$$

and minimizing $|\alpha|$ with respect to r and x_0, one finds

$$|\alpha| = \sqrt{1 - \frac{2}{\pi}}\,\frac{e\hbar H}{mc} = 0.60\,\frac{e\hbar H}{mc}$$

The nucleation field, which we call H_{c3}, is given by (6-73). By making use of (6-64), it can be written as

$$H_{c3} = 2.4\kappa H_c = 1.7\,H_{c2} \tag{6-74}$$

Discussion of Eq. (6-74).

(1) Materials of the second type. The above calculation shows us that above H_{c2}, in the interval $H_{c2} < H < H_{c3}$, superconductivity is not entirely destroyed. The bulk of the material is normal, but a *superconducting sheath* subsists in certain surface regions. For a long cylinder placed in an axial field, this sheath occupies the entire lateral surface of the cylinder. For a sphere, the situation is quite different. When H is slightly less than H_{c3}, the sheath is limited to a narrow band around the equatorial circle (Fig. 6-6b). As H decreases, this band broadens and finally reaches the poles at $H = H_{c2}$.

A number of methods have been used to detect the surface sheath: *resistance measurements* (at low currents, the sheath acts as a short circuit); *inductive measurements* at low frequencies—a typical geometry is shown in Fig. 6-6c. When the sheath is present on the cylinder surface, flux lines cannot pass it and the self-inductance of the coil is low. When H reaches H_{c3}, the sheath disappears and the full self-inductance is restored (Fig. 6-7a). Similar but more complicated experiments can also be carried out in the microwave region.

In films [much thicker than $\xi(T)$], various sensitive techniques can be applied: (a) tunneling experiments display a density of states that deviates significantly from the normal state value; (b) the sheath acts as a small diamagnetic region that can be detected by torque measurements.

All these experiments have shown that for clean sample surfaces [radius of curvature $\gg \xi(T)$] there is, indeed, a sheath in an interval $H_{c2} < H < H_{c3}(\theta)$, where θ is the angle between the field and the surface. For $\theta = 0$ the experimental value $H_{c3}(0)$ is close to the theoretical value (6-74). For $\theta \neq 0$, $H_{c3}(\theta)$ varies as shown on Fig. 6-7b. For $\theta = \frac{1}{2}\pi$, $H_{c3}(\theta) \to H_{c2}$.

(2) Type I superconductors with $0.42 < \kappa < 1/\sqrt{2}$. Here $H_{c2} < H_c$, but $H_{c3} > H_c$. Thus in the field range $H_c < H < H_{c3}$ there still exists a superconducting

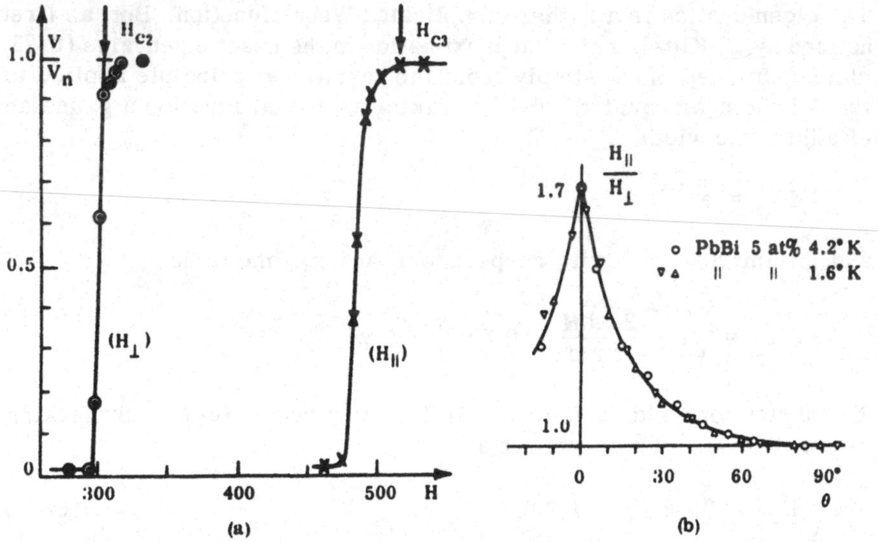

Figure 6-7

(a) Results of induction measurements (with the coil of Fig. 6-6c) on hollow cylinders of a SnIn alloy. When the field H is normal to the cylinder axis ($H_\perp$), the voltage V (proportional to the self-inductance of the coil) increases for $H = H_{c2}$. When H is parallel to the axis ($H_\parallel$), the increase occurs only when $H = H_{c3}$ (after Burger et al.). (b) Angular dependence of the critical field in PbBi alloys at various temperatures. [After P. Burger, G. Deutscher, E. Guyon, and A. Martinet, *Phys. Rev.*, **137A**, 853 (1965).]

sheath (in the regions where H is parallel or nearly parallel to the sample surface). This sheath in a Type I superconductor has been observed in a few cases such as lead and dilute lead alloys.

(3) Type I superconductors with $\kappa < 0.42$. Here $H_{C3} < H_C$ and it is not possible to observe the sheath in a state of thermodynamic equilibrium. It is still possible, however, to measure H_{C3} in the following way: In very good samples it is possible to lower the field below H_C while still retaining the sample in the normal (metastable) state. In aluminum, for instance, the field can be lowered to $< H_C/10$ with no transition occurring. However, the metastable normal phase never extends down to 0 applied field; the reason is that, when we reach the field H_{C3}, it becomes possible to create weakly superconducting regions near the sample surface, with no energy expense. The normal phase becomes strictly unstable when $H \leq H_{C3}$. The corresponding nucleation process has, in fact, been studied very carefully by Faber on various Type I superconductors. He showed experimentally that nucleation took place near the sample surface (however, without realizing at the time that this was a fundamental property) and he measured the nucleation field we now identify with H_{C3}. From his results one could, in principle, derive the corresponding κ values by Eq. (6-74). If we do this blindly, we find

$$\kappa = \begin{array}{ll} 0.015 & \text{Al} \\ 0.07 & \text{In} \\ 0.10 & \text{Sn} \end{array}$$

This procedure is probably not quite correct for these pure metals where the Landau-Ginsburg equations are valid only for temperatures T very close to T_0, while the supercooling fields are usually measured at lower temperatures. For the sake of completeness, we quote the "theorctical" values of κ derived from the definition (6-27) where $\lambda_L(0)$ and ξ_0 are determined from Fermi surface measurements in the normal state (plus the experimental value of T_0). They are

$$\kappa = \begin{array}{ll} 0.010 & \text{Al} \\ 0.05 & \text{In} \\ 0.15 & \text{Sn} \end{array}$$

6–7 STRUCTURE OF THE VORTEX PHASE IN STRONG FIELDS ($H \sim H_{c2}$)

Consider a cylindrical sample of a second type material ($\kappa > 1/\sqrt{2}$) placed in an exterior field H parallel to the axis of the cylinder. Let H decrease. When H becomes equal to H_{c2}, nucleation of the superconducting state begins in the volume of the sample. The order parameter $|\psi|$ is small and the form of ψ on nucleation is obtained from the linearized Landau-Ginsburg equations (6-63). If H still decreases, the nucleation region develops, $|\psi|$ becomes larger and in order to determine the nature of the order it is necessary to solve the complete nonlinearized Landau-Ginsburg equations. This represents a formidable numerical problem.

Things become simpler if H is taken to be only slightly less than H_{c2} (A. Abrikosov, 1956). By continuity, the solution ψ of the complete Landau-Ginsburg equations must have a strong resemblance to a certain solution ψ_L of the linearized equation. This does not mean that the problem is immediately solved. We only know that the ψ_L satisfies

$$\frac{1}{2m}\left(-i\hbar\nabla - \frac{2eA_0}{c}\right)^2 \psi_L = -\alpha\psi_L \tag{6-75a}$$

with

$$\text{curl } \mathbf{A}_0 = (0, 0, H_{c2}) \tag{6-75b}$$

However, the eigenvalues of (6-75a) are strongly degenerate, that is, there exist many independent solutions describing nucleation in one or another part of the sample. For example, in the gauge ($A_x = A_z = 0$, $A_y = H_{c2}x$) we have seen that the solutions are of the form

$$\psi_k = e^{iky} e^{-(\frac{1}{2})(x - x_0)^2/\xi^2(T)}$$
$$x_0 = \frac{\hbar ck}{2eH_{c2}} \tag{6-76}$$

where k is an arbitrary parameter. The nucleation region for such a situation is a band of width $\xi(T)$ perpendicular to the x axis. Its average abscissa x_0 depends on the value chosen for k. We know that ψ_L must be a linear combination of the ψ_k. In fact, from our simplified discussion in Chapter 3, we expect to find $|\psi|$ to have a periodic structure in x and y. If $2\pi/q$ is the spatial period in the y direction, such a function ψ_L will be the combination

$$\psi_L = \sum_n C_n e^{inqy} \exp[-\tfrac{1}{2}(x - x_n)^2/\xi^2(T)]$$
$$x_n = \frac{n\hbar cq}{2eH_{c2}} \tag{6-77}$$

In order that $|\psi_L|$ also be periodic in x, it is necessary to impose a periodicity on the C_n's. The form of condition chosen by Abrikosov was $C_{n+\nu} = C_n$, where ν is a fixed integer. Then

$$\psi_L\left(x + \frac{\nu\hbar cq}{2eH_{c_2}}, y\right) = e^{i\nu qy}\psi_L(x, y)$$

However, independent of the detailed structure of ψ_L postulated in (6-77), it is possible to establish certain general theorems:

(1) Normalization of ψ_L: As the starting equations (6-11) and (6-12) are nonlinear in ψ, the normalization adopted for ψ_L plays a nontrivial role. Let us assume that the free energy (6-9) is stationary if the trial function ψ_L is changed to $(1 + \epsilon)\psi_L$ where ϵ is small and independent of $\mathbf{r}$. To first order in ϵ the variation in the free energy is

$$\delta\mathfrak{F} = 2\epsilon \int d\mathbf{r} \left[\alpha |\psi_L|^2 + \beta |\psi_L|^4 + \frac{1}{2m}\left|\left(-i\hbar\nabla - \frac{2e\mathbf{A}}{c}\right)\psi_L\right|^2\right] \tag{6-78}$$

In order to simplify the notation, we shall write an integral such as $\int |\psi_L|^2 d\mathbf{r}$ extended over a macroscopic volume V in the form $V\overline{|\psi_L|^2}$. From (6-78), the condition $\delta\mathfrak{F} = 0$ can be written

$$\alpha\overline{|\psi_L|^2} + \beta\overline{|\psi_L|^4} + \frac{1}{2m}\overline{\left|\left(-i\hbar\nabla - \frac{2e\mathbf{A}}{c}\right)\psi_L\right|^2} = 0 \tag{6-79}$$

This equation can be simplified by using (6-75a), which is satisfied by ψ_L. Set $\mathbf{A} = \mathbf{A}_0 + \mathbf{A}_1$ where $\mathbf{A}_0$ (defined by 6-75b) is the vector potential which would exist in the presence of the field H_{c2} and where $\mathbf{A}_1$ represents the modifications due to the facts that (a) the applied field is slightly smaller than H_{c2}; and (b) there exist supercurrents, which also contribute to the fields. Expanding (6-79) to first order in $\mathbf{A}_1$ and using (6-75a), we obtain

$$\beta\overline{|\psi_L|^4} - \frac{1}{c}\overline{\mathbf{A}_1 \cdot \mathbf{j}_L} = 0 \tag{6-80}$$

where

$$\mathbf{j}_L = \frac{e}{m}\left[\psi_L^*\left(-i\hbar\nabla - \frac{2e\mathbf{A}_0}{c}\right)\psi_L + \text{C.C.}\right] \tag{6-81}$$

$\mathbf{j}_L$ represents the current associated with the unperturbed solution. If we integrate the second term in (6-80) by parts and set curl $\mathbf{A}_1 = \mathbf{h}_1$ and curl $\mathbf{h}_s = (4\pi/c)\mathbf{j}_L$, we obtain

$$\beta\overline{|\psi_L|^4} - \frac{1}{4\pi}\overline{\mathbf{h}_1 \cdot \mathbf{h}_s} = 0 \tag{6-82}$$

(2) Relation between field and order parameter: In order to make

the condition (6-82) explicit, we must calculate the fields h_1 and h_s. (From the results of Chapter 3, we predict that h_1 and h_s are everywhere parallel to the z axis.) We can write

$$h_1(r) = H - H_{c2} + h_s(r) \tag{6-83}$$

The term $H - H_{c2}$ represents the contribution (α) and h_s gives the effect of the supercurrents.[6]

When ψ_L is a solution of (6-75a) corresponding to the lowest eigenvalue $-\alpha = \frac{1}{2}\hbar\omega_c$, *the lines of current* associated with j_L *coincide with the lines* $|\psi_L| = const.$ This property allows us to calculate h_s simply.

Proof.

Set $\Pi = (2m\hbar\omega_c)^{-1/2}\left(-i\hbar\nabla - \frac{2eA_0}{c}\right)$

and $\Pi^{\pm} = \Pi_x \pm i\Pi_y$. The commutator $[\Pi^+, \Pi^-]$ is one and (6-75a) can be written in the form

$$\Pi^+\Pi^- \psi_L = 0$$

whose bound state solutions correspond to

$$\Pi^-\psi_L = 0 \tag{6-84}$$

We still set

$$\begin{aligned} \psi_L &= |\psi_L| e^{i\theta} \\ j_L &= \frac{2e}{m}|\psi_L|^2\left(\hbar\nabla\theta - \frac{2e}{c}A_0\right) \end{aligned} \tag{6-85}$$

On equating the real and the imaginary parts of $\Pi^-\psi_L$ to zero and using (6-85), we find

$$\begin{aligned} j_{Lx} &= -\frac{e\hbar}{m}\frac{\partial}{\partial y}|\psi_L|^2 \\ j_{Ly} &= \frac{e\hbar}{m}\frac{\partial}{\partial x}|\psi_L|^2 \end{aligned} \tag{6-86}$$

[6] $j_L = (c/4\pi)$ curl h_s is not exactly the supercurrent since it is A_0 and not A_1 which enters in (6.81). However we shall see that $H_{c2} - H$ and $|\psi|^2$ are of the same order and small, and therefore $\overline{(A_0 - A)|\psi|^2}$ is of order $|\psi|^4$ and negligible at this stage

We see that the current lines are the lines of constant $|\psi_L|$.

By comparing (6-86) to the definition curl $\mathbf{h}_s = (4\pi/c)\mathbf{j}_L$, we find explicitly that the field $\mathbf{h}_s$ due to the supercurrents is given by

$$\mathbf{h}_s = -4\pi \frac{e\hbar}{mc} |\psi_L|^2 \tag{6-87}$$

The constant of integration is such that, for $\overline{|\psi_L|^2} = 0$, h_s disappears. By regrouping (6-82), (6-83), and (6-87), we obtain the explicit condition

$$\beta \overline{|\psi_L|^4} + \frac{e\hbar}{mc} \overline{|\psi_L|^2 \left(H - H_{c2} - \frac{4\pi e\hbar}{mc} |\psi_L|^2\right)} = 0 \tag{6-88a}$$

It is again useful to set $|\psi_L| = \psi_0 f$, to use the definition (6-26) of κ, and (6-64) for H_{c2}, to find

$$\overline{f^4}\left(1 - \frac{1}{2\kappa^2}\right) - \overline{f^2}\left(1 - \frac{H}{H_{c2}}\right) = 0 \tag{6-88b}$$

The relation (6-88b) is important because it is independent of the detailed form of the hypothesis for ψ_L, that is, independent of the nature of the lattice formed by the flux lines. If we now select a given lattice, that is, fix q and the periodicity of the coefficients C_n, we can explicitly calculate the ratio

$$\frac{\overline{f^4}}{(\overline{f^2})^2} = \beta_A \tag{6-89}$$

which corresponds to this lattice. If β_A is known, (6-88b) and (6-89) can be solved for $\overline{f^4}$ and $\overline{f^2}$. This is all that is necessary to calculate the induction and free energy.

(1) Induction **B**: This is the average value of the field $\mathbf{H} + \mathbf{h}_s$ and is found from (6-87) to be

$$B = H + \overline{h_s} = H - \frac{4\pi e\hbar}{mc} \overline{|\psi_L|^2}$$

$$= H - \frac{H_c}{\kappa\sqrt{2}} \overline{f^2} \tag{6-90}$$

(2) The *free energy* per unit volume F is calculated using (6-11) to simplify (6-9)

$$F = -\frac{\beta}{2} \overline{|\psi_L|^4} + \frac{\overline{h^2}}{8\pi} = -\frac{H_c^2}{8\pi}\overline{f^4} + \frac{\overline{h^2}}{8\pi} \tag{6-91}$$

By returning to (6-87) we can express h^2 as a function of H, f^2, and f^4. We then eliminate H, f^2, f^4 by means of (6-88b), (6-89), and (6-90) to obtain the free energy only as a function of B:

$$F = \frac{B^2}{8\pi} - \frac{(H_{c2} - B)^2}{1 + (2\kappa^2 - 1)\beta_A}\,\frac{1}{8\pi} \tag{6-92}$$

Conclusions.

(1) For fixed B, F is an increasing function of β_A (since $\kappa > 1/\sqrt{2}$). The most favorable "lattice" will correspond to the *smallest* β_A. [Note that from (6-89) $\beta_A \geq 1$.]

The choice of lattice corresponds algebraically to a certain choice of the periodicity conditions satisfied by the C_n's in Eq. (6-77). Two lattices have been studied in practice: a *square* lattice of vortex lines is obtained when all C_n's are taken as being equal; a *triangular* lattice is obtained with the conditions

$$C_{n+2} = C_n$$

$$C_1 = iC_0$$

(after minimization of β_A with respect to q). It is found that $\beta_A = 1.18$ for the square lattice, and $\beta_A = 1.16$ for the triangular lattice. The latter is thus slightly more stable. The contour diagram of $|\psi|^2$ for the triangular solution is shown on Fig. 3-5, p. 67.

(2) Once β_A is known, the usual thermodynamic calculations can be performed with the free energy (6-92). We can first verify that

$$4\pi \frac{dF}{dB} = H$$

as expected. Then the magnetization $M = (B - H)/4\pi$ can be calculated. M is of order $|\psi|^2$ as expected, and given explicitly by

$$M = \frac{B - H}{4\pi} = \frac{H - H_{c2}}{4\pi\beta_A(2\kappa^2 - 1)}, \qquad \frac{H_{c2} - H}{H_{c2}} \ll 1 \tag{6-93}$$

The magnetization vanishes at $H = H_{c2}$—the transition at $H = H_{c2}$ is of second order. (However, if κ is only slightly greater than $1/\sqrt{2}$, the slope $|dM/dH|$ becomes very large.) A detailed test of Eq. (6-93) has been carried out by Kinsel, Lynton, and Serin on an alloy InBi with 2.5% Bi, an example of a Type II superconductor with a nearly reversible magnetization curve. The phase diagram deduced from their magnetic measurements on this alloy is shown in Fig. 6-8. From the experimental values of H_{c2} near T_0, one finds

$$\kappa = \frac{1}{\sqrt{2}}\,\frac{H_{c2}}{H_c} = 1.80$$

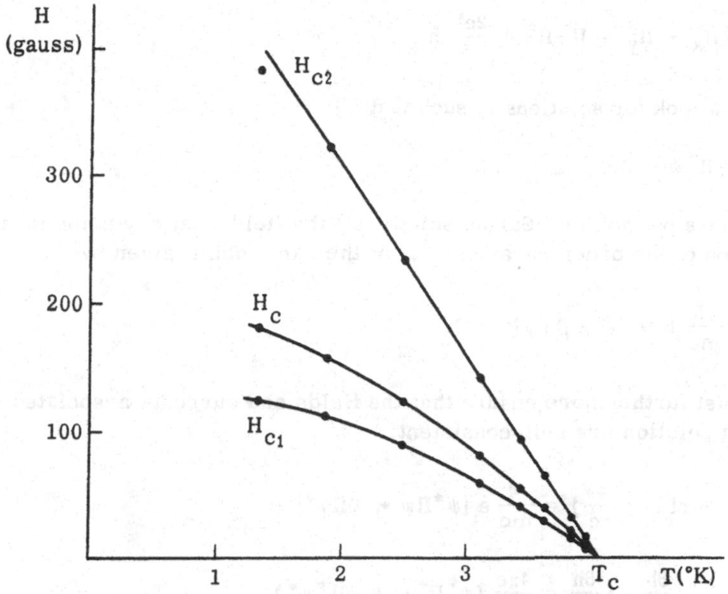

Figure 6-8
The phase diagram of a Type II superconductor (InBi alloy). [After T. Kinsel, E. A. Lynton, and B. Serin, *Phys. Letters*, **3**, 30 (1962).]

From (6-93), taking $\beta_A = 1.16$, one finds from the magnetization curves

$$\kappa = 1.81$$

This very good agreement[7] is further confirmed by a theoretical calculation of κ described later.

Problem. Show that the wall energy in a Type I superconductor vanishes for $\kappa = 1/\sqrt{2}$.

Solution. We follow a method due to G. Sarma. By taking the fields **h** to be along the z direction, we first reduce the Landau-Ginsburg equation to a two-dimensional form

$$\frac{1}{2m}(\Pi_x^2 + \Pi_y^2)\psi + \alpha\psi + \beta|\psi|^2\psi = 0$$

where $\Pi = \mathbf{p} - 2e\mathbf{A}/c$. By putting $\Pi^{\pm} = \Pi_x \pm i\Pi_y$, we have

[7] The agreement is even surprising since the experimental magnetization curves are not perfectly reversible, which implies a certain uncertainty in $(dM/dH)_{H_{c_2}}$.

$$\Pi_x^2 + \Pi_y^2 = \Pi^- \Pi^+ + \frac{2e\hbar}{c} h$$

We now look for solutions ψ such that

$$\Pi^+ \psi = 0$$

For these particular "Sarma solutions" the field h at any point is a known function of the order parameter ψ at the same point, given by

$$\frac{e\hbar}{mc} h + \alpha + \beta |\psi|^2 = 0$$

We must furthermore ensure that the fields and currents associated with the Sarma solution are self-consistent:

$$\text{curl } \mathbf{h} = \frac{4\pi}{c} \mathbf{j} = \frac{4}{mc} e [\psi^* \mathbf{\Pi} \psi + \psi \mathbf{\Pi} \psi^*]$$

$$\frac{\partial h}{\partial y} - i \frac{\partial h}{\partial x} = \frac{4\pi e}{mc} (\psi^* \Pi^+ \psi + \psi \Pi^+ \psi^*)$$

$$= \frac{4\pi e}{mc} \psi \Pi^+ \psi^*$$

This relation must agree with the $h(\psi)$ relation obtained earlier. Differentiation of the $h(\psi)$ relation gives

$$\frac{e\hbar}{mc}\left(\frac{\partial h}{\partial y} - i \frac{\partial h}{\partial x}\right) + \beta\psi^*\left(\frac{\partial \psi}{\partial y} - i \frac{\partial \psi}{\partial x}\right) + \beta\psi\left(\frac{\partial \psi^*}{\partial y} - i \frac{\partial \psi^*}{\partial x}\right) = 0$$

This may be transformed in the following way: The condition $\Pi^+ \psi = 0$ may be written as

$$\hbar\left(\frac{\partial \psi}{\partial y} - i \frac{\partial \psi}{\partial x}\right) = \frac{2e}{c} (A_x + iA_y) \psi$$

from which we obtain

$$\frac{\partial h}{\partial y} - i \frac{\partial h}{\partial x} = \frac{\beta mc}{e\hbar^2} \psi \Pi^+ \psi^*$$

This does agree with the equation for the current provided that

$$\frac{4\pi e}{mc} = \frac{\beta mc}{e\hbar^2} \quad \text{or} \quad \kappa = \frac{1}{\sqrt{2}}$$

Consider now more specifically a wall in the yz plane. For $x \to -\infty$ (normal side), $h = H_c$ and $\psi = 0$. For $x \to +\infty$ (superconducting side), $h = 0$ and

$|\psi|^2 = -\alpha/\beta$. These two conditions are in agreement with the $h(\psi)$ relation for the Sarma solution—thus the Sarma solution does apply to the present problem. The thermodynamic potential (per unit area in the yz plane) $\mathcal{G}$ is given by

$$\mathcal{G} = \int dx \left[\alpha |\psi|^2 + \tfrac{1}{2}\beta |\psi|^4 + \frac{1}{2m} |\Pi\psi|^2 + \frac{h^2}{8\pi} - \frac{hH_c}{4\pi} \right]$$

The last term is the microscopic analog of the $-BH/4\pi$ term, the thermodynamic field H being here equal to H_c. Since ψ satisfies the Landau-Ginsburg equation, $\mathcal{G}$ can be integrated by parts to give

$$\mathcal{G} = \int dx \left[-\tfrac{1}{2}\beta |\psi|^4 + \frac{h^2 - 2H_c h}{8\pi} \right]$$

To obtain the wall energy $\mathcal{G}_{\text{wall}}$, we subtract from $\mathcal{G}$ the term $\int (H_c^2/8\pi)\, dx$, which corresponds to the potential in the normal (or in the superconducting) phase:

$$\mathcal{G}_{\text{wall}} = \int dx \left[-\tfrac{1}{2}\beta |\psi|^4 + \frac{(h - H_c)^2}{8\pi} \right]$$

The $h(\psi)$ relation can be rewritten as

$$h = H_c - \frac{mc}{e\hbar} \beta |\psi|^2$$

and

$$\frac{(h - H_c)^2}{8\pi} = \frac{1}{8\pi} \beta^2 \left(\frac{mc}{e\hbar}\right)^2 |\psi|^4 = \tfrac{1}{2}\beta |\psi|^4 \quad \text{when } \kappa = \frac{1}{\sqrt{2}}$$

Thus $\mathcal{G}_{\text{wall}} = 0$ for $\kappa = 1/\sqrt{2}$.

A similar calculation in terms of Sarma solutions can be done for the energy of an isolated vortex line, when $\kappa = 1/\sqrt{2}$. The conclusion there is that the first penetration field H_{c1} is equal to H_c. For that particular value of κ, H_{c1}, H_{c2}, and H_c coincide.

REFERENCES

A much more detailed study of the various applications of the Landau-Ginsburg equation is to be published by Saint-James, Sarma, Thomas (Pergamon).

7

MICROSCOPIC ANALYSIS OF THE LANDAU-GINSBURG EQUATIONS

7–1 LINEARIZED SELF-CONSISTENCY EQUATION

In the preceding chapter we constructed the Landau-Ginsburg equations from a postulated form of the free energy F, which introduces an unknown coefficient [for example, the coefficient C of (6-7)] giving the energy associated with spatial variations in the order parameter. Gorkov (1959) has shown that it is possible to establish the Landau-Ginsburg equations from the microscopic theory and to calculate in particular the coefficient C. Here we give a simplified version of this calculation.

Δ Treated as a Perturbation

Our starting point will be the self-consistency equation

$$\Delta(\mathbf{r}) = V \sum_n v_n^*(\mathbf{r}) u_n(\mathbf{r}) [1 - 2f(\epsilon_n)] \tag{7-1}$$

where u_n and v_n are solutions corresponding to positive eigenvalues of the Bogolubov system

$$\begin{aligned} \epsilon_n u_n &= \left[\frac{1}{2m}\left(\mathbf{p} - \frac{e}{c}\mathbf{A}\right)^2 + U - E_F\right] u_n + \Delta v_n \\ \epsilon_n v_n &= -\left[\frac{1}{2m}\left(\mathbf{p} + \frac{e}{c}\mathbf{A}\right)^2 + U - E_F\right] v_n + \Delta^* u_n \end{aligned} \tag{7-2}$$

As usual, the effects of impurities, and so on, are included in U. We expand the right-hand side of (7-1) as a power series in Δ and set

$$u_n = u_n^0 + u_n^1 + \cdots$$
$$v_n = v_n^0 + v_n^1 + \cdots \tag{7-3}$$

From (7-2), we see that u_n^0 and v_n^0 are proportional to the eigenfunctions ϕ_n of one electron in a normal metal, defined by

$$\xi_n \phi_n = \left[\frac{1}{2m}\left(p - \frac{eA}{c}\right)^2 + U - E_F\right]\phi_n \tag{7-4}$$

To zeroth order, we have

$$u_n^0 = \phi_n, \quad v_n^0 = 0 \qquad (\xi_n > 0)$$
$$u_n^0 = 0, \qquad v_n^0 = \phi_n^* \qquad (\xi_n < 0) \tag{7-5}$$

and $\epsilon_n^0 = |\xi_n|$. Since $u_n^0 v_n^0 = 0$, there is no term of lowest order on the right-hand side of (7-1). In order to determine the first-order corrections u_n^1 and v_n^1, we set

$$u_n^1 = \sum_m e_{nm}\phi_m$$
$$v_n^1 = \sum_m d_{nm}\phi_m^* \tag{7-6}$$

If we insert these values into (7-2), multiply the first equation by ϕ_m^*, the second by ϕ_m, and integrate, we find

$$(|\xi_n| - \xi_m)\, e_{nm} = \int \Delta(\mathbf{r})\,\phi_m^*(\mathbf{r}) v_n^0(\mathbf{r})\, d\mathbf{r}$$
$$(|\xi_n| + \xi_m)\, d_{nm} = \int \Delta^*(\mathbf{r})\,\phi_m(\mathbf{r}) u_n^0(\mathbf{r})\, d\mathbf{r} \tag{7-7}$$

As usual, the diagonal terms such as e_{nn} are taken to be zero to preserve the normalization of $\binom{u}{v}$.

We can now insert u_n^1 and v_n^1 defined by (7-6) and (7-7) into the self-consistency equations (7-1). We obtain to first order:

$$\Delta(s) = \int K(s,r)\, \Delta(r)\, dr \tag{7-8}$$

where

$$K(s,r) = V \sum_{n,m} [1 - 2f(|\xi_n|)]$$

$$\times \left[\frac{u_n^{0*}(r) u_n^0(s)}{|\xi_n| + \xi_m} + \frac{v_n^{0*}(s) v_n^0(r)}{|\xi_n| - \xi_m} \right] \phi_m(s) \phi_m^*(r)$$

u_n^0 is nonzero only for $\xi_n > 0$, while v_n^0 is nonzero for $\xi_n < 0$. Also the function $1 - 2f(\xi_n) = \tanh(\beta \xi_n/2)$ is odd in ξ_n. These remarks enable us to collect the uu and vv terms into one single component. Finally, we can symmetrize the expression with respect to the indices n and m (since it is to be summed over n and m) and obtain

$$K(s,r) = \frac{V}{2} \sum_{nm} \frac{\tanh\left(\frac{\beta \xi_n}{2}\right) + \tanh\left(\frac{\beta \xi_m}{2}\right)}{\xi_n + \xi_m}$$

$$\times\, \phi_n^*(r) \phi_m^*(r) \phi_n(s) \phi_m(s) \tag{7-9}$$

It is sometimes useful to transform this result in the following manner: Let us write

$$\tanh \frac{\beta \xi}{2} = 2k_B T \sum_\omega \frac{1}{\xi - i\hbar\omega} \tag{7-10}$$

where $\hbar\omega = 2\pi k_B T(\nu + \frac{1}{2})$ and Σ represents a sum over all positive or negative integers ν. Equation (7-10) can be verified by comparing the poles and residues of both sides of (7-10) in the complex ξ plane. (Also note that we can use either $\pm i$ in the denominator of (7-10) since both ω and $-\omega$ contribute.)

Then

$$\frac{\tanh\left(\frac{\beta\xi}{2}\right) + \tanh\left(\frac{\beta\xi'}{2}\right)}{\xi + \xi'} = 2k_B T \sum_\omega \frac{1}{\xi + \xi'} \times \left(\frac{1}{\xi - i\hbar\omega} + \frac{1}{\xi' + i\hbar\omega}\right)$$

$$= 2k_B T \sum_\omega \frac{1}{(\xi - i\hbar\omega)(\xi' + i\hbar\omega)}$$

and

$$K(s,r) = Vk_B T \sum_\omega \sum_{n,m} \frac{\phi_n^*(r)\phi_m^*(r)\phi_n(s)\phi_m(s)}{(\xi_n - i\hbar\omega)(\xi_m + i\hbar\omega)} \tag{7-11}$$

Equation (7-8) with the explicit form (7-11) for the kernel constitutes the linearized form of the self-consistency condition (7-7). The enormous advantage over (7-1) is that the functions u and v are eliminated, and $\Delta(r)$ is then the only unknown in (7-8). However we must remember that the linearized form only applies for very small Δ, that is, in the immediate neighborhood of a second-order transition.

Separation of Magnetic Effects

We now assume that the spatial variations in the vector potential are small. Then the eigenfunctions ϕ_n in the normal metal in the presence of **A** differ from the eigenfunctions w_n in the absence of **A** by only a phase factor,

$$\phi_n^*(r)\phi_n(s) \rightarrow w_n(r)w_n(s)\, e^{ie\mathbf{A}\cdot(s-r)/\hbar c} \tag{7-12}$$

where the w_n are taken to be real. It is easy to verify that (7-12) is compatible with (7-4) and the definition of the w_n if the spatial variations in **A** are systematically neglected. What is the range of validity of this approximation?

(1) We see later that the range of the kernel K(s,r) in a *pure metal* is of the order of $\xi_0 = 0.18\, \hbar v_F / k_B T_0$. The spatial variations in the vector potential must therefore be small on this scale. A condition is thus obtained by requiring the field h to vary slowly, which implies that the penetration depth be large with respect to ξ_0

$$\lambda_L(T) \gg \xi_0 \tag{7-13}$$

(2) A slow variation of h = curl **A** does not guarantee that **A** varies slowly. Over the distance $|\mathbf{s} - \mathbf{r}| \sim \xi_0$, **A** can vary by $\sim \xi_0 h$. This results in an uncertainty in the phase

$$\frac{e}{\hbar c}\,\xi_0 h \xi_0 \sim \frac{\hbar\omega_c E_F}{(k_B T_0)^2}$$

This must be small compared to one, and we are then led to

$$\frac{\left(k_B T_0\right)^2}{E_F} \gg \hbar\omega_c \tag{7-14}$$

where ω_c = eh/mc is the cyclotron frequency of electrons in the normal metal in the field h. For a bulk, first-kind superconductor, h is at most equal to the thermodynamic field $H_c(T)$. From Eqs. (6-25) and (6-19) we have, for a pure metal,

$$H_c \sim \frac{\phi_0}{\kappa\xi_0^2}\,\frac{T_0 - T}{T_0} \qquad (T \to T_0)$$

where κ is the Landau–Ginsburg parameter and ϕ_0 = hc/2e is the quantum of flux. The condition (7-14) then becomes $(T_0 - T)/T_0 < \kappa$. This is less restrictive than condition (7-13) which can be written as $\kappa > [(T_0 - T)/T_0]^{1/2}$. For a second-kind superconductor, h is at most of the order of $H_{c2} \sim (\phi_0/\xi_0^2)[(T_0 - T)/T_0]$ and (7-14) becomes $(T_0 - T)/T_0 \ll 1$.

(3) For (7-12) to be correct, it is evident that the radius $R = mcv_F/eh$ of the electronic orbits in the field h must be large with respect to the range of the kernel K(s,r). This requirement assures that all effects related to Landau diamagnetism are negligible. The condition $R \gg \xi_0$ can also be written

$$\hbar\omega_c \ll k_B T_0 \tag{7-15}$$

again less restrictive than (7-14). We conclude that for pure metals the substitution (7-12) is valid provided that T is close enough to T_0. This, as we shall see, leads directly to the Landau–Ginsburg equations. For "dirty" alloys, we show later that the situation is even more favorable, and that the linear integral equation (7-8) can be replaced at *all temperatures* by a second-order differential equation of the Landau-Ginsburg form. Finally we are led to

$$\Delta(\mathbf{s}) = \int d\mathbf{r}\; K_0(\mathbf{s},\mathbf{r}) \exp\left[-\frac{2ie}{\hbar c}\,\mathbf{A}\cdot(\mathbf{s} - \mathbf{r})\right]\Delta(\mathbf{r}) \tag{7-16}$$

$$K_0(\mathbf{s},\mathbf{r}) = Vk_B T \sum_{\omega} \sum_{n,m} \frac{w_n(\mathbf{r})\, w_m(\mathbf{r})\, w_n(\mathbf{s})\, w_m(\mathbf{s})}{(\xi_n - i\hbar\omega)(\xi_m + i\hbar\omega)} \tag{7-17}$$

Infinite Homogeneous Medium

We usually consider the kernel K_0 in an infinite homogeneous medium. Then, if this is a pure metal, it is clear that $K_0(\mathbf{s},\mathbf{r})$ depends only on $\mathbf{s} - \mathbf{r}$. On the contrary, for an alloy, this translational invariance is lost. In this case, the translational invariance is restored by making an additional approximation. In (7-16), the average is taken over all impurity configurations. On the right-hand side of (7-16) appears the average $\overline{K_0(\mathbf{s},\mathbf{r})\Delta(\mathbf{r})}$. We approximate this in the following way:

$$\overline{K_0(\mathbf{s},\mathbf{r})\Delta(\mathbf{r})} \rightarrow \overline{K_0(\mathbf{s},\mathbf{r})}\;\overline{\Delta(\mathbf{r})} \tag{7-18}$$

Equation (7-18) is not rigorous because it neglects certain distortions of the pair potential in the immediate neighborhood of each impurity. However, detailed calculations (C. Caroli, 1962) have shown that this approximation is reasonable when the impurity potentials can be treated as weak perturbations. We therefore limit our discussion to alloys whose constituents are not too different chemically. Then (7-18) is applicable and the integral equation for $\overline{\Delta(\mathbf{r})}$ can be written completely in terms of the average kernel $\overline{K_0(\mathbf{s},\mathbf{r})}$, which only depends on $\mathbf{s} - \mathbf{r}$.

Relation between K_0 and a Correlation Function

A priori, the product of the four functions w_n that occurs in (7-17) is rather discouraging. However, it can be shown that this product is related to rather simple physical concepts. Taking, for simplicity, an infinite homogeneous medium, we study the Fourier transform

$$K_0(\mathbf{q}) = L^{-3} \int d\mathbf{r}\, d\mathbf{s}\, K_0(\mathbf{s} - \mathbf{r})\, e^{i\mathbf{q}\cdot(\mathbf{s}-\mathbf{r})}$$

$$= VL^{-3} k_B T \sum_{\omega,n,m} \frac{\langle n | e^{iqx} | m \rangle \langle m | e^{-iqx} | n \rangle}{(\xi_n - i\hbar\omega)(\xi_m + i\hbar\omega)} \tag{7-19}$$

where L^3 is the volume of the sample. We have taken $\mathbf{q}$ along the x direction and set

$$\langle n | e^{iqx} | m \rangle = \int w_n(\mathbf{r})\, e^{i\mathbf{q}\cdot\mathbf{r}}\, w_m(\mathbf{r})\, d\mathbf{r}$$

It is useful to first discuss the real function

$$g(q,\Omega) = \overline{\sum_m \langle n|e^{iqx}|m\rangle\langle m|e^{-iqx}|n\rangle\ \delta(\xi_m - \xi_n - \hbar\Omega)} \tag{7-20}$$

sometimes called the *spectral density* of the one-electron operator e^{iqx}. The symbol represents an average over all states of fixed energy ξ_n (for example, $\xi_n = 0$ corresponds to the Fermi level). In practice, $g(q,\Omega)$ depends strongly on Ω but only slightly on ξ_n and the average will be taken at the Fermi level. If g is known, then $K_0(q)$ is immediately given by

$$K_0(q) = N(0)Vk_B T \sum_\omega \int \frac{d\xi\, d\xi'\, g\left(q, \frac{\xi - \xi'}{\hbar}\right)}{(\xi - i\hbar\omega)(\xi' + i\hbar\omega)} \tag{7-21}$$

Now $g(q,\Omega)$ has a simple physical significance. Introduce the Heisenberg operator

$$e^{iqx(t)} = \exp(i\mathcal{H}_e t/\hbar)\ e^{iqx}\ \exp(-i\mathcal{H}_e t/\hbar) \tag{7-22}$$

which describes the evolution of e^{iqx} in time for an electron in the pure normal metal described by the Hamiltonian $\mathcal{H}_e = (p^2/2m) + U(\mathbf{r})$.

In terms of the operator $e^{iqx(t)}$, the spectral density takes a very simple form:

$$g(q,\Omega) = \frac{1}{2\pi\hbar} \int dt\ e^{i\Omega t}\ \overline{\langle n|e^{-iqx(0)}\ e^{iqx(t)}|n\rangle} \tag{7-23}$$

Equation (7-23) can be verified by writing explicitly the matrix elements of $e^{iqx(t)}$ between states $|n\rangle$ and $|m\rangle$.

In order to determine g, it is only necessary to determine the correlation function of e^{iqx},

$$\langle e^{-iqx(0)}\ e^{iqx(t)}\rangle_{E_F} = \overline{\langle n|e^{-iqx(0)}\ e^{iqx(t)}|n\rangle} \tag{7-24}$$

for an electron at the Fermi energy in a normal metal.

(1) First for a *pure* metal, we can assume that at $t = 0$ the electron

has abscissa x_0 and velocity $v_F \cos\theta$ along the x axis (θ is the angle between **q** and the electron velocity vector).

$$e^{-iqx(0)} = e^{-iqx_0}$$

$$e^{iqx(t)} = \exp[iq(x_0 + v_F t \cos\theta)]$$

Thus

$$\langle e^{-iqx(0)} e^{iqx(t)} \rangle_{E_F} = \tfrac{1}{2}\int_0^\pi \sin\theta \, d\theta \exp(iqv_F \cos\theta t) \quad (7\text{-}25)$$

and

$$g(q,\Omega) = \frac{1}{2\hbar}\int_0^\pi \sin\theta\, d\theta\, \delta(\Omega - qv_F \cos\theta)$$

$$= \begin{cases} (2qv_F\hbar)^{-1} & |\Omega| < qv_F \\ 0 & |\Omega| > qv_F \end{cases} \quad (7\text{-}26)$$

(2) For an *impure* metal where the mean free path ℓ is small compared to the wavelength studied q^{-1}, $e^{iqx(t)}$ is controlled by a diffusion (random walk) process. If $D = v_F \ell/3$ is the diffusion coefficient, we have

$$\langle e^{-iqx(0)} e^{iqx(t)} \rangle_{E_F} = e^{-Dq^2|t|} \qquad q\ell \ll 1 \quad (7\text{-}27)$$

$$g(q,\Omega) = \frac{1}{\pi\hbar}\frac{Dq^2}{\Omega^2 + D^2q^4}$$

Explicit Calculation of the Kernel K_0

Using the forms (7-26) and (7-27) for $g(q,\Omega)$, one can explicitly calculate the kernel $K_0(q)$ from (7-21). The variables of integration are taken to be ξ and $\hbar\Omega = \xi' - \xi$ and the first integration is performed by the theory of residues over ξ. For the pure metal, one finds

$$K_0(q) = \frac{N(0)Vk_B T\pi}{qv_F\hbar}\sum_\omega \int_{-qv_F}^{qv_F} \frac{d\Omega}{2\omega - i\Omega}$$

$$= \frac{2\pi N(0)Vk_B T}{\hbar qv_F}\sum_\omega \tan^{-1}\left(\frac{qv_F}{2|\omega|}\right) \quad (7\text{-}28a)$$

and for the impure metal

$$K_0(q) = N(0)Vk_B T \frac{2i}{\hbar} \sum_{\omega} \int d\Omega \frac{Dq^2}{(Dq^2 + i\Omega)(2i\omega + \Omega)}$$

$$= \frac{N(0)Vk_B T}{\hbar} \sum_{\omega} \frac{1}{Dq^2 + 2|\omega|} \qquad (q\ell \ll 1) \qquad (7\text{-}28b)$$

Discussion of these results

(1) Precautions related to the frequency cut-off in the interaction V. Until now we have neglected the fact that the interaction V in the BCS approximation only couples those states of energy $|\xi| < \hbar\omega_D$. This causes the sums in (7-28) to diverge. However, this is easily remedied by writing

$$K_0(q) = K_0(0) + [K_0(q) - K_0(0)] \qquad (7\text{-}29)$$

The only divergent term is $K_0(0)$, which can be calculated directly from (7-9).

$$K_0(0) = V \sum_n \frac{\tanh(\beta\xi/2)|w_n(\mathbf{r})|^2}{2\xi_n}$$

$$= N(0)V \int_{-h\omega_D}^{h\omega_D} \frac{d\xi}{2\xi} \tanh\left(\frac{\beta\xi}{2}\right)$$

$$= N(0)V \ln \frac{1.14\hbar\omega_D}{k_B T} \qquad (7\text{-}30)$$

where the $\mathbf{s}$ integration was performed by using the orthogonality of the function $w_n(\mathbf{s})$.[1]

(2) Spatial form of the kernel K_0. On taking the inverse Fourier transform of (7-28), we find

$$K_0(\mathbf{s}, \mathbf{r}) = K_0(\mathbf{R}) = \begin{cases} \dfrac{N(0)Vk_B T}{2\hbar v_F} \displaystyle\sum_{\omega} \frac{1}{R^2} \exp[(-2|\omega|R)/v_F] & (R \ll \ell) \quad (7\text{-}31a) \\[2ex] \dfrac{N(0)Vk_B T}{2\hbar D} \displaystyle\sum_{\omega} \frac{1}{R} \exp\{-(2|\omega|/D)^{1/2} R\} & (R \gg \ell) \quad (7\text{-}31b) \end{cases}$$

$$\mathbf{R} = \mathbf{s} - \mathbf{r}$$

---

[1]We had already obtained the result (7-30) in the simpler case of an infinite homogeneous medium without fields or currents where we predicted that $\Delta(\mathbf{r})$ is constant in space. Then the linearized self-consistency equation is simply written $\Delta = K_0(0)\Delta$. This has a nonzero solution only if $K_0(0) = 1$. This condition gives the temperature T_0 at which a nonzero order parameter can appear, that is, the transition temperature.

For $R \neq 0$, the sums converge. It is particularly interesting to study the asymptotic form of $K_0(R)$ at large distances. The only important terms are then $\hbar\omega = \pm\pi k_B T (\simeq \pm\pi k_B T_0)$. Then we find for $T = T_0$

$$\frac{K_0(R)}{N(0)V} = \frac{k_B T_0}{2\hbar v_F} \frac{1}{R^2} \exp\left[-\frac{2\pi k_B T_0 R}{\hbar v_F}\right] = \frac{k_B T_0}{2\hbar v_F} \frac{1}{R^2} \exp(-1.13R/\xi_0)$$

$$(R \ll \ell) \qquad (7\text{-}32a)$$

$$= \frac{k_B T_0}{\hbar D} \frac{1}{R} \exp\left[-\left(\frac{6\pi k_B T_0}{v_F \ell}\right)^{1/2} R\right]$$

$$= \frac{k_B T_0}{\hbar D R} \exp(-1.8R/\sqrt{\xi_0 \ell}) \qquad (R \gg \ell) \qquad (7\text{-}32b)$$

Conclusion. For a pure metal and $T \sim T_0$, the range of $K_0(R)$ is of the order of $\xi_0 = 0.18\hbar v_F/kT_0$ as we previously stated. On the contrary when ℓ is small, (7-32b) gives a range $\sqrt{\xi_0 \ell}$. The range of validity of (7-32b) is limited by $R \gg \ell$, which implies $(\xi_0 \ell)^{1/2} \gg \ell$ or

$$\xi_0 \gg \ell \quad \text{(dirty metal)} \qquad (7\text{-}33)$$

[When (7-33) is satisfied, the alloy is said to be *dirty* following the terminology introduced by Anderson.] If the matrix is a nontransition metal, ξ_0 is generally high (for example, $\xi_0 = 16 \times 10^3$ Å for Al), and a small fraction of impurities (10^{-3}) is sufficient to make the metal dirty.

It is interesting to compare K_0 with the kernel $S_{\mu\nu}$, studied in Chapter 5, which gives the current responds of a superconductor in the presence of a vector potential **A**. We have seen that in a dirty alloy the range of $S_{\mu\nu}$ is ℓ, but the range of K_0 is $\sqrt{\xi_0 \ell}$. What is the reason for this difference?

Answer. The kernels $S_{\mu\nu}$ and K_0 can both be expressed as average one-electron correlation functions in a normal metal in the absence of fields:

$$S_{\mu\nu} = \frac{1}{4} \int d\xi\, d\xi' \int \frac{dt}{2\pi} L(\xi, \xi')\, e^{i(\xi-\xi')t/\hbar} \langle j_\mu(\mathbf{r}_1, 0)\, j_\nu(\mathbf{r}_2, t)\rangle_{E_F}$$

$$(7\text{-}34)$$

$$K_0 = \frac{1}{2} N(0)V \int d\xi\, d\xi' \left[\frac{\tanh(\beta\xi/2) + \tanh(\beta\xi'/2)}{\xi + \xi'}\right]$$

$$\times \int \frac{dt}{2\pi} e^{i(\xi-\xi')t/\hbar} \langle \delta(\mathbf{r}(0) - \mathbf{r}_1)\, \delta(\mathbf{r}(t) - \mathbf{r}_2)\rangle_{E_F} \qquad (7\text{-}35)$$

where $\langle j_\mu \; j_\nu \rangle$ is a transverse current correlation function and $\langle \delta \, \delta \rangle$ is a density correlation function. Equations (7-34) and (7-35) can easily be verified by explicitly calculating the matrix elements that appear in products such as $j_\mu \; j_\nu$. For $T \sim T_0$, the integrations over ξ and ξ' in (7-34) and (7-35) limit the useful interval to $t \sim h/k_B T_0$. If there are no collisions, the distance traveled by an electron in the time t is $v_F t \sim \xi_0$. Then both $S_{\mu\nu}$ and K_0 have the range ξ_0. If the alloy is dirty, the velocity correlation function falls to zero after the electron travels a distance ℓ: the range of $S_{\mu\nu}$ is ℓ. However the *density* correlation function is not destroyed by a collision. It obeys a diffusion equation. The average distance traveled in time t is $\sqrt{Dt}$, where D is the diffusion coefficient $D = \frac{1}{3} v_F \ell$. Thus the range of the kernel K_0 becomes $\sqrt{Dt} \sim \sqrt{\xi_0 \ell}$.

7—2 LANDAU-GINSBURG EQUATIONS

Addition of Nonlinear Terms to the Self-Consistency Equations

The linearized self-consistency equation (7-8) for the pair potential applies when Δ is infinitesimal, that is, at the transition point (provided that the transition of interest is of *second order*, that is, Δ goes to 0 continuously when the temperature is raised). Call T^* the highest temperature for which (7-8) has a nonzero solution $\Delta(\mathbf{r})$. T^* is the ordering temperature in the conditions of the experiment: For instance, in a bulk Type II material, if we impose an external field H, T^* is such that $H_{c2}(T^*) = H$. Later in this section we make extended use of this property. For the present time, our aim is slightly different. We want to extend our analysis to temperatures (slightly) below T^*. Then, it is not sufficient to consider only a first-order approximation in Δ in the right-hand side of (7-1). To extend the calculation, u and v must be calculated to higher order in Δ. The result is of the form

$$\Delta(\mathbf{s}) = \int K_0(\mathbf{s},\mathbf{r})\, e^{-[2ie\mathbf{A}\cdot(\mathbf{s}-\mathbf{r})]/hc}\, \Delta(\mathbf{r})\, d\mathbf{r} + \int R(\mathbf{s},\mathbf{r},\boldsymbol{\ell},\mathbf{m})\, \Delta^*(\mathbf{r})\, \Delta(\boldsymbol{\ell})\, \Delta(\mathbf{m})\, d\mathbf{r}\, d\boldsymbol{\ell}\, d\mathbf{m} \qquad (7\text{-}36)$$

Notice in particular that there are no terms of order Δ^2. The kernel $R(\mathbf{s},\mathbf{r},\boldsymbol{\ell},\mathbf{m})$ can be calculated but its only important property to remember is that in a pure metal its range is of the order of ξ_0. The only contributions to the integral $d\mathbf{r}\, d\boldsymbol{\ell}\, d\mathbf{m}$ come from regions where the three points $\mathbf{r},\boldsymbol{\ell},\mathbf{m}$ are simultaneously near $\mathbf{s}$. In an alloy, the ranges are further reduced. These remarks allow us to avoid a complete calculation of the kernel R.

Hypothesis of Slow Variation

We now assume that quantities such as $\Delta(\mathbf{r})e^{[2ie\mathbf{A}\cdot(\mathbf{r}-\mathbf{s})]/\hbar c}$ vary slowly with respect to the range of K_0 or of R. This restriction defines the Landau-Ginsburg domain. We can then expand $\Delta(\mathbf{r})$ in a Taylor series about the point $\mathbf{s}$:

$$\Delta(\mathbf{r})e^{-2i\phi(\mathbf{r})} = \Delta(\mathbf{s})e^{-2i\phi(\mathbf{s})} + (\mathbf{r}-\mathbf{s})\cdot\nabla(\Delta e^{-2i\phi})$$

$$+ \tfrac{1}{2}\sum_{\alpha\beta}(\mathbf{r}-\mathbf{s})_\alpha(\mathbf{r}-\mathbf{s})_\beta\frac{\partial^2}{\partial r_\alpha\,\partial r_\beta}(\Delta e^{-2i\phi})$$

where we have introduced the phase parameter

$$\phi_s(\mathbf{r}) = e[\mathbf{A}(\mathbf{s})\cdot\mathbf{r}]/\hbar c$$

By performing the differentiations, we get

$$\Delta(\mathbf{r})e^{-2i\phi(\mathbf{r})} = e^{-2i\phi(\mathbf{s})}\{\Delta(\mathbf{s}) + \sum_\alpha(\mathbf{r}-\mathbf{s})_\alpha\,\delta_\alpha\,\Delta$$

$$+ \tfrac{1}{2}\sum_{\alpha\beta}(\mathbf{r}-\mathbf{s})_\alpha(\mathbf{r}-\mathbf{s})_\beta\,\delta_\alpha\,\delta_\beta\,\Delta\} \tag{7-37}$$

$$\delta_\alpha = \frac{\partial}{\partial s_\alpha} - \frac{2ieA_\alpha(\mathbf{s})}{\hbar c}$$

We insert the expansion (7-37) into the linear part of (7-36). For the correction term of order Δ^3 in Eq. (7-36), it will be sufficiently accurate to neglect completely the spatial variations of Δ, and to set $\Delta(\mathbf{r}) = \Delta(\boldsymbol{\ell}) = \Delta(\mathbf{m}) = \Delta(\mathbf{s})$. Finally, in an infinite, homogeneous metal, terms such as $\int K_0(\mathbf{r},\mathbf{s})(\mathbf{r}-\mathbf{s})\,d\mathbf{r}$ vanish by symmetry. This finally gives

$$\Delta(\mathbf{s}) = Q\Delta(\mathbf{s}) + \tfrac{1}{2}\sum_{\alpha,\beta}L_{\alpha\beta}\,\delta_\alpha\,\delta_\beta\,\Delta(\mathbf{s}) + R\,|\Delta(\mathbf{s})|^2\,\Delta(\mathbf{s}) \tag{7-38}$$

$$Q = \int K_0(\mathbf{s},\mathbf{r})\,d\mathbf{r}$$

$$L_{\alpha\beta} = \int K_0(\mathbf{s},\mathbf{r})(s_\alpha - r_\alpha)(s_\beta - r_\beta)\,d\mathbf{r} \tag{7-39}$$

$$R = \int R(\mathbf{s},\mathbf{r},\boldsymbol{\ell},\mathbf{m})\,d\mathbf{r}\,d\boldsymbol{\ell}\,d\mathbf{m}$$

For a cubic crystal ($L_{\alpha\beta} = L\,\delta_{\alpha\beta}$), Eq. (7-38) has exactly the form predicted by Landau and Ginsburg (6-11). The coefficients Q, L, R introduced here are related to the coefficients A, B, C in (6-7) by

$$\frac{-Q+1}{A} = \frac{R}{B} = \frac{1}{2}\frac{L}{C} \tag{7-40}$$

Discussion of the Coefficients

We have already determined A and B for the case $A = 0$, $\Delta =$ const (6-3). Numerical values for Q and R can also be found by the same method. We know that for $A = 0$, $\Delta =$ const, and the complete self-consistency equation (7-1) reduces to

$$\Delta = N(0)V \int_0^{\hbar\omega_D} \Delta \frac{\tanh(\beta\epsilon/2)}{\epsilon}\, d\xi$$
$$\epsilon = (\xi^2 + \Delta^2)^{1/2} \tag{7-41}$$

Anderson's theorem (Chapter 5) shows that (7-41) is valid for an alloy as well as for a pure metal. Expanding the right-hand side as a power series in Δ and performing the integration, we find

$$Q = N(0)V\left[\ln\frac{1.14\hbar\omega_D}{k_B T}\right] \sim 1 + N(0)V\,\frac{T_0 - T}{T_0}$$
$$R = -0.1066\,\frac{N(0)V}{(k_B T_0)^2} \tag{7-42}$$

The most interesting coefficient is L, which describes the effects of spatial variations of Δ. From the explicit form of the kernel K_0 (7-31), we find for a pure metal

$$L = \frac{\pi}{6} N(0)Vv_F^2\, k_B T_0 \sum_\omega \frac{1}{|\omega|^3}$$
$$= 0.033\, N(0)V\left(\frac{\hbar v_F}{k_B T_0}\right)^2 \tag{7-43a}$$

and for a dirty metal ($\ell \ll \xi_0$)

$$L = \pi N(0) V D \frac{k_B T_0}{\hbar} \sum_{\omega} \frac{1}{|\omega|^2}$$

$$= \frac{\pi}{12} N(0) V \frac{\hbar v_F \ell}{k_B T_0} \tag{7-43b}$$

(The sum $\Sigma_\omega 1/|\omega|^3$ must be evaluated numerically while $\Sigma_\omega 1/|\omega|^2$ determined by differentiating (7-10) with respect to ξ and then setting ξ equal to zero.) These results can be expressed in various ways:

(1) The coefficient C of the free energy (6-7) becomes

$$C = \frac{L}{2} \frac{A}{1 - Q} = \frac{1}{2} \frac{L}{V} \tag{7-44}$$

(2) The characteristic length $\xi(T)$ that appears in the Landau-Ginsburg equation in the absence of a field is obtained by comparing (6-17) to (6-8)

$$\xi(T) = \left[\frac{L}{2(Q - 1)}\right]^{1/2} \tag{7-45}$$

It diverges as $[T_0/(T_0 - T)]^{1/2}$ as predicted in (6-19).

(3) The Landau-Ginsburg parameter κ is given by Eq. (6-26)

$$\kappa = \frac{1}{4}\left(\frac{B}{2\pi}\right)^{1/2} \frac{\hbar c}{eC} \tag{7-46}$$

For a pure metal, this formula leads to the previous result $\kappa = 0.96\, \lambda_L(0)/\xi_0$. For a dirty metal, we have

$$\kappa = 0.75 \frac{\lambda_L(0)}{\ell} \tag{7-47}$$

where $\lambda_L(0) = (mc^2/4\pi n e^2)^{1/2}$ is the London penetration depth in the pure metal at absolute 0.

This result can be expressed in another way that involves directly measurable quantities in the normal state instead of ℓ and λ_L. The mean free path ℓ is related to the resistivity ρ by the equation[2]

[2]Proof of (7-48): In the normal state, the current j is given by $j = -eD\nabla n + E/\rho$ where n is the number of electrons per cm^3. At equilibrium, $n = n_0 - 2N(0)\, eV$ where V is the potential such that $E = \nabla V$. On setting $j = 0$ we obtain (7-48).

$$\rho^{-1} = 2e^2 N(0)D = \tfrac{2}{3} N(0)e^2 v_F \ell \qquad (7\text{-}48)$$

The London penetration depth is given by[3]

$$\lambda_L(0) = \left(\frac{3c^2}{8\pi N(0)v_F^2 e^2}\right)^{1/2} \qquad (7\text{-}49)$$

Finally, N(0) can be expressed as a function of the electronic specific heat at low temperatures C_e (per cm^3):

$$C_e = \gamma T$$

$$\gamma = \frac{2\pi^3}{3} N(0)k_B^2$$

Finally this gives

$$\frac{\lambda_L(0)}{\ell} = \frac{1}{2\pi\sqrt{\pi}} \frac{ce}{k_B} \rho\gamma^{1/2} \qquad (7\text{-}50)$$

If ρ is expressed in ohm.cm instead of cgs units, we find

$$\kappa = 7.5 \times 10^3 \rho\gamma^{1/2} \qquad (\xi_0 >> \ell) \qquad (7\text{-}51)$$

Remarks about Dirty Alloys.

(1) From (7-47), κ only depends on ℓ, that is, the transport properties of the alloy in the normal state.

(2) If the pure metal is a superconductor of the first kind ($\kappa_{pure} << 1$), then the alloy is of the second type ($\kappa > 1/\sqrt{2}$) when ℓ is less than a critical value $\ell_c = 1.06\lambda_L(0)$. This law has been verified by Seraphim for various impurities in indium. The critical concentrations vary a great deal according to the impurity used (0.8% for Bi, 7% for Tl). However, the value of ℓ_c is nearly independent of the nature of the impurity ($\ell_c = 440 \pm 100$ Å for Bi, Pb, Su, Cd, Tl, Hg). (The London penetration depth $\lambda_L(0)$ of indium is estimated to be 400 Å.)

(3) For the alloy InBi (2.5% Bi), studied by Kinsel, Lynton, and Serin, the magnetic measurements give $\kappa = 1.79$. The theoretical value deduced from (7-51) is of the order of 1.7.

---

[3]Proof of (7-49): In the presence of a static vector potential **A**, the entire Fermi surface is displaced by a momentum $\delta p = eA/c$. The corresponding current is $\mathbf{j} = e\Sigma[(\partial f/\partial_p)\cdot\delta p]v_p$ where f is the distribution function. This gives $\mathbf{j} = -A(2e^2/3c)N(0)v_F^2$ and thus (7-49).

Table 7-1
Characteristic Lengths

	Pure Metal	Dirty Metal
Range of the kernel $S_{\mu\gamma}$ (relation between current and vector potential)	$\xi_0 = 0.18 \frac{\hbar v_F}{k_B T_0}$	ℓ
Range of the kernel K_0 [self-consistency equation for $\Delta(\mathbf{r})$]	$\sim \xi_0$	$\sim \sqrt{\xi_0 \ell}$
Scale of the spatial variations of $\|\Delta\|$ in the Landau-Ginsburg domain	$\xi(T) = 0.74\xi_0 \left(\frac{T_0}{T_0 - T}\right)^{1/2}$	$\xi(T) = 0.85 \left(\frac{\xi_0 \ell T_0}{T_0 - T}\right)^{1/2}$
Penetration depth in the Landau-Ginsburg domain	$\lambda(T) = \frac{1}{\sqrt{2}} \lambda_L(0) \left(\frac{T_0}{T_0 - T}\right)^{1/2}$	$\lambda(T) = 0.64\lambda_L(0) \left(\frac{\xi_0}{\ell} \frac{T_0}{T_0 - T}\right)^{1/2}$

Final Remarks.
Our discussion has only dealt with the first Landau-Ginsburg equation (6-11), that is, the self-consistency equation for $\Delta(\mathbf{r})$ [or $\psi(\mathbf{r})$]. In order to complete the microscopic analysis, it is also necessary to calculate the current $\mathbf{j}(\mathbf{r})$ at each point and show that it is given by a formula of the type (6-12). This calculation is performed just as the preceding one, by expressing $\mathbf{j}(\mathbf{r})$ as a function of the Bogolubov u's and v's, and then calculating $\mathbf{j}$ to second order in Δ. We do not do this here since it is very analogous to the preceding calculation and gives no new information. The three coefficients A, B, C of the free energy (6-7) have already been determined. We could also have directly treated the free energy itself. All these approaches are equivalent. We chose to discuss the self-consistency equation for Δ, in view of other applications to be discussed later.

Problem. Construct a Landau-Ginsburg equation for a thin film ($d \ll \xi_0$) with diffuse reflection at the boundaries of the film.

Solution. In general, there is no solution. No Landau-Ginsburg equation applies to all cases. For instance the effective κ would depend on the angle between the magnetic field and the film plane. Even more seriously, it may depend on the ratio $d/\lambda(T)$ which is strongly temperature dependent.

However, in one case we can construct a well-defined (two-dimensional) Landau-Ginsburg equation. We need two conditions: (1) The amplitude of Δ must be constant in the film thickness. This will be true when $d < \xi_0$, and implies

$$\left(\frac{\partial}{\partial z} - \frac{2ie}{\hbar c} A_z\right)\Delta \to 0$$

(z is the direction normal to the film). (2) The variations of $\mathbf{A}$ in the film thickness are negligible. This requires that the field be normal to the film plane, and also that the film be thinner than some effective penetration depth (we come back to that condition later).

On the other hand, Δ and A shall be allowed to vary slowly in the plane of the film (xy plane). The free energy is then of the form

$$F = A|\Delta|^2 + \frac{B}{2}|\Delta|^4 + C\left\{\left|\left(\frac{\partial}{\partial x} - \frac{2ie}{\hbar c}A_x\right)\Delta\right|^2 + \left|\left(\frac{\partial}{\partial y} - \frac{2ie}{\hbar c}A_y\right)\Delta\right|^2\right\}$$

In order to determine C, we start from the linearized self-consistency equation (7-16) and study the Fourier transform $K_0(\mathbf{q})$ of the kernel, where $\mathbf{q}$ is a vector in the xy plane. When d is small ($d \ll \xi_0$), most of the electrons experience several collisions with the walls during the time interval $\hbar/k_B T_0$ and the function $g(\mathbf{q}, \Omega)$ is governed by a diffusion equation just as for a dirty superconductor.[4] The calculations are quite analogous to those giving rise to (7-43) and (7-44) and give

$$C = \frac{\pi}{8} N(0) \frac{\hbar D}{k_B T_0}$$

The diffusion coefficient D is no longer equal to $\frac{1}{3} v_F \ell$ but rather of the order of $v_F d$. The coefficient D can be related to the experimental conductivity $\bar{\sigma}$ of the film in the normal state by means of (7-48)

$$\bar{\sigma} = 2e^2 N(0) D$$

$$C = \frac{\pi}{16} \frac{\hbar\bar{\sigma}}{2k_B T_0}$$

Detailed calculations of $\bar{\sigma}$ can be found for instance in the book by Olsen, *Electron Transport in Metals* (New York: Interscience, 1962), p. 80. In practice of course it is better to extract $\bar{\sigma}$ from a measurement in the normal state on the same film.

Discussion.

(1) The order of magnitude of $\bar{\sigma}$ is $ne^2 d/mv_F$. Thus $C \sim N(0)\xi_0 d$ and $\kappa \sim \lambda_L(0)/d$.

(2) Once C is determined, we can write the equation for the current—in the x direction

$$j_x = \frac{2eC}{i\hbar}\left(\Delta^* \frac{\partial \Delta}{\partial x} - \Delta \frac{\partial \Delta^*}{\partial x}\right) - \frac{8e^2 C}{\hbar c} A_x$$

[4]There are a few electrons whose velocity is just in the plane of the film and which thus suffer no collisions. One can take account of them exactly. For simplification we here assume that there also exist volume impurities and therefore a bulk mean free path $\ell \gtrsim \xi_0$ (but $\ell > d$). Under this condition there is no significant contribution from these "anomalous electrons."

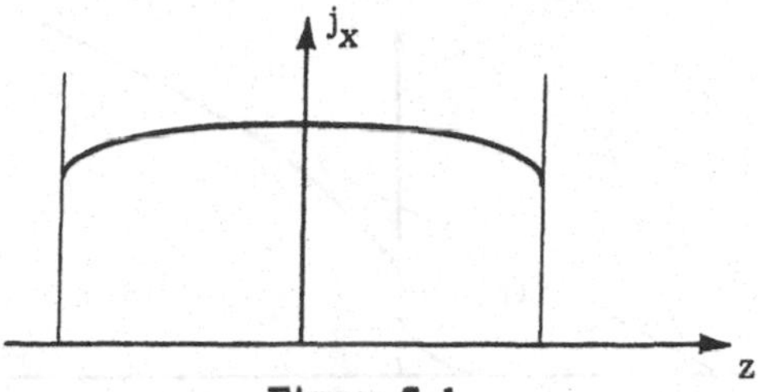

Figure 7-1
Current distribution across a thin film ($d < \xi_0$) with diffuse reflection at the surface.

However it is important to realize that, even if d is small, the current distribution $j_x(z)$ throughout the thickness of the film is not uniform (Fig. 7-1). The current j_x given by the preceding formula is just the average of j_x over the thickness.

(3) Although j is not independent of z, **A** is to a good approximation independent of z: using the relation $\partial^2 A_x / \partial z^2 = (4\pi/c) j_x$, it is easy to show that, when A_x is an even function of z, the relative variations of A_x through the thickness are of the order $d^3/\xi_0 \lambda^2 << 1$.

7–3 SURFACE PROBLEMS IN THE LANDAU-GINSBURG REGION

Boundary Conditions at an Interface

In the preceding section we have reconstructed the Landau-Ginsburg equations from the microscopic theory, for an infinite homogeneous medium. We shall now consider a superconducting specimen occupying the half-space $x > 0$. The region $x < 0$ is either an insulator or normal metal. In all cases, we shall assume that there is no current flowing through the boundary.[5]

Over a width ξ_0 at the surface,[6] $K_0(\mathbf{r},\mathbf{r}')$ and $S_{\mu\nu}(\mathbf{r},\mathbf{r}')$ are modified. On the microscopic scale it is necessary to calculate these modifications and then to determine the spatial distribution of the currents j and of the pair potential Δ in the perturbed layer. The result of such calculations is represented in (Fig. 7-2a) for Δ. On the other hand, on the Landau-Ginsburg scale [lengths of the order of $\xi(T), \lambda(T) \gg \xi_0$] the effect of the surface can be simply described by means of a boundary condition for $x = 0$ (Fig. 7-2b). What will be the form of this

[5]When normal currents and supercurrents coexist, the implied dissipative effects must be calculated by a dynamic equation more general than the Landau-Ginsburg equations.

[6]We quote here the orders of magnitude that apply for a pure metal. In a dirty alloy, of course, the "transition width" would decrease to $\sqrt{\xi_0 \ell}$.

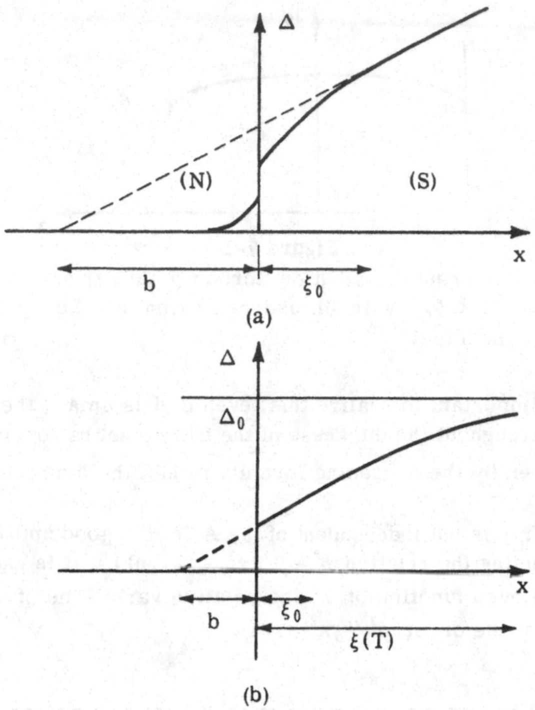

Figure 7-2

(a) The microscopic variation of the order parameter near a superconducting-normal metal boundary. (b) "Macroscopic" description of the same situation, at temperatures where $\xi(T) \gg b$.

condition? Let us first consider the case **A** = 0, Δ = real. The exact boundary condition will be of the form

$$\left(\frac{d\Delta}{dx}\right)_{x=0} = \frac{1}{b}\Delta_{x=0} + \frac{1}{c}\frac{\Delta^3_{x=0}}{(k_B T_0)^2} + \cdots \tag{7-52}$$

Justification of (7-52):

(1) From the general form of the self-consistency equation, only the odd powers of Δ occur in the expansion. Also the lengths b, c, and so on, are real.

(2) Higher derivatives do not occur in (7-52) because they can be expressed in terms of the first and zeroth derivatives by the Landau-Ginsburg equations.

(3) We expect nonlinear effects to become important when $\Delta \sim k_B T_0$. Thus the quantities 1/b, 1/c, and so on, are of the same order

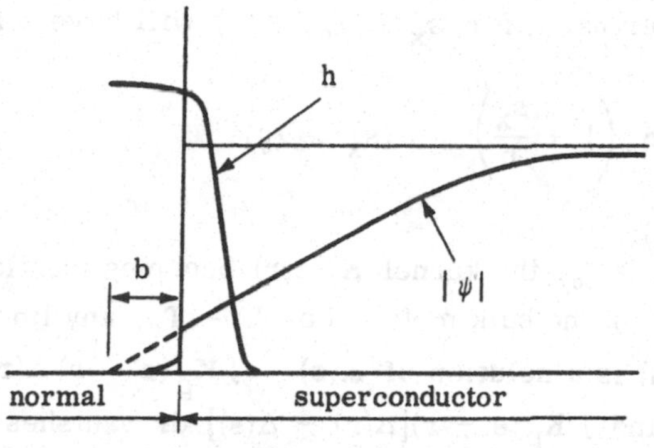

Figure 7-3
The magnetic field distribution at a normal-superconducting boundary.

of magnitude ($\gtrsim 1/\xi_0$). The Δ^3 term in (7-52), smaller than the Δ term by $\sim(\Delta/k_B T_0)^2 \ll 1$, can be neglected. In the Landau-Ginsburg region, the boundary condition is *linear*, and the coefficient b can be obtained by solving the linearized self-consistency equation (7-8) in the neighborhood of the surface (for $T = T_0$, $A = 0$). In the presence of a magnetic field, we write, instead of (7-52),

$$\left(\frac{d\Delta}{dx} - \frac{2ie}{\hbar c} A_x \Delta\right)_{x=0} = \frac{(\Delta)_{x=0}}{b} \tag{7-53}$$

The form of (7-53) assures the gauge invariance dictated by the general formulas of Chapter 5. In (7-53), Δ may be replaced by ψ to which it is proportional. Using (7-53), it is easy to verify that for real b, the current (6-12) crossing the boundary is zero, as desired.

Contact between Superconductor and Insulator (or Vacuum)

We shall now show that 1/b is very small when the boundary separates the superconductor from an insulator. The equation to be solved (for $A = 0$ and $T = T_0$) is

$$\Delta(\mathbf{s}) = \int K_0(\mathbf{s},\mathbf{r})\Delta(\mathbf{r})\, d\mathbf{r} \tag{7-54}$$

$K_0(\mathbf{s},\mathbf{r})$ is defined by (7-17) and therefore $\Delta(\mathbf{s})$ contains in particular a product of two functions $w_n(\mathbf{s})$ and $w_m(\mathbf{s})$, which decrease expotentially in the insulating region. Therefore $\Delta(\mathbf{s})$ vanishes rapidly for $s_x < 0$.

On the other hand, for $s_x > \xi_0$, $\Delta(\mathbf{s})$ will have a linear form,[7]

$$\Delta(\mathbf{s}) = \Delta_0\left(1 + \frac{s_x}{b}\right) \qquad (s_x > \xi_0) \tag{7-55}$$

Proof of (7-55):

For $s_x, r_x > \xi_0$, the kernel $K_0(\mathbf{s},\mathbf{r})$ becomes identical to the kernel $K_p(\mathbf{s} - \mathbf{r})$ in the bulk metal. For $T = T_0$, any linear function of the type (7-55) is a solution of $\Delta(\mathbf{s}) = \int K_p(\mathbf{s} - \mathbf{r})\Delta(\mathbf{r})\, d\mathbf{r}$ because $\int K_p\, d\mathbf{r} = 1$ and $\int K_p(\mathbf{s} - \mathbf{r})[\Delta(\mathbf{r}) - \Delta(\mathbf{s})]\, d\mathbf{r}$ vanishes by symmetry.

In order to determine the length b in which we are interested, (7-54) is rewritten in the form

$$\Delta(\mathbf{s}) - \int K_p(\mathbf{r} - \mathbf{s})\Delta(\mathbf{r})\, d\mathbf{r} = \int [K_0(\mathbf{s},\mathbf{r}) - K_p(\mathbf{r} - \mathbf{s})]\,\Delta(\mathbf{r})\, d\mathbf{r}$$
$$= -H(\mathbf{s}) \tag{7-56}$$

The function $H(\mathbf{s})$ vanishes for $s_x > \xi_0$ (because $K_0 \to K_p$) and also vanishes for $s_x < -\xi_0$ because $\Delta(\mathbf{r})$ is zero for $r_x < 0$ and the range of K_p is ξ_0. The advantage of (7-56) is that a localized perturbation appears on the right. For the moment we assume $H(\mathbf{s})$ as known and solve for Δ by a Laplace transformation

$$\Delta(p) = \int \Delta(s_x)\exp(-ps_x)\, d\mathbf{s}$$

$$H(p) = \int H(s_x)\exp(-ps_x)\, d\mathbf{s} \tag{7-57}$$

$$K_p(p) = \int K_p(\mathbf{r})\exp(-pr_x)\, d\mathbf{r}$$

$$= 1 + \frac{p^2}{2}L \cdots$$

where $L = \int K_p(\mathbf{r}) r_x^2\, d\mathbf{r}$ is given explicitly by (7-39). The transform of (7-56) is

---

[7]The linear dependence (7-55) might seem physically absurd since it leads to very large Δ's at large distances from the boundary. In fact, when T is slightly below T_0, the correct $\Delta(x)$ has a negative curvature and reaches the BCS values deep in the superconductor $[x > \xi(T)]$. But here we are interested in $x \sim \xi_0 \ll \xi(T)$ and the curvature is negligible.

$$\Delta(p)[1 - K_p(p)] = -H(p) \tag{7-58}$$

Taking the limit $p \to 0$, from (7-57)

$$-\frac{L}{2} p^2 \Delta(p) \to -H(0) = -\int H(s_x)\, ds$$

As $p \to 0$, the dominant term in (7-55) is the linear part,

$$\Delta(p) \to \frac{\Delta_0}{b} \int_0^\infty s_x \exp(-ps_x)\, ds = \frac{\Delta_0 \Sigma}{bp^2} \tag{7-59}$$

where Σ is the surface area of the boundary. Combining (7-58) and (7-59), we find

$$\frac{1}{b} = \frac{2}{L\Sigma\, \Delta_0} \int H(s_x)\, ds = \frac{2}{L\Sigma\Delta_0} \int [K_p(s - r) - K_0(s,r)] \times \Delta(r)\, dr\, ds \tag{7-60}$$

In order to simplify (7-60), we shall assume that the interaction V is constant. Then $\int K_0(r,s)\, ds$ can be integrated by using the orthogonality of the functions $w_n(s)$ and we find

$$\int K_0(r,s)\, ds = N(r) V \ln \frac{1.14\hbar\omega_D}{k_B T_0} = \frac{N(r)}{N(0)} \tag{7-61}$$

where $N(r) = \Sigma_n |w_n(r)|^2 \delta(\xi_n)$ is the local density of states at the Fermi surface and N(0) is its usual value for the bulk metal. Finally

$$\frac{1}{b} = \frac{2}{L} \int_{-\infty}^{\infty} dx\, \frac{\Delta(x)}{\Delta_0} \left[1 - \frac{N(x)}{N(0)}\right] \tag{7-62}$$

where $\Delta(x)/\Delta_0$ approaches zero in the insulating region and is of the order of 1 in the metallic region. $N(x)/N(0)$ passes from $0 \to 1$ in a few interatomic distances and the integrand is nonvanishing only in a width of order of the interatomic distance a. The effective distance b is given by

$$b \sim \frac{L}{a} \tag{7-63}$$

This is generally enormous: for a pure metal $L \sim \xi_0^2$ with $\xi_0 = 10^{-4}$ cm, which gives $b \sim 1$ cm. Therefore for a boundary separating a superconductor from an insulator we can set $1/b = 0$. This leads to the boundary condition (6-13). More precisely, the criterion for the validity of (6-13) is that $b \gg \xi(T)$, which is equivalent to $a/\xi_0 \ll [(T_0 - T)/T_0]^{1/2}$.[8]

Contact between Superconductor and Normal Metal

We now briefly discuss the case of the boundary between a super-conductor and normal metal. If there is a good electrical contact between the N and S regions, the Cooper pairs can diffuse into the normal region. Algebraically this means that $\Delta(s)$ extends up to a rather long distance ξ_N into the N regions $(x < 0)$ $(\xi_N \sim \hbar v_{FN}/k_B T$ if N is a pure metal).

Equation (7-60) remains valid. It is, however, necessary to determine the exact form of $\Delta(x)/\Delta_0$ and therefore solve the integral equation (7-54) completely. Qualitatively the results can be represented in the form

$$\Delta \sim \Delta_0 \frac{N_N}{N} T_j \exp(x/\xi_N) \qquad (x < 0) \qquad (7\text{-}64)$$

where N_N/N is the density of states factor occuring in (7-61) and T_j is the transmission coefficient at the boundary for an electron at the Fermi surface. By substituting (7-64) into (7-62), we find

$$b \sim \frac{N}{N_N} \frac{1}{T_j} \frac{L}{\xi_N} \qquad (7\text{-}65)$$

If $(N_N/N) T_j \sim 1$, we can have $b \sim \xi_N$, b is then short when compared to $\xi(T)$ and the boundary conditions are drastically modified from the insulator case.

Problem. How is the penetration depth modified in a superconductor S when it is covered by a normal metal layer N? (P.G. de Gennes and J. Matricon, 1965).

Solution. Consider the Landau-Ginsburg domain (see Fig. 7-3). Assume (as is often the case) that the N metal does not contribute to the screening of magnetic

---

[8]Remark: Notice the difference in this behavior with that of liquid He[4], which is also a superfluid satisfying, at least approximately, a Landau-Ginsburg equation in the neighborhood of T_c. In He[4], ξ_0 is of the order of a and $b \sim a \gg \xi(T)$. The correct boundary condition is then completely different and closer to $\psi_{x=0} = 0$.

fields. Then the field at the NS boundary is the applied field. For the order parameter, the presence of N imposes a boundary condition on the type (7-53) at the surface ($x = 0$) of the superconductor.

(1) In zero field, $\psi(x)$ is given by

$$f(x) = \frac{\psi(x)}{\psi_0} = \tanh \frac{x - x_0}{\sqrt{2}\,\xi(T)}$$

where x_0 is a parameter to be determined by the condition (7-53),

$$\left(\frac{1}{\psi}\frac{d\psi}{dx}\right)_{x=0} = \frac{1}{b}$$

If the electrical contact between N and S is good, if the density of states of N is comparable to S, and if N is thick compared to $\xi_0 \simeq \hbar v_F / k_B T_0$, b is of the order of ξ_0. Therefore we can make $b \ll \xi(T)$ and decrease strongly the order parameter at the surface $f(0) \sim b/[\sqrt{2}\,\xi(T)] \ll 1$.

(2) If we now apply a weak magnetic field $h_z(x) = \partial A_y(x)/\partial x$ along the z axis, its distribution is determined by

$$\frac{\partial^2 A_y}{\partial x^2} = \frac{16\pi e^2}{mc^2} \,|\psi_0|^2 f^2 A_y = \frac{f^2}{\lambda^2(T)} A_y$$

In a weak field, we can use the unperturbed form f given above. It happens that the equation for A_y is then soluble exactly in terms of hypergeometric functions (for a similar problem, see Landau and Lifshitz, *Quantum Mechanics*, London: Pergamon Press, 1959, p. 69). We only quote here the result for the limiting case $b/\xi(T) \to 0\,[x_0 \to 0,\ f(0) \to 0]$. Then

$$\frac{\lambda}{\lambda(T)} = \frac{1}{\kappa\sqrt{2}} \frac{\Gamma\left(-\frac{S}{2} + \frac{1}{\kappa\sqrt{2}}\right)\Gamma\left(\frac{1}{2} + \frac{S}{2} + \frac{1}{\kappa\sqrt{2}}\right)}{\Gamma\left(\frac{1}{2} - \frac{S}{2} + \frac{1}{\kappa\sqrt{2}}\right)\Gamma\left(1 + \frac{S}{2} + \frac{1}{\kappa\sqrt{2}}\right)}$$

For small κ, $\lambda/\lambda(T) \to 1.75\,\kappa^{-1/2}$. This could have been guessed by the following argument: If the penetration depth is small, we can replace the exact expression for f (with $x_0 = 0$) by the approximate form $f = x/[\sqrt{2}\,\xi(T)]$. This approximation will be correct in the region where the field penetrates. Then the equation for A_y reduces to

$$\frac{\partial^2 A_y}{\partial x^2} = \frac{x^2}{2\lambda^2\xi^2} A_y$$

and the penetration depth is of the order of $(\lambda\xi)^{1/2}$.

The opposite limit $\kappa \to \infty$ leads to $[\lambda/\lambda(T)] \to 1$. This was also expected since, when $\xi(T) \ll \lambda(T)$, the distortion of $\psi(x)$ near the surface, due to the presence of the N layer, is present only in a small fraction of $\lambda(T)$.

(3) In strong fields, the penetration depth increases spectacularly if $\kappa \sim 1/\sqrt{2}$ and $b \ll \xi(T)$. This can be shown from the complete nonlinear Landau-Ginsburg equations, but has not yet been verified experimentally.

S-N-S′ Junction

Consider now the system of three layers S-N-S′ represented in Fig. (7-4); two superconductors S and S′ separated by either an insulating film N of thickness d (~20–50Å), or a normal metal with $d \sim 10^3$ Å. When d is not too large, the Cooper pairs can tunnel from S to S′ and vice versa. We shall analyze this effect under the following assumptions:

(1) The conditions for the validity of the Landau-Ginsburg equations are satisfied in S and S′.

(2) The thickness d is small with respect to $\lambda(T)$ and $\xi(T)$.

The effect of the film N is simply to impose boundary conditions relating the values of Δ and $d\Delta/dx$ at one surface of N to the values at the other surface. As previously, the boundary conditions can be reduced to be linear. Consider first the case when the vector potential $\mathbf{A} = 0$. Then

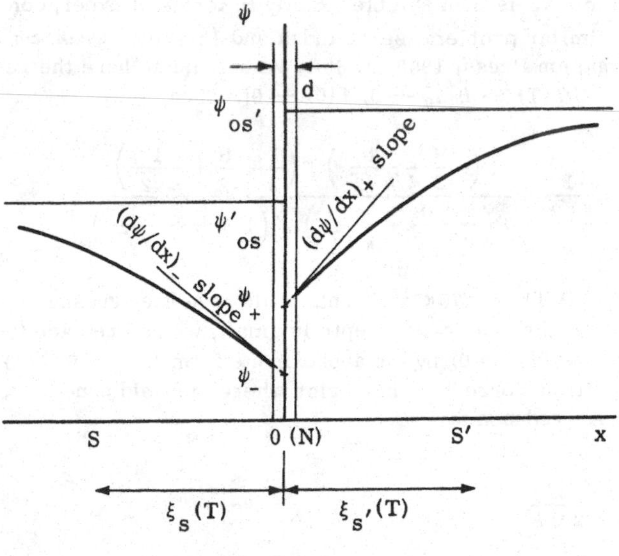

Figure 7-4

An S-N-S′ junction where the normal layer may either be an insulator (d ~ 50Å) or a metal (d ~ 10^3 Å).

$$\Delta_+ = L_{11}\,\Delta_- + L_{12}\left(\frac{d\Delta}{dx}\right)_- \qquad (\mathbf{A} = 0) \qquad (7\text{-}66)$$
$$\left(\frac{\partial\Delta}{\partial x}\right)_+ = L_{21}\,\Delta_- + L_{22}\left(\frac{\partial\Delta}{\partial x}\right)_-$$

Here the x axis is normal to the film; Δ_+ represents the value of $\Delta(x,y,z)$ extrapolated to the point $(0,y,z)$ from the solution of the Landau-Ginsburg equations in the S region as is shown in Fig. (7-4), and so on. The coefficients L_{ij} are obtained by completely solving the linearized self-consistency equation (7-8) (for $\mathbf{A} = 0$, $T = T_0$) in and around the N region. For $\mathbf{A} = 0$, the kernel $K_0(\mathbf{s},\mathbf{r})$ of their equation is real, therefore, the L_{ij} are real. Sometimes it is more useful to deal with ψ instead of Δ, and then (7-66) can be rewritten

$$\psi_+ = M_{11}\,\psi_- + M_{12}\left(\frac{\partial\psi}{\partial x}\right)_- \qquad (\mathbf{A} = 0) \qquad (7\text{-}66')$$
$$\left(\frac{\partial\psi}{\partial x}\right)_+ = M_{21}\,\psi_- + M_{22}\left(\frac{\partial\psi}{\partial x}\right)_-$$

[If the metals S and S′ are different, the ratio ψ/Δ defined by (6-8) is not the same in S as in S′. Thus the coefficients M_{ij} are not identical to the L_{ij}.] The M_{ij} are real. They are not all independent, but are related to each other by one relation expressing current conservation through the boundary. The supercurrent I_x crossing a unit area of the wall is given by

$$I_x = \frac{-ie\hbar}{m}\left(\psi^*\frac{\partial\psi}{\partial x} - \text{C.C.}\right)_+ = \frac{-ie\hbar}{m}\left(\psi^*\frac{\partial\psi}{\partial x} - \text{C.C.}\right)_- \qquad (\mathbf{A} = 0) \qquad (7\text{-}67)$$

By comparing (7-67) and (7-66′), we find

$$M_{11}M_{22} - M_{12}M_{21} = 1 \qquad (7\text{-}68)$$

and the current I_x can be written in the form

$$I_x = \frac{-ie\hbar}{m}\left\{\psi_-^*\left(\frac{1}{M_{12}}\psi_+ - \frac{M_{11}}{M_{12}}\psi_-\right) - \text{C.C.}\right\}$$
$$= \frac{-ie\hbar}{M_{12}\,m}\left(\psi_-^*\psi_+ - \text{C.C.}\right) \qquad (7\text{-}69)$$

Therefore *a supercurrent must flow through the junction* (B. D. Josephson, 1962). The junction cannot be too thick because then the coefficient M_{12} would become too large. Such currents have first been clearly observed by Anderson and Rowell (1963).

These results are easily generalized to the case $\mathbf{A} \neq 0$. We always assume that, on the Landau-Ginsburg scale, the vector potential $\mathbf{A}$ is continuous at the boundary.[9] Then, putting $A_x = A_x(0,yz)$, we replace Eq. (7-66′) by

$$\psi_+ = M_{11}\psi_- + M_{12}\left(\frac{\partial}{\partial x} - \frac{2ie}{\hbar c}A_x\right)\psi_-$$
$$\left(\frac{\partial}{\partial x} - \frac{2ieA_x}{\hbar c}\right)\psi_+ = M_{21}\psi_- + M_{22}\left(\frac{\partial}{\partial x} - \frac{2ie}{\hbar c}A_x\right)\psi_- \qquad (7\text{-}70)$$

The equation (7-69) for the current is not changed.[10]

In practice, the use of the current formula (7-69) is particularly simple when the region N is only slightly transparent to pairs, that is, when I_x is small relative to the critical currents in S or S′ in their bulk states. With this assumption, the absolute values $|\psi_+|$ and $|\psi_-|$ are only slightly modified by the presence of the current and they can be replaced by their values $\psi_{S'}$ and ψ_S in the state without current:

$$\psi_+ \to \psi_{S'}\exp(i\phi_{S'}) \qquad (\psi_S, \psi_{S'} \text{ are real})$$
$$\psi_- \to \psi_S\exp(i\phi_S)$$

The phases ϕ_S and $\phi_{S'}$, on the other hand, will, in general, be very sensitive to the presence of currents and magnetic fields and nothing can be said about them *a priori*. With the substitution of these values into (7-69), the current takes the form

$$I_x = \frac{2e\hbar}{mM_{12}}\psi_S\psi_{S'}\sin(\phi_S - \phi_{S'}) \qquad (7\text{-}71)$$

Its maximum value is

---

[9] $\mathbf{A}$ could be discontinuous if a different choice of gauge was made on the two sides.

[10] A particularly simple case occurs when the junction is symmetric. Then $M_{11} = M_{22}$ and, by taking account of (7-68), we see that the M_{ij} depend on only two independent parameters.

$$I_m = \frac{2e\hbar}{mM_{12}} \psi_S \psi_{S'} \tag{7-72}$$

To determine ψ_S and $\psi_{S'}$, it is only necessary to solve the Landau-Ginsburg equation in S and S′ in the simple case **A** = 0 and ψ real. We have already encountered the form of these solutions in one dimension

$$\psi(x) = \begin{cases} \psi_{oS'} \tanh\left(\dfrac{x + x_{S'}}{\sqrt{2}\,\xi_{S'}(T)}\right) & x > 0 \\[2ex] \psi_{oS} \tanh\left(\dfrac{-x + x_{S'}}{\sqrt{2}\,\xi_S(T)}\right) & x < 0 \end{cases} \tag{7-73}$$

where ψ_{oS} and $\xi_S(T)$ are, respectively, the order parameter in the bulk S metal and the coherence length at the temperature T in S. The lengths x_S and $x_{S'}$ are related to ψ_S and $\psi_{S'}$ by

$$\psi_S = \psi_{oS} \tanh\left(\frac{x_S}{\sqrt{2}\,\xi_S(T)}\right) \qquad \psi_{S'} = \psi_{oS'} \tanh\left(\frac{x_{S'}}{\sqrt{2}\,\xi_{S'}(T)}\right) \tag{7-74}$$

They can be determined explicitly by demanding that (7-73) satisfy the boundary (7-66′), which gives

$$\psi_{S'} = M_{11}\psi_S - M_{12} \frac{\psi_{oS}}{\sqrt{2}\,\xi_S(T)}\left(1 - \frac{\psi_S^2}{\psi_{oS}^2}\right)$$

$$\frac{\psi_{oS'}}{\sqrt{2}\,\xi_{S'}(T)}\left(1 - \frac{\psi_{S'}^2}{\psi_{oS'}^2}\right) = M_{21}\psi_S - M_{22} \frac{\psi_{oS}}{\sqrt{2}\,\xi_S}\left(1 - \frac{\psi_S^2}{\psi_{oS}^2}\right) \tag{7-75}$$

The system of equations (7-75) can be solved numerically to obtain ψ_S and $\psi_{S'}$. Certain limiting cases are simple:

(1) If N is an insulator, $\psi(x)$ is only slightly modified by the presence of N($\psi_S \sim \psi_{oS}$). Then M_{12} and M_{21} are small, and M_{11} is nearly 1. (See the microscopic calculation that follows later.)

(2) If N is metallic [sufficiently thick for I_x to be small, but thin compared to $\lambda(T)$ and $\xi(T)$], we predict that

$$\frac{\psi_s}{\psi_{os}} \sim \frac{\psi_{s'}}{\psi_{os'}} \sim \frac{b}{\xi(T)} \ll 1$$

by analogy with the NS boundary studied previously. In this case, (7-75) can be linearized.

Conclusions.
(1) If N is an insulating layer, the maximum current becomes

$$I_m = \frac{2e\hbar}{mM_{12}} \psi_{os}\psi_{os'}$$

M_{12} is finite as $T \to T_0$; ψ_{os} varies as $(T_0 - T)^{1/2}$ from (6-15) and consequently the maximum current varies as $T_0 - T$.

(2) If N is metallic, $I_m \sim (e\hbar/mM)[b/\xi(T)]^2 \psi_{os}\psi_{os'}$, which varies as $(T_0 - T)^2$. (This latter result has not yet been experimentally verified.)

MICROSCOPIC CALCULATION OF M_{ij} FOR AN INSULATING JUNCTION

We consider a symmetric situation for $d \ll \xi_0$. We start from the linearized self-consistency equation (7-8) for $\mathbf{A} = 0$ and $T = T_0$.

From (7-35), the kernel $K_0(\mathbf{r},\mathbf{r}')$ is essentially related to the function $\langle \delta[\mathbf{r} - \mathbf{r}(0)]\ \delta[\mathbf{r}' - \mathbf{r}(t)]\rangle_{E_F}$, which represents the probability of finding an electron at $\mathbf{r}'$ at the time t (in the normal state) if it has been injected at $\mathbf{r}$ at time 0, with the Fermi energy. This allows us to write $K_0(\mathbf{r},\mathbf{r}')$ directly in terms of $K_{pure}(\mathbf{r} - \mathbf{r}')$ which we have already studied for the infinite metal S. Equation (7-8) becomes

$$\Delta(\mathbf{r}) = \int_I K_{pure}(\mathbf{r} - \mathbf{r}')\Delta(\mathbf{r}')\, d\mathbf{r}' + \int_{II} K_{pure}(\mathbf{r} - \mathbf{r}') \times [T_j\, \Delta(\mathbf{r}') + R_j\, \Delta(\overline{\mathbf{r}'})\, d\mathbf{r}'] \tag{7-76}$$

where the first integral (I) corresponds to a direct propagation from $\mathbf{r}$ to $\mathbf{r}'$ and the region of integration is limited by the condition that $\mathbf{r}$ and $\mathbf{r}'$ be on the same side of the junction. The integral (II) gives the effects of transmission and reflection arising from the presence of the junction. The coefficient $T_j = 1 - R_j$ is the transmission coefficient of the function for an electron of energy E_F in the normal metal. The point $\overline{\mathbf{r}'} = (-x', y', z')$ is the image point of $\mathbf{r}'$ with respect to the junction. For simplification we assume that the reflection is specular and that the reflection coefficient is independent of the angle of incidence.

Equation (7-76) has an even solution $\Delta(x) = \Delta(-x) = \text{constant}$ by virtue of the property $\int K_{pure}(\mathbf{r} - \mathbf{s})\, d\mathbf{r} = 1$ at $T = T_0$. In order to determine M_{12}, it is necessary to construct a second independent solution, for example, the odd solution $\Delta(x) = -\Delta(-x)$.

If we set $\Delta^+(x) = \Delta(s)S(x)$ where

$$S(x) = \begin{cases} 1 & x > 0 \\ 0 & x < 0 \end{cases}$$

Eq. (7-76) can be put into the form

$$\Delta^+(x) = S(x) \int \Delta^+(x')[K_0(\mathbf{r},\mathbf{r}') + (1 - 2T_j)K_0(\mathbf{r},\mathbf{r}')]\, d\mathbf{r}'$$

or further

$$\begin{aligned} \Delta^+(\mathbf{r}) - \int K_0(\mathbf{r},\mathbf{r}')\Delta^+(\mathbf{r}')\, d\mathbf{r}' &= \int d\mathbf{r}\; \Delta^+(\mathbf{r}')[-S(-x)K_0(\mathbf{r},\mathbf{r}') \\ &\quad + S(x)(1 - 2T_j)K_0(\mathbf{r},\mathbf{r}')] \\ &= -H(x) \end{aligned} \tag{7-77}$$

The function H(x) represents a source localized in the region $|x| < \xi_0$. The equation is very similar to (7-56) and can be solved in the same way,

$$\left(\frac{L}{2}\right)\left(\frac{d\Delta}{dx}\right)_{x \gg \xi_0} = \int H(x)\, dx = 2T_j \int_\infty^0 dx \int_\infty^0 dx'$$

$$\times \int dy'\, dz'\; \Delta^+(x')K_0(\mathbf{r},\mathbf{r}') \tag{7-78}$$

As the junction is insulating, $\Delta^+(x')$ does not vary much over the region of integration ($x' \sim \xi_0$), $\Delta^+(x) \to \Delta^+$. The integral that remains can be calculated by recalling the definition of K_{pure},

$$\int_{-\infty}^0 dx \int_0^\infty dx' \int dy'\, dz'\, K_{pure}(\mathbf{r} - \mathbf{r}') = \xi_0 \tag{7-79}$$

Finally, we find

$$\frac{L}{2}\left(\frac{d\Delta}{dx}\right)_+ = T_j\, \xi_0\, \Delta_+$$

and comparing this with (7-66),

$$L_{12} = M_{12} = \frac{L}{T_j \, \xi_0} \tag{7-80}$$

The coefficient M_{12} is of the order of ξ_0/T_j since $L \sim \xi_0^2$. In practice the transmission coefficient T_j across an insulating wall is very small when d is larger than a few angströms. This implies that the maximum supercurrent flowing across the junction is small

$$I_m \sim \frac{eh}{m\xi_0} |\psi_0|^2 T_j \tag{7-81}$$

DISTRIBUTION OF FIELDS AND CURRENTS NEAR A JUNCTION

A typical situation is shown on Fig. 7-5. The field H is applied in the plane of the junction. Surface currents appear on S and S′ to screen out the field. How will the currents manage to go through the junction? Because the maximum current density I_m through the junction is small, the currents must go deep in between S and S′. Typically, for an insulating junction, they will penetrate on a length $\delta \sim 1$ mm. At higher fields, things will become even more dramatic, and vortex lines will pile up in the junction (Fig. 7-6).

The behavior of the junction in this type of situation is governed by a simple equation due to Ferrell and Prange. In order to obtain this equation, we consider a junction occupying the entire xy plane; the magnetic fields are directed along the z axis and it is convenient to describe them by a vector potential directed along the y axis:

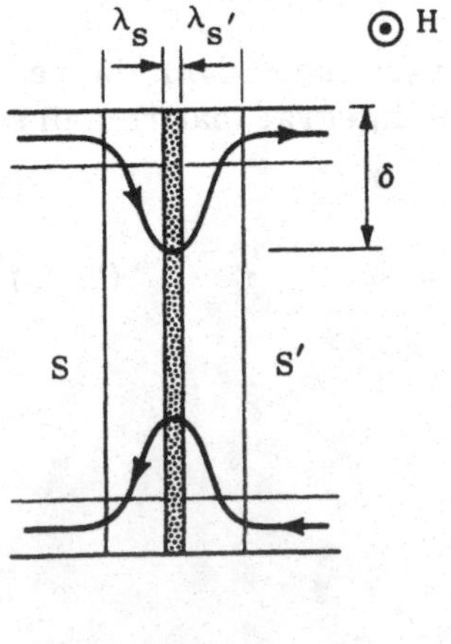

Figure 7-5

Current streamlines around an S-N-S′ junction under a very weak applied field. Because of the weak current capacity of the junction, the field penetrates up to a large distance δ.

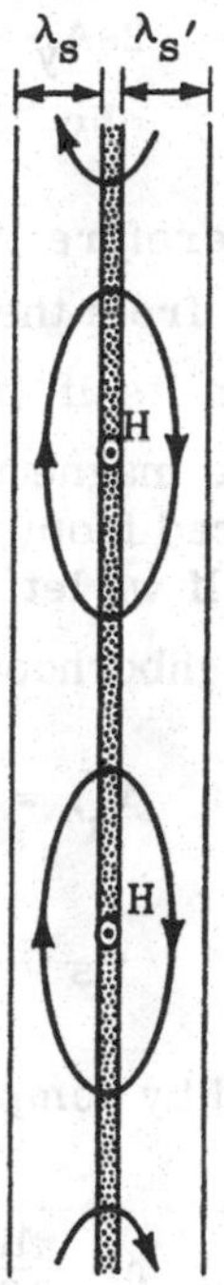

Figure 7-6
Current distribution in a junction at higher fields: Vortex lines are stacked along the junction.

$$h(x,y) = \frac{\partial}{\partial x} A_y(x,y)$$

We assume a solution of the Landau-Ginsburg equations of the form

$$\psi(x,y) = \begin{cases} \psi(x) \exp[i\phi_{s'}(y)] & x > 0 \\ \psi(x) \exp[i\phi_s(y)] & x < 0 \end{cases} \tag{7-82}$$

As we shall see, the variations of ϕ_s (or $\phi_{s'}$) take place over distances (δ) much larger than $\xi(T)$. Then $\psi(x)$ is essentially the solution (7-73).

Let us first consider the region $x > 0$ far from the junction ($x \gg \lambda_{s'}$). Here the Meissner effect is complete, no current flows and we find from Eq. (6-12)

$$\frac{2e}{\hbar c} A_y = \frac{\partial \phi_{s'}(y)}{\partial y} \qquad (x \gg \lambda_{s'}) \tag{7-83a}$$

and similarly

$$\frac{2eA_y}{\hbar c} = \frac{\partial \phi_s(y)}{\partial y} \qquad (x \ll -\lambda_s) \tag{7-83b}$$

Therefore $A_y(x,y)$ has well-defined limiting values, on either side far from the junction. Call these values $A_s(y)$, $A_{s'}(y)$. At the junction itself $(x = 0)$, we can, for convenience, take $A_y(x = 0, y) = 0$. The magnetic field $H(x = 0, y) = h_0(y)$ at the junction is simply deduced from the following considerations.

If we let λ_s and $\lambda_{s'}$ be the penetration depths in S and S′ in the neighborhood of the junction,[11] then by definition

$$\begin{aligned} A_{s'} &= \int_0^{\infty} h\,dx = \lambda_{s'} h_0 \\ A_s &= \int_0^{-\infty} h\,dx = \lambda_s h_0 \end{aligned} \tag{7-84}$$

and by comparison with (7-83)

$$h_0 = \frac{\hbar c}{2e} \frac{1}{\lambda_{s'}} \frac{\partial \phi_{s'}}{\partial y} = \frac{\hbar c}{2e} \frac{1}{\lambda_s} \frac{\partial \phi_s}{\partial y} \tag{7-85}$$

We now write the relation between field and current at the junction $\text{curl}\,\mathbf{h} = (4\pi/c)\,\mathbf{j}$,

$$\frac{\partial h_0}{\partial y} = \frac{4\pi}{c} I_x = \frac{4\pi}{c} I_m \sin(\phi_s - \phi_{s'}) \tag{7-86}$$

where we have used (7-71). By combining (7-85) and (7-86), we obtain the Ferrell-Prange equation

$$\frac{\partial^2 \phi}{\partial y^2} = \frac{1}{\delta^2} \sin \phi \tag{7-87}$$

where $\phi = \phi_s - \phi_{s'}$

$$\delta^2 = \frac{\hbar c^2}{8\pi e I_m (\lambda_s + \lambda_{s'})} \tag{7-88}$$

---

[11]When the N region is metallic, $\psi(x)$ becomes small near the junction and the penetration depths must be calculated taking this into account, as we have previously seen for an NS boundary.

We now estimate the characteristic length δ for an insulating junction. The order of magnitude of I_m is given by (7-81) and

$$\delta^2 \sim \frac{mc^2}{16\pi e^2 |\psi_0|^2} \frac{\xi_0}{\lambda(T)} \frac{1}{T_j} \sim \frac{\lambda(T)\xi_0}{T_j} \tag{7-89}$$

For an insulator of thickness $d \sim 20\text{Å}$, the transmission coefficient is typically of the order of 10^{-8}. If $\lambda(T) \sim \xi_0 \sim 10^3\text{Å}$, then $\delta \sim 1$ mm.

Equation (7-87) can be completely discussed by noticing that it is formally identical to the equation of motion of a pendulum.

Two limiting cases are particularly interesting.

(1) Very weak fields. Then (7-87) can be linearized to give solutions of the form

$$\phi = \phi_0 e^{-y/\delta}, \quad h_0 = H_0 e^{-y/\delta} \tag{7-90}$$

where $y = 0$ corresponds to the intersection of the junction and the sample surface, and H_0 is the field just outside of the sample. Thus the penetration depth of the junction in weak fields is δ. This regime holds when $\Phi \ll 1$ or $H \ll \phi_0/(\lambda_s + \lambda_{s'})\,\delta$, that is, the flux that penetrates into the junction must be much less than the flux quantum ϕ_0.

(2) Domain of complete penetration. If $H \gg \phi_0/(\lambda_s + \lambda_{s'})\,\delta$, the field becomes nearly uniform along the junction and $h_0(y) \to H$. ϕ can be determined from (7-85),

$$\phi \approx \frac{2\pi y}{L} + \text{const} \qquad L = \frac{\phi_0}{(\lambda_s + \lambda_{s'})H} \tag{7-91}$$

The structure of the currents is then periodic with period L

$$I_x = I_m \sin\phi \cong I_m \sin\left(\frac{2\pi(y - y_0)}{L}\right) \tag{7-92}$$

By placing (7-91) into the right-hand side of (7-87), it is easy to verify that $\partial^2\phi/\partial y^2$ and $\partial h_0/\partial y$ are essentially negligible for $L \ll \delta$.

The current structure is shown on Fig. (7-6). Each period carries one quantum of flux. It resembles a series of vortex lines stacked in the plane of the junction. Because the penetration depth δ is very large ($\delta \gg \xi$), the situation is similar to a Type II superconductor where

$\lambda \gg \xi$. In the limit $L \ll \delta$, it can be shown, by a complete calculation of the thermodynamic potential, that the field h_0 is very nearly equal to the applied field.

It is also interesting to calculate the total current that flows from S to S′ when the junction has a finite length $0 < y < D$. Per cm of length along the z axis, we have

$$I_{\text{total}} = \int^{D} I_m \sin\phi \, dy$$

$$\cong \frac{I_m L}{2\pi} \int d\phi \sin\phi \qquad (L \ll \delta)$$

$$= \frac{I_m L}{2\pi}\left[\cos\phi(0) - \cos\left(\phi(0) + \frac{2\pi D}{L}\right)\right]$$

The maximum value of I_{total} is obtained when $\phi(0) = \pi/2 - \pi D/L$,

$$I_{\text{total max}} = \frac{I_m L}{2\pi}\left|\sin\frac{\pi D}{L}\right| \qquad (L \ll \delta) \tag{7-93}$$

The factor D/L is the number of flux quanta in the junction. The total current vanishes when D/L is an integer, and is maximum when D/L is half-integral. Rather detailed verifications of (7-93) have recently been obtained by Fiske and by Rowell.

Problem. Discuss the maximum superfluid current I of a system of two SNS junctions in parallel, as a function of the magnetic field H ("quantum interferometer"). The (schèmatic) arrangement and the notation are shown on Fig. 7-7. A flux Φ is enclosed in the ring between the two junctions. For simplicity assume that (a) the junctions A and B are small, the flux trapped in each of the junctions is negligible ($2\, d\lambda H \ll \phi_0$); (b) the superconducting pieces (1) and (2) are much thicker than the penetration depth λ.

Solution. The overall supercurrent flowing from (1) to (2) is

$$I = I_m d\,[\sin(\phi_{2A} - \phi_{1A}) + \sin(\phi_{2B} - \phi_{1B})]$$

The different phases ϕ_{1A}, and so on, are coupled by a relation involving the flux ϕ, which we now derive. At any point in superconductor I, for instance, the current density j_s is proportional to $\hbar\nabla\phi - 2eA/c$ (where ϕ is the local phase of the pair potential). Consider a path C_1 in the superconductor distant enough from the surface, so that $j_s \equiv 0$ on it. Then, on C_1, $\hbar\nabla\phi = 2eA/c$. Integrating this relation we obtain

$$\phi_{1B} - \phi_{1A} = \frac{2e}{\hbar c}\int_{C_1} \mathbf{A}\cdot d\boldsymbol{\ell}$$

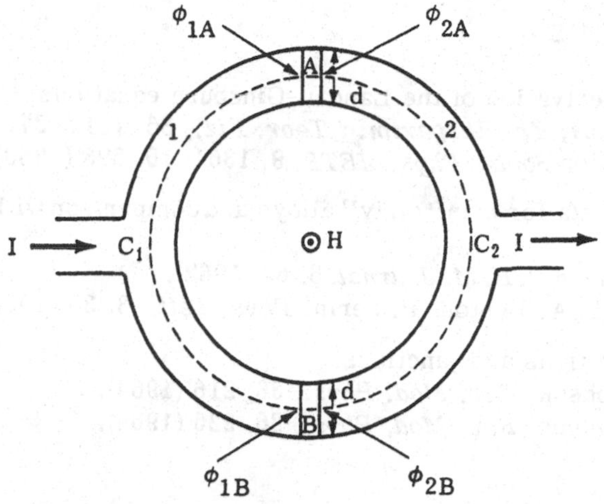

Figure 7-7
Principle of a "quantum interferometer." A and B are two Josephson junctions. The maximum superfluid current I is measured as a function of the field H in the loop.

and similarly

$$\phi_{2A} - \phi_{2B} = \frac{2e}{\hbar c} \int_{C_2} \mathbf{A} \cdot d\boldsymbol{\ell}$$

Adding these equations we arrive at

$$\phi_{1B} - \phi_{2B} + \phi_{2A} - \phi_{1A} = \frac{2e}{\hbar c} \oint \mathbf{A} \cdot d\boldsymbol{\ell} = 2\pi \frac{\Phi}{\Phi_0}$$

where Φ is the total flux in the loop.

Apart from this relation, the phases, ϕ_{1A}, and so on, are arbitrary. The current I is maximum when we choose

$$\phi_{2A} - \phi_{1A} = -\left(\phi_{2B} - \phi_{1B}\right) = \pi \frac{\Phi}{\Phi_0}$$

$$I = 2I_m d \left| \sin\left(\pi \frac{\Phi}{\Phi_0}\right) \right|$$

The maximum superfluid current is a periodic function of Φ or H. To observe the periodicity in field the surface S of the loop must be small. The experiment has been achieved by Jaklevic, Lambe, Silver, and Mercereau. A and B can be two Josephson junctions, or, even more simply, two mechanical contacts between superconducting wires.

REFERENCES

Microscopic derivation of the Landau-Ginsburg equations:

L. P. Gor'kov, *Zh. Eksperim. i Teor. Fiz.*, **36**, 1918; **37**, 833 (1959); Translation *Soviet Phys. JETP*, **9**, 1364; **10**, 593 (1960).

Experimental data for κ in "dirty" alloys and comparison with Gor'kov's formula:

B. B. Goodman, *I. B. M. Journal*, **6**, 62 (1962).

T. Kinsel, E. A. Lynton, B. Serin, *Phys. Lett.*, **3**, 30 (1962).

Boundary conditions and junctions:

B. D. Josephson, *Rev. Mod. Phys.*, **36**, 216 (1964).

P. G. de Gennes, *Rev. Mod. Phys.*, **36**, 225 (1964).

8

EFFECTS OF STRONG MAGNETIC FIELDS AND OF MAGNETIC IMPURITIES

We shall discuss two apparently very different physical situations:

(1) dirty superconductors and small superconducting samples in high magnetic fields;

(2) superconductors containing magnetic impurities.

There are in fact some crucial features shared by (1) and (2).

(a) In both cases the perturbation acts with opposite signs on the two members of a Cooper pair. In case (1) there is a term $(1/2m)(\mathbf{p}\cdot\mathbf{A}+\mathbf{A}\cdot\mathbf{p})$ if the one-electron Hamiltonian, and this changes sign when going from $\mathbf{p}$ to $-\mathbf{p}$. In case (2) we may often describe the effect of the magnetic impurity (i) of spin $\mathbf{S}_i$ on a conduction electron (of coordinate $\mathbf{r}_e$ and spin $\mathbf{S}_e$) by an interaction of the form

$$\Gamma(\mathbf{r}_e - \mathbf{r}_i)\mathbf{S}_e\cdot\mathbf{S}_i$$

(Γ is called an exchange coupling constant, but in fact its origin is complex.) Γ has an arbitrary sign, and is typically of order 0.1 eV = 10^3 °K.[1] Since the two electrons of a Cooper pair have opposite spins, this Γ coupling acts with opposite signs on each.

(b) Experimentally, such "antisymmetric" perturbations lead to a

---

[1] Note that Γ has nothing to do with the dipolar field created by the magnetic impurity. The dipolar coupling energy is of order $\mu_B^2/a^3 \sim 1°\mathrm{K}$ (μ_B = Bohr magneton, a = atomic distance) and much smaller than Γ (C. Herring, 1958).

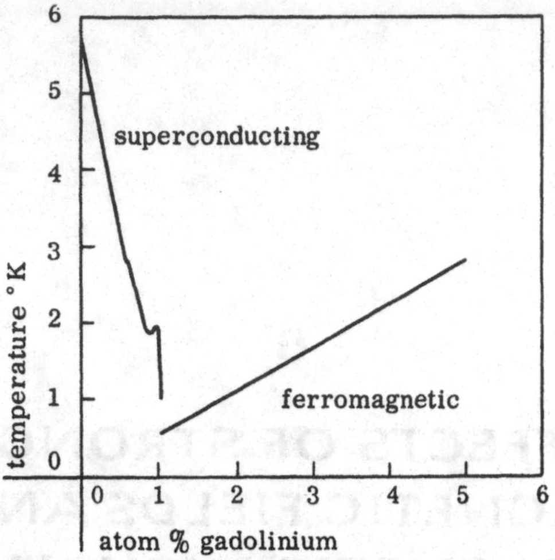

Figure 8-1

Superconducting and ferromagnetic transition temperatures of La Gd alloys [after B. T. Matthias, *IBM J.*, **6**, 250 (1962)]. Note the sharp drop of the superconducting transition point when the (magnetic) gadolinium atoms are added to lanthanum.

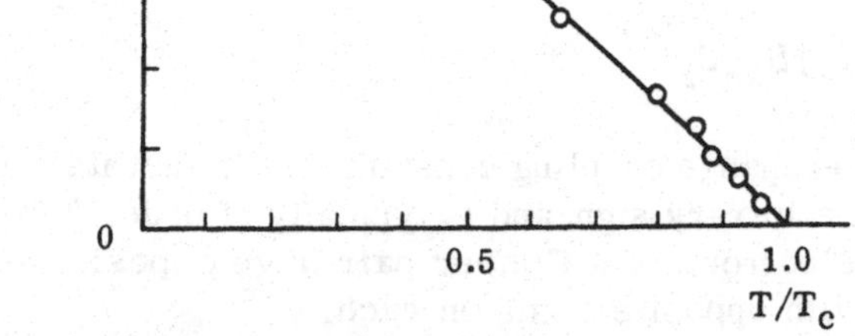

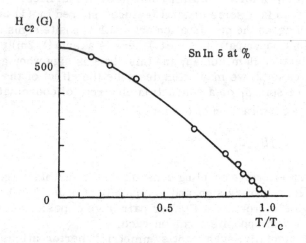

Figure 8-2

Relation between temperature and field at the upper transition point of a "dirty" Type II superconductor. Experimental points by E. Guyon and A. Martinet. The theoretical curve is derived from Eq. (8-40).

strong decrease of the transition temperature. This is shown on examples in Figs. 8-1 and 8-2. Note that the "antisymmetry" is crucial. Recall that nonmagnetic impurities (which give rise only to "symmetric" perturbations) have very little effect on the transition point (Anderson's theorem).

(c) Another important property is shared by classes (1) and (2). Electrons are randomly scattered by impurities or on the sample surface. Thus the "antisymmetric" perturbation seen by one given Cooper pair is modulated in time. This, as we shall see, can lead to a number of strange properties, and in particular to gapless superconductivity (A. Abrikosov and L. P. Gorkov, 1961).

8–1 RELATION BETWEEN TRANSITION TEMPERATURE AND TIME-REVERSAL PROPERTIES

Our problem is to translate the qualitative asymmetry property of the perturbation into more numerical terms. Consider for instance the calculation of a transition point, in a given field (case 1) or for a given concentration of magnetic impurities (case 2). Experimentally for both cases the transition is found to be of second order. $\Delta(\mathbf{r})$ is small near the transition and a linearized self-consistency equation like (7-8) can be used.

Let us first restrict our attention to case (1), where spin indices play no role, and where (7-8) thus applies directly. We rewrite the kernel $K(\mathbf{r}_1\mathbf{r}_2)$ of (7-8) under a form equivalent to (7-11).

$$K(\mathbf{r}_1\mathbf{r}_2) = k_B T \sum_{\omega} \mathcal{V} N(0) V \int \frac{d\xi\, d\xi'}{(\xi - i\omega\hbar)(\xi' + i\omega\hbar)} \times f\left(\mathbf{r}_1\mathbf{r}_2, \frac{\xi - \xi'}{\hbar}\right) \tag{8-1}$$

$$f(\mathbf{r}_1\mathbf{r}_2\Omega) = \sum_m \langle \phi_n^*(\mathbf{r}_1)\phi_m^*(\mathbf{r}_1)\phi_m(\mathbf{r}_2)\phi_n(\mathbf{r}_2) \times \delta(\xi_m - \xi_n - \hbar\Omega)\rangle \tag{8-2}$$

[$\mathcal{V}$ is the sample volume and N(0) the density of states per unit volume in the normal state so that $\mathcal{V}N(0)$ is the total density of states of the sample when it is normal. We have made the substitution $\Sigma_n \to \int \mathcal{V}N(0)\, d\xi_n$.] As usual, the sum Σ_ω runs over the values $\pi(k_B T/\hbar) \times (2\nu + 1)$. The average in Eq. (8-2) is taken over all states n at a fixed energy ξ_n (in practice, at the Fermi energy $\xi_n = 0$). We would like to express the product of two ϕ's and two ϕ^*'s in terms of matrix elements between one-electron states n and m. Obviously, the * symbols are not located as desired. We circumvent this by introducing an ad hoc operator K, which by definition changes a function into its complex conjugate

$$K\phi_m = \phi_m^* \qquad \text{(case 1)} \tag{8-3}$$

We call K the *time-reversal operator.* In terms of K we may write

$$\begin{aligned}
\phi_n^*(\mathbf{r}_1)\,\phi_m^*(\mathbf{r}_1) &= (n\,|\,\delta(\mathbf{r}-\mathbf{r}_1)K\,|\,m) \\
\phi_n(\mathbf{r}_2)\,\phi_m(\mathbf{r}_2) &= (m\,|\,K^+\,\delta(\mathbf{r}-\mathbf{r}_2)\,|\,n) \\
f(\mathbf{r}_1\mathbf{r}_2\Omega) &= \sum_m \langle (n\,|\,\delta(\mathbf{r}-\mathbf{r}_1)K\,|\,m)(m\,|\,K^+\delta(\mathbf{r}-\mathbf{r}_2)\,|\,n) \\
&\qquad \times\, \delta(\xi_n - \xi_m - \hbar\Omega)\rangle
\end{aligned} \tag{8-4}$$

At this stage, as usual, we find it convenient to use the Fourier transform of f

$$\begin{aligned}
f(\mathbf{r}_1\mathbf{r}_2 t) &= \int d\Omega\, e^{-i\Omega t} f(\mathbf{r}_1\mathbf{r}_2\Omega) \\
&= \langle \delta[\mathbf{r}(t)-\mathbf{r}_1]\, K(t)K^+(0)\, \delta[\mathbf{r}(0)-\mathbf{r}_2]\rangle
\end{aligned} \tag{8-5}$$

where K(t) and $\mathbf{r}(t)$ are Heisenberg operators describing the time evolution of K or $\mathbf{r}$ for one electron submitted to the forces described by the Hamiltonian

$$\begin{aligned}
\mathcal{H}_e &= \frac{1}{2m}\left(\mathbf{p} - \frac{e\mathbf{A}}{c}\right)^2 + U(\mathbf{r}) - E_F \qquad \text{(case 1)} \\
K(t) &= \exp\left(i\,\frac{\mathcal{H}_e t}{\hbar}\right) K \exp\left(+i\,\frac{\mathcal{H}_e t}{\hbar}\right)
\end{aligned} \tag{8-6}$$

(Note the unusual sign in the last exponent, due to the fact that Ki = −iK.) The rate of change of K is given by

$$\frac{dK}{dt} = \frac{i}{\hbar}[\mathcal{H}_e, K] = \frac{-ie}{m\hbar c}(\mathbf{p}\cdot\mathbf{A} + \mathbf{A}\cdot\mathbf{p})K \qquad \text{(case 1)} \tag{8-7}$$

The motion of K is due only to the "antisymmetric" part of the magnetic perturbation in $\mathcal{H}_e$. (Rather than "antisymmetric," we now say: the part that changes sign under time reversal.)

Finally, our result is the following: The kernel $K(\mathbf{r}_1\mathbf{r}_2)$ of the integral equation giving the transition temperature is known in terms of a correlation function $f(\mathbf{r}_1\mathbf{r}_2 t)$ defined in (8-5). This correlation function is useful because, as in many cases discussed before, it is closely related to the transport properties in the normal state. All the complications due to "antisymmetric" perturbations are described in terms of the time-reversal operator K(t).

These conclusions can be extended directly to cases where there are magnetic impurities, that is, where spin indices must be taken into account. We now briefly describe this extension. The complete Hamiltonian is

$$\mathcal{H} = \sum_{\sigma} \int d\mathbf{r} \Bigg\{ \psi^{+}(\mathbf{r},\sigma) \left[\frac{\left(\mathbf{p} - \frac{e}{c}\mathbf{A}\right)^2}{2m} - E_F \right] \psi(\mathbf{r},\sigma) + \sum_{\sigma\nu} \psi^{+}(\mathbf{r},\sigma)\, U_{\sigma\nu}(\mathbf{r})\, \psi(\mathbf{r},\nu) - V \sum_{\sigma\nu} \psi^{+}(\mathbf{r},\sigma)\, \psi^{+}(\mathbf{r},\nu)\, \psi(\mathbf{r},\nu)\, \psi(\mathbf{r},\sigma) \Bigg\} \tag{8-8}$$

where $\psi^{+}(\mathbf{r},\sigma)$ creates an electron of spin σ at the point $\mathbf{r}$. $U_{\sigma\nu}$ is the potential acting upon the electrons, which for magnetic alloys is spin dependent. (In order to simplify the notation, we assume that U be a local operator coupling ψ^{+} and ψ at the same point.) The main assumption made about the potential U is that it is *static*. We do not take into account possible motions of the impurity spins. (This, as we later see, is correct only for very dilute alloys. In more concentrated systems, the impurities are coupled among themselves and precess in the exchange field of their neighbors.) The pair potential $\Delta(\mathbf{r})$ is still defined by

$$\Delta(\mathbf{r}) = V \langle \psi(\mathbf{r}\uparrow)\, \psi(\mathbf{r}\downarrow) \rangle = -V \langle \psi(\mathbf{r}\downarrow)\, \psi(\mathbf{r}\uparrow) \rangle$$

which can be rewritten in the condensed form

$$\Delta = \frac{V}{2} \sum_{\sigma\mu} \rho_{\sigma\mu} \langle \psi(\mathbf{r},\sigma)\, \psi(\mathbf{r},\mu) \rangle \tag{8-9}$$

where

$$\rho = \begin{pmatrix} 0 & 1 \\ -1 & 0 \end{pmatrix}$$

As in Chapter 5, we write the equations of motion of ψ and ψ^{+} and linearize them, allowing for a nonzero Δ. The linearized equation of motion for ψ is

$$i\hbar \frac{\partial\psi}{\partial t}(\mathbf{r},\sigma) = [\mathcal{H},\psi(\mathbf{r},\sigma)]$$

$$= \left(\frac{p^2}{2m} - E_F\right)\psi(\mathbf{r},\sigma)$$

$$+ \sum_\mu V_{\sigma\mu}\,\psi(\mathbf{r},\mu) + \sum_\mu \rho_{\sigma\mu}\,\Delta(\mathbf{r})\psi^*(\mathbf{r},\mu) \qquad (8\text{-}10)$$

(As usual, we have subtracted the chemical potential E_F.) The elementary excitations are found by performing the transformation

$$\psi(\mathbf{r},\sigma) = \sum_n [u_n(\mathbf{r},\sigma)\gamma_n + v_n^*(\mathbf{r},\sigma)\gamma_n^+]$$

and imposing the condition $[\mathcal{H},\gamma_n] = -E_n\gamma_n$. The $\binom{u}{v}$ are eigenfunctions of the system

$$\epsilon u(\mathbf{r},\sigma) = \left(\frac{\left(\mathbf{p} - \frac{e\mathbf{A}}{c}\right)^2}{2m} - E_F\right) u(\mathbf{r},\sigma)$$

$$+ \sum_\mu \left[V_{\sigma\mu}(\mathbf{r})u(\mathbf{r},\mu) + \Delta(\mathbf{r})\rho_{\sigma\mu}\,v(\mathbf{r},\mu)\right]$$

$$-\epsilon v(\mathbf{r},\sigma) = \left[\left(\mathbf{p} + \frac{e\mathbf{A}}{c}\right)^2 \frac{1}{2m} - E_F\right] v(\mathbf{r},\sigma)$$

$$+ \sum_\mu \left[V_{\sigma\mu}^*(\mathbf{r})v(\mathbf{r},\mu) + \Delta^*(\mathbf{r})\rho_{\sigma\mu}\,u(\mathbf{r},\mu)\right]$$

We are interested in the limit $\Delta \to 0$. To zeroth order in Δ, $\binom{u}{v}$ are related to the electronic eigenfunctions ϕ_n in the normal metal, defined by

$$\xi\phi(\mathbf{r},\sigma) = \left[\frac{1}{2m}\left(\mathbf{p} - \frac{e\mathbf{A}}{c}\right)^2 - E_F\right]\phi(\mathbf{r},\sigma) + \sum_\mu U_{\sigma\mu}(\mathbf{r})\phi(\mathbf{r},\mu)$$

$$\equiv \mathcal{H}_e\,\phi(\mathbf{r},\sigma)$$

$$u_n^0 = \begin{cases}\phi_n & \xi_n > 0\\ 0 & \xi_n < 0\end{cases}$$

$$v_n^0 = \begin{cases}0 & \xi_n > 0\\ \phi_n^* & \xi_n < 0\end{cases} \qquad \epsilon_n^0 = |\xi_n|$$

To first order in Δ we write

$$u_n = u_n^0 + \sum_{m \neq n} a_{nm} \phi_m$$

$$v_n = v_n^0 + \sum_{m \neq n} b_{nm} \phi_m^*$$

The coefficients a_{nm} and b_{nm} are obtained by the usual perturbation method:

$$a_{nm} = \begin{cases} 0 & (\xi_n > 0) \\ \dfrac{-1}{\xi_n + \xi_m} \displaystyle\sum_{\sigma\mu} \int d\mathbf{r}_2\, \phi_m^*(\mathbf{r}_2\sigma) \rho_{\sigma\mu}\, \phi_n^*(\mathbf{r}_2\mu) \Delta(\mathbf{r}_2) & (\xi_n < 0) \end{cases}$$

$$b_{nm} = \begin{cases} \dfrac{-1}{\xi_n + \xi_m} \displaystyle\sum_{\sigma\mu} \int d\mathbf{r}_2\, \phi_n(\mathbf{r}_2\sigma) \rho_{\mu\sigma}\, \phi_n(\mathbf{r}_2\mu) \Delta(\mathbf{r}_2) & (\xi_n > 0) \\ 0 & (\xi_n < 0) \end{cases}$$

We now insert these corrected values for u and v into the self-consistency equation for Δ. The latter, when spin indices are taken into account, takes the form

$$\Delta(\mathbf{r}_1) = \tfrac{1}{2} V \sum_{\sigma\mu} \sum_{n} \rho_{\sigma\mu}\, v_n^*(\mathbf{r},\sigma)\, u_n(\mathbf{r},\mu)[1 - 2f(\xi_n)]$$

To first order in Δ this again has the form

$$\Delta(\mathbf{r}_1) = \int K(\mathbf{r}_1\mathbf{r}_2)\Delta(\mathbf{r}_2)\, d\mathbf{r}_2 \qquad (8\text{-}11)$$

and the kernel can still be expressed in terms of a correlation function by Eq. (8-1) and (8-5). The only change is that the definition of the time-reversal operator in (8-5) now involves the spin indices. Explicitly we have

$$\begin{aligned} K\phi(\mathbf{r}\uparrow) &= \phi^*(\mathbf{r}\downarrow) \\ K\phi(\mathbf{r}\downarrow) &= -\phi^*(\mathbf{r}\uparrow) \end{aligned} \qquad (8\text{-}3')$$

In terms of Pauli spin matrices, one often writes

$$K = i\sigma_y C$$

where $C\phi = \phi^*$ and $i\sigma_y$ coincides with our earlier defined $\rho = \begin{pmatrix} 0 & 1 \\ -1 & 0 \end{pmatrix}$. Thus in all cases of interest we have reduced our problem to a study of the motion of the operator K for one electron in the normal metal.

At first sight, K is a rather unfamiliar object, but with some practice you can get a very physical understanding of it. Take for instance the case of orbital magnetic fields, where the equation of motion for K is (8-7). In Eq. (8-1) the order of magnitude of $|\xi - \xi'|$ is $k_B T$. Thus in $f(\mathbf{r}_1 \mathbf{r}_2 t)$ the time scale of interest is $t \sim \hbar/k_B T$. The spatial distances traveled by an electron during such times are very large when compared to the Fermi wavelength. Then we can treat the motion of the operator K *classically* and write (8-7) in the form

$$\frac{dK(t)}{dt} = -i \frac{d\phi}{dt} K$$

$$K(t) = e^{-i\phi(t)} K(0) \tag{8-12}$$

$$\phi(t) = \frac{2e}{\hbar c m} \oint_0^t \mathbf{A} \cdot \mathbf{p} \, dt$$

In this last equation, the integral is to be taken along a classical one-electron trajectory in the normal metal. Replacing $\mathbf{p}$ by $m\mathbf{v}$ (the difference $e\mathbf{A}/c$ between $\mathbf{p}$ and $m\mathbf{v}$ being very small when compared to the Fermi momentum), we may rewrite the phase ϕ in the form

$$\phi = \frac{2e}{\hbar c} \int \mathbf{A} \cdot \mathbf{v} \, dt = \frac{2e}{\hbar c} \int \mathbf{A} \cdot d\boldsymbol{\ell} \tag{8-13}$$

In this case the correlation function $f(\mathbf{r}_1 \mathbf{r}_2 t)$ may be understood as follows: Consider all one-electron trajectories linking $\mathbf{r}_1$ and $\mathbf{r}_2$ in a time interval t. For a pure metal, there is only one such trajectory —essentially a straight line (the cyclotron radius being very large on our scale $v_F t$). For more "dirty" systems, there are many wiggly trajectories allowed. For each of these we compute $e^{i\phi}$. Finally we sum over the different allowed trajectories, each of them being weighted. according to its probability to occur in a succession of scattering processes: As announced, the calculation of f is reduced to a study of transport properties.

In case 2 (magnetic impurities), the equation of motion of K is derived in the following way: the one-electron Hamiltonian $\mathcal{H}_e$ is of the form

$$\mathcal{H}_e = A + \mathbf{B} \cdot \mathbf{S}_e \tag{8-14}$$

where **A** includes the kinetic energy and the spin-independent forces, while **B** stems from the Γ interaction with all impurities. Using the definition (8-3′) of K and the corresponding equations for the adjoint K^+, we can show that

$$K^+\mathcal{H}_e K = A^* - \mathbf{B}\cdot\mathbf{S}_e \tag{8-15}$$

that is, $K^+\mathcal{H}_e K$ is obtained from $\mathcal{H}_e$ by changing A into A* (this amounts to changing **p** into −**p** in the magnetic coupling term) and $\mathbf{S}_e$ into $-\mathbf{S}_e$. (This could be reached by changing the direction of time for the conduction electrons and justifies the name "time-reversal operator.") We also notice that $K^+K = KK^+ = 1$, and write

$$[\mathcal{H}_e, K] = \mathcal{H}_e K - K\mathcal{H}_e = K(K^+\mathcal{H}_e K - \mathcal{H}_e) \tag{8-16}$$

Consider for instance pure case 2 (no orbital magnetic fields A = A*). Then, by (8-15),

$$\frac{dK}{dt} = \frac{i}{\hbar}[\mathcal{H}_e, K] = -\frac{2i}{\hbar} K\mathbf{B}\cdot\mathbf{S}_e \tag{8-17}$$

Here we might also say that K acquires a time-dependent phase (due to the Γ coupling). However, we cannot treat the phase classically, because the potential **B** varies strongly in one atomic distance, or in one Fermi wavelength. We see later how the motion of K is analyzed in such situations. Before going into such details, we now discuss some general qualitative properties of K(t) [in both cases (1) and (2)] related to the shape of the one-electron trajectories.

8–2 ERGODIC VERSUS NONERGODIC BEHAVIOR—GAPLESS SUPERCONDUCTIVITY

For the moment, let us restrict our attention to systems where the pair potential Δ(**r**) has an amplitude |Δ| constant in space. This greatly simplifies the discussion of the linear equations (8-8) or (8-11) and allows us to bring out directly the most important features. Also, it is not an academic case. It is found in the following situations:

(1) small superconducting colloids;

(2) thin superconducting films in fields parallel to the film surface;

(3) superconducting alloys, with magnetic impurities, in zero external field. In colloids, |Δ| cannot vary very much if all dimensions of the grains are small compared with ξ(T). In thin films, we have to be more careful, since only one of the dimensions is small. In particular, if the field is normal to the film, it often penetrates by means

of vortices as in Type II superconductors (M. Tinkham, 1962), and then $|\Delta|$ is strongly varying in space. To avoid this, we only consider fields parallel to the surface. There is also another requirement on the colloids and films, related to the fact that we discuss second-order transition. In Type I materials, the transition is indeed observed to be of second order only if the sample thickness is smaller than one or two penetration depths.

Finally, for magnetic alloys (in zero field), it is also a reasonable starting point to take $|\Delta| = \text{const.}$ (This neglects only some rather weak local modulations of $\Delta(\mathbf{r})$ in the vicinity of each impurity.)

We now write down explicitly the equation for the transition temperature. For a singly connected specimen, we can choose a particular gauge for the vector potential $\mathbf{A}$, such that $\Delta(\mathbf{r})$ is real, and constant in space (since $|\Delta|$ is constant). This enables us to integrate the self-consistency equation (8-11) on $\mathbf{r}_2$ and $\mathbf{r}_1$, obtaining

$$\Delta \mathcal{V} = \int d\mathbf{r}_1\, d\mathbf{r}_2\, K(\mathbf{r}_1 \mathbf{r}_2)\Delta$$

By making use of (8-1), (8-5), and canceling out a factor $\Delta\mathcal{V}$ on both sides, we reach the condition

$$1 = N(0)V \int d\xi\, d\xi' \sum_{\omega} \frac{k_B T}{(\xi - i\omega\hbar)(\xi' + i\omega\hbar)} \frac{1}{\hbar} g\left[\frac{(\xi - \xi')}{\hbar}\right] \tag{8-18}$$

$$g(\Omega) = \int \frac{dt}{2\pi} \langle K^+(0)\, K(t)\rangle\, e^{-i\Omega t}$$

$g(\Omega)$ is the *power spectrum of the operator K.* Instead of using the summation Σ_ω we may also, if we wish, write (8-18) in terms of Fermi functions $f(\xi)$

$$1 = N(0)V \int d\xi\, d\xi' \frac{1 - f(\xi) - f(\xi')}{\xi + \xi'} g\left[\frac{(\xi - \xi')}{\hbar}\right] \tag{8-18'}$$

[Some features are more striking on (8-18′), but (8-18) is more convenient for computational purposes.]

Discussion of this equation: First we notice that for all cases where the one-electron Hamiltonian $\mathcal{H}_e$ is invariant by time reversal ($[K, \mathcal{H}_e] = 0$), $K(t) = K(0)$:

$$\langle K^+(0)\, K(t)\rangle \equiv 1$$

$$g(\Omega) = \delta(\Omega)$$

The power spectrum of K has a single line at zero frequency. Then (8-18′) reduces to

$$1 = N(0)V \int d\xi \, \frac{1 - 2f(\xi)}{2\xi}$$

This is the BCS condition for the transition temperature in conventional superconductors.

When $[K, \mathcal{H}_e] \neq 0$, we must study g(t) in more detail. As pointed out, we are interested in $\Omega \sim k_B T/\hbar$ and $t \sim \hbar/k_B T$, that is, a very long time scale. Thus we consider first the limiting behavior of $\langle K^+(0) K(t) \rangle$ at very large times. There can be many types of limiting behavior. Two of them are particularly interesting:

$$\text{(I)} \quad \lim_{t \to \infty} \langle K^+(0) K(t) \rangle = \eta \neq 0 \tag{8-19}$$

$$\text{(II)} \quad \lim_{t \to \infty} \langle K^+(0) K(t) \rangle = \exp\left(-t/\tau_K\right) \tag{8-20}$$

where η and τ_K are certain functions of the strength of the perturbation (the magnetic field, or the magnetic impurity concentration).

If our system is of class I, we say it is nonergodic.
If our system is of class II, we say it is ergodic and Markoffian.

Nonergodic Systems

We may write

$$\langle K^+(0) K(t) \rangle = \eta + R(t) \tag{8-21}$$

where R(t) is a rapidly decreasing function of time [for instance, in small samples of class I, the time constant of R(t) is of order d/v_F where d is the sample dimension, and here $d \ll \xi_0$ or $d/v_F \ll \hbar/k_B T$]. Thus we may, in a first approximation, neglect completely R(t) at the times of interest, and Eq. (8-18′) takes again the BCS form

$$1 = N(0)V\eta \int_{-\hbar\omega_D}^{\hbar\omega_D} d\xi \, \frac{1 - 2f(\xi)}{2\xi} \tag{8-22}$$

leading to a transition temperature

$$k_B T = 1.14 \hbar\omega_D \exp\left(-\frac{1}{N(0)V\eta}\right) \tag{8-23}$$

The behavior is close to that of a conventional superconductor in zero field, but the coupling constant $N(0)V\eta$ is now dependent on the strength of the perturbation (through η). The following problem discusses a physical system of class I.

Problem. Discuss the behavior of $\langle K^+(0)K(t)\rangle$ at large times for a thin film (thickness d) with diffuse reflection on the boundaries but no volume defects, in a parallel field H (P. G. de Gennes and M. Tinkham, 1964).

Solution. Consider an electron starting at point $\mathbf{r}_0$ at time 0 (Fig. 8-3), then suffering collisions alternatively on both sides of the film at times $t_1, t_2, \ldots, t_n$ and finally arriving at $\mathbf{r}_t$ at time t. The phase ϕ defined by Eq. (8-13) is a sum of increments:

$$\phi = \Delta\phi_{01} + \Delta\phi_{12} + \cdots + \Delta\phi_{n-1\,n} + \Delta\phi_{nt}$$

The vector potential, in the gauge where Δ is real, has one nonzero component $A_y = Hx$. Then each of these increments (except $\Delta\phi_{01}$ and $\Delta\phi_{nt}$) is in fact equal to 0, as shown in Fig. 8-3.

$$\Delta\phi_{p\,p+1} = \frac{2eH}{\hbar c}\int_{x=-d/2}^{x=d/2} x\,dy = 0$$

The trajectory between two successive collisions is essentially a straight line; y is a linear function of x. Thus the above integral, taken between even limits, vanishes and $\phi = \Delta\phi_{01} + \Delta\phi_{nt}$. Furthermore, when $t \gg d/v_F$, the number of collisions n is large and the relevant time intervals $0 \to t_1$ and $t_n \to t$ are uncorrelated. We may thus write

$$\eta = \langle \exp(i\Delta\phi_{01})\rangle\langle\exp(i\Delta\phi_{nt})\rangle \qquad \eta^{1/2} = \langle\exp(i\Delta\phi_{01})\rangle$$

η is nonvanishing: in spite of the numerous collisions, the system is *not ergodic* in our sense. To compute η we consider an electron starting from an arbitrary point $\mathbf{r}$, with a randomly oriented velocity of length v_F, the projection of this velocity on the xy plane making an angle ψ with the x axis.

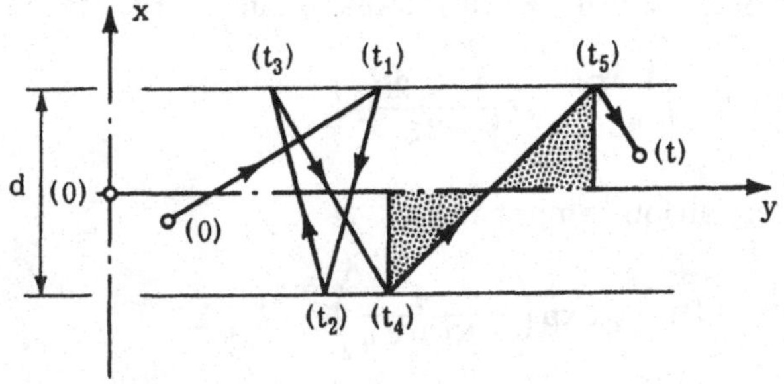

Figure 8-3
Successive collisions of one electron in a thin film with diffuse scattering on the boundaries.

$$\eta^{1/2} = \frac{1}{\pi d} \int_{-\pi/2}^{\pi/2} d\psi \int_{-d/2}^{d/2} dx \, \exp\left[i\pi \frac{H}{\phi_0} \left(\frac{d^2}{4} - x^2 \right) \tan\psi \right]$$

$$= \frac{2}{d} \int_0^{d/2} dx \, \exp\left[\frac{\pi H}{\phi_0} \left(\frac{x^2 - d^2}{4} \right) \right]$$

where $\phi_0 = ch/2e$ is the flux quantum.

$$\eta = \begin{cases} 1 - \dfrac{\pi}{3} \dfrac{H}{\phi_0} d^2 & H \ll \dfrac{\phi_0}{d^2} \\ \left(\dfrac{2\phi_0}{\pi H d^2} \right)^2 & H \gg \dfrac{\phi_0}{d^2} \end{cases}$$

Ergodic Markoffian Systems

If the exponential decay law (8-20) is obeyed at the times t of interest, we can write

$$g(\Omega) = \frac{1}{\pi} \frac{\tau_K}{1 + \Omega^2 \tau_K^2} \tag{8-24}$$

and perform the integrations $\int d\xi \, d\xi'$ in Eq. (8-18). We arrive at the condition

$$1 = N(0)V \sum_\omega 2\pi \frac{k_B T}{\hbar} \frac{1}{2|\omega| + \dfrac{1}{\tau_K}} \tag{8-25}$$

The sum Σ_ω diverges. This is because we did not take into account the frequency cutoff ω_D of the interaction. A similar complication has been discussed after Eq. (7-28), p. 218. When the cutoff is taken into account, (8-25) becomes

$$1 = N(0)V \left[\ln\left(\frac{1.14\hbar\omega_D}{k_B T} \right) + \frac{2\pi k_B T}{\hbar} \times \sum_\omega \left(\frac{1}{2|\omega| + \dfrac{1}{\tau_K}} - \frac{1}{2|\omega|} \right) \right] \tag{8-25'}$$

The sum is now convergent and can be expressed in terms of a tabulated function $\psi(z)$.

$$\psi(z) = \frac{\Gamma'(z)}{\Gamma(z)} = -0.577 - \frac{1}{z} + \sum_{\nu=1}^{\nu=\infty} \left(\frac{1}{\nu} - \frac{1}{\nu + z}\right)$$

$$\frac{2\pi k_B T}{\hbar} \sum_{\omega} \left(\frac{1}{2|\omega| + \frac{1}{\tau_K}} - \frac{1}{2|\omega|} \right)$$

$$= \left(\frac{1}{2}\right) - \psi\left(\frac{1}{2} + \frac{\hbar}{4\pi \tau_K k_B T}\right)$$

Finally the relation between $1/\tau_K$ and T takes the implicit form

$$\ln \frac{T_0}{T} = -\psi(\tfrac{1}{2}) + \psi\left(\frac{1}{2} + \frac{\hbar}{4\pi \tau_K k_B T}\right) \tag{8-26}$$

Since a relation of this form occurs very often in the theory of superconductivity, we give a special name to it. We say that

$$\frac{\hbar}{\tau_K} = k_B T_0 \,\mathrm{Un}\left(\frac{T}{T_0}\right) \tag{8-26$'$}$$

The "universal function" $\mathrm{Un}(x)$ is shown in Fig. 8-4. For $x \to 0$,

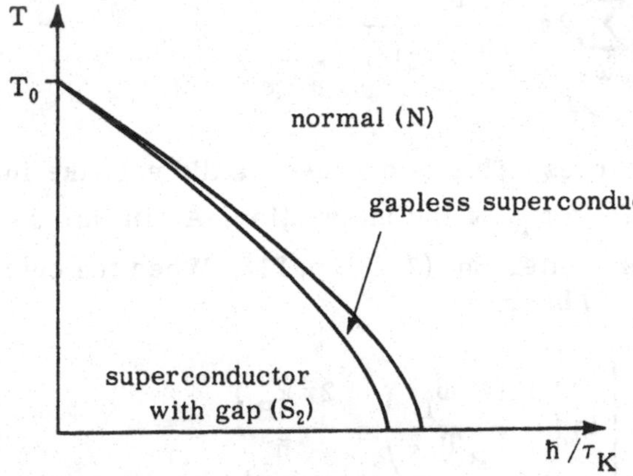

Figure 8-4

Phase diagram for a superconductor with "asymmetric" perturbations on the Cooper pairs + ergodicity. $\hbar/\tau_K$ measures the strength of the perturbation.

$Un(x) \to 1.76$ and $\hbar/\tau_K \to 1.76 k_B T_0 = \Delta_0$ [the BCS gap parameter in the unperturbed superconductor at $(T = 0)$]. For $x \to 1$

$$\begin{aligned} Un(x) &\to \frac{\pi}{8}(1 - x) \\ \frac{\hbar}{\tau_K} &\cong \frac{\pi}{8} k_B (T_0 - T) \qquad (T \to T_0) \end{aligned} \tag{8-27}$$

EXAMPLES OF "ERGODIC" SYSTEMS

(1) In a field H:

(a) small colloidal particles,

(b) thin films with both surface and volume scattering (for films of well controled thickness, surface scattering is not enough to bring in the ergodicity, as mentioned above) in parallel fields,

(c) thin films of irregular thickness.

(2) Alloys with magnetic impurities.

In the following problems, specific calculations of τ_K are described for Markoffian systems. In case (1) for instance, once τ_K is known as a function of magnetic field, we insert the result in Eq. (8-26) and obtain the critical field versus temperature curve.

Problem. Compute τ_K for a small "dirty" sample of arbitrary shape under a field (K. Maki, 1963).

Solution. We consider a sample of dimensions d much larger than the mean free path ℓ. It is important to realize, however, that there is an upper bound on d, related to our assumption $|\Delta| =$ const. At larger d's, we know in fact that the order parameter is not constant in space: The superconducting phase nucleates at a field H_{c2} or H_{c3}, the nucleus having a size

$$\xi(T) \sim \sqrt{\xi_0 \ell \frac{T_0}{T_0 - T}}$$

where $\xi_0 = 0.18(\hbar v_F / k_B T_0)$. Thus, to be sure that $|\Delta|$ is constant, we must have

$$d < \sqrt{\xi_0 \ell \frac{T_0}{T_0 - T}}$$

and the inequalities

$$\ell < d < \sqrt{\xi_0 \ell \frac{T_0}{T_0 - T}}$$

allow only for a rather narrow range of d values. We discuss this region, however, since it is a simple example of ergodic behavior. From the point of view of the one-electron trajectories, the condition $\sqrt{\xi_0 \ell} > d$ means that the diffusion length $\sqrt{Dt}$ (where $D = \frac{1}{3} v_F \ell$ and $t \sim \hbar / k_B T_0$) is larger than d, and consequently that at time t the electron has explored all regions of the sample. Then the phase $\phi(t)$ defined in Eq. (8-13) is a sum of many uncorrelated increments, and has a gaussian distribution. This implies that

$$\langle e^{i\phi(t)} \rangle = e^{-1/2 \langle \phi^2(t) \rangle}$$

We write $\phi(t) = \int_0^t dt' \omega(t')$ with $\omega = 2e\mathbf{v} \cdot \mathbf{A}/\hbar c$. At times t much larger than the collision time $\tau = \ell / v_F$, we have

$$\langle \phi^2 \rangle \to 2t \int_0^\infty \langle \omega(0)\, \omega(t') \rangle \, dt' \qquad (t \gg \tau)$$

For trajectories originating from point $\mathbf{r}$, we may write

$$\langle \omega(0)\, \omega(t') \rangle = \left(\frac{2e}{\hbar c}\right)^2 \sum_{\alpha\beta} A_\alpha(\mathbf{r}) A_\beta(\mathbf{r}') \langle v_\alpha(0) v_\beta(t') \rangle$$

$$= \left(\frac{2e}{\hbar c}\right)^2 A^2(\mathbf{r}) \frac{v_F^2}{3} e^{-t'/\tau} \qquad \alpha, \beta = x, y, z$$

since at times of order τ the change of $\mathbf{r}$ along the trajectory is $\sim \ell$ and negligible on the scale of variation of $\mathbf{A}(\sim d)$. Then

$$\langle \exp[i\phi(t)] \rangle = \exp(-t/\tau_K) \qquad (t \gg \tau)$$

$$\frac{1}{\tau_K} = \frac{1}{3}\tau \left(\frac{2ev_F}{\hbar c}\right)^2 \langle A^2(\mathbf{r}) \rangle$$

The average of $\mathbf{A}^2$ is to be taken over the sample volume.

It is very important to realize that the choice of gauge for $\mathbf{A}$ is imposed by the condition: Δ = const.

For simple geometries, this gauge can be found by inspection:

(1) Thin film (of thickness d along Ox) in uniform field (along Oz), carrying no over-all current.

$$A_x = A_z = 0$$

$$A_y = Hx \qquad \text{(where x is measured from the midplane of the film)}$$

$$\langle A^2 \rangle = \frac{H^2 d^2}{12}$$

(2) Spherical grain of radius R in uniform field

$$\mathbf{A} = \tfrac{1}{2} \mathbf{r} \times \mathbf{H}$$

$$\langle A^2 \rangle = \frac{H^2 R^2}{10}$$

Problem. Compute τ_K for a magnetic alloy (A. Abrikosov and L. P. Gorkov, 1961).

<u>Solution</u>. The following is a "lowbrow" derivation. Start from the equation

$$g(\Omega) = \overline{\sum_m |(n|K|m)|^2 \, \delta\left(\frac{\xi_m - \xi_n}{\hbar} - \Omega\right)}$$

$$= \frac{1}{\pi} \frac{\tau_k}{1 + \Omega^2 \tau_k^2}$$

the average being taken over one-electron states at a fixed energy ξ_n (and in practice at the Fermi level). Choose a value $\Omega \gg 1/\tau_k$ (although $\Omega \ll E_F/\hbar$). Then $g(\Omega) \sim 1/\pi \Omega^2 \tau_K$ and we can write

$$\frac{1}{\tau_K} = \pi\hbar \sum_m \overline{|\langle n|K|m\rangle|^2} \, \Omega^2 \, \delta(\xi_n - \xi_m - \hbar\Omega)$$

$$= \frac{\pi}{\hbar} \sum_m \overline{|\langle n|K|m\rangle|^2 \, (\xi_n - \xi_m)^2} \, \delta(\xi_n - \xi_m - \hbar\Omega)$$

$$= \frac{\pi}{\hbar} \sum_m \overline{|\langle n|[\mathcal{H}_e, K]|m\rangle|^2} \, \delta(\xi_n - \xi_m - \hbar\Omega)$$

The commutator $[\mathcal{H}_e, K]$ has been derived in Eq. (8-17)

$$[\mathcal{H}_e, K] = -2K\mathbf{B} \cdot \mathbf{S}_e$$

where $\mathbf{B} = \Sigma_i \Gamma(\mathbf{r} - \mathbf{r}_i) \mathbf{S}_i$.

To *second order* in the impurity potentials, we may now replace the exact one-electron states $|n\rangle, |m\rangle$ by plane waves $\phi_n \rightarrow L^{-3/2} e^{i\mathbf{k}\cdot\mathbf{r}}$

$$\frac{1}{\tau_K} = \frac{\pi}{2\hbar} \sum_{\sigma,\mu} \sum_{\mathbf{k}'} (\sigma\mathbf{k}| 2\mathbf{B}\cdot\mathbf{S}\rho |\mu\mathbf{k}') \, \delta(\xi_k - \xi_{k'} - \hbar\Omega)$$

The summation on spin indices (σ, μ) is performed according to the rules

$$\frac{1}{2}\sum_{\sigma\mu} \langle\sigma| S_e^x |\mu\rangle \langle\mu| S_e^x |\sigma\rangle = \frac{1}{2}\sum_{\sigma} \langle\sigma| (S_e^x)^2 |\sigma\rangle = \frac{1}{4}$$

$$\frac{1}{2}\sum_{\sigma\mu} \langle\sigma| S_e^x |\mu\rangle \langle\mu| S_e^y |\sigma\rangle = 0, \text{ and so on}$$

With the Γ interaction, the orbital part of the matrix element is given by

$$\langle k|\mathbf{B}|k'\rangle = L^{-3}\sum_i \exp[i(\mathbf{k}-\mathbf{k}')\cdot\mathbf{r}_i]\,S_i\,\Gamma(\mathbf{k}-\mathbf{k}')$$

where

$$\Gamma(\mathbf{q}) = \int e^{iqr}\,\Gamma(\mathbf{r})\,dr$$

On squaring and only keeping terms of the form $S_i S_i$, not $S_i S_j$ (no correlation between impurities), we obtain

$$\frac{1}{\tau_K} = \frac{\pi}{\hbar}\frac{n}{L^3}\frac{S_i(S_i+1)}{3}\sum_{k'} |\Gamma(\mathbf{k}-\mathbf{k}')|^2\,\delta(\xi_k - \xi_{k'} - \hbar\Omega)$$

where n is the number of impurities per cm^3.

Finally we may simplify this by taking $\hbar\Omega \to 0$, because $\Sigma_{k'}$ is nearly independent of $\hbar\Omega$ (when $\hbar\Omega \ll E_F$) and $\xi_k \to 0$, that is, we calculate $1/\tau_K$ just at the Fermi surface.[2] Then

$$\frac{1}{\tau_K} = \frac{\pi}{\hbar}\,n\,\frac{S_i(S_i+1)}{3}\,N(0)\int\frac{d\Omega_{k'}}{4\pi}\,|\Gamma(\mathbf{k}-\mathbf{k}')|^2$$

where $\int d\Omega_{k'}$ is an integration over the polar angles of $\mathbf{k}'$. The lengths are $k = k' = k_F$. If x is the impurity fraction and a^3 the atomic volume ($n = x/a^3$):

$$N(0) = \frac{1}{a^3 E_F}, \qquad \Gamma(q) \sim \Gamma\cdot a^3$$

and thus

$$\frac{\hbar}{\tau_K} \sim x\,\frac{\Gamma^2}{E_F}$$

Comments on this formula:

(1) At small x, $1/\tau_K$ is small and Eq. (8-27) may be used to derive the transition point T. Qualitatively, this gives

$$k_B(T_0 - T) \sim x\,\frac{\Gamma^2}{E_F}$$

T decreases very rapidly when magnetic impurities are added. Typically $\Gamma \sim 0.2$ eV, $E_F \sim 2$ eV, $\Gamma^2/E_F \sim 2.10^3$ °K. For $x = 10^{-3}$, T has dropped by 2°. Such rapid drops are indeed observed in practice. They are now, in fact,

[2] We can pass to the limit $\Omega \to 0$ in the final expression even though in the initial expression we took $\Omega \gg 1/\tau_K$. This simplification is introduced when the true states $|n\rangle$ are replaced by the unperturbed states $|k\rangle$.

one of the best methods to ascertain whether an impurity is, or is not, magnetic in a given matrix.

(2) The whole calculation is valid only when $x(\Gamma^2/E_F) \ll k_B T_0$. This comes from the following effect: In the presence of the Γ coupling, an indirect interaction appears between the impurities, of the form

$$H_{ij} = \mathbf{S}_i \cdot \mathbf{S}_j \frac{\Gamma^2}{E_F} f(k_F R_{ij})$$

where $E_F = \hbar^2 k_F^2 / 2m$ is the Fermi energy of the conduction electrons and $f(k_F R_{ij})$ is a dimensionless function which behaves at large distances as $(k_F R)^{-3} \cos 2k_F R$ (M. A. Ruderman and C. Kittel, 1954). Because of H_{ij}, the impurities order magneticallv (ferro- or antiferromagnetically) at low temperatures. The magnetic transition point is roughly given by $k_B T_c = x(\Gamma^2/E_F)$. We need $T_c \ll T$ where T_c is the superconducting transition temperature of the alloy. Under these conditions:

(a) Each impurity is only weakly coupled to the others and its spin direction is nearly independent of time (static impurities).

(b) The spin correlation between two impurities is of the order of T_c/T and is thus negligible (independent impurities).

(3) Note that if a spin-orbit coupling term is added to $\mathcal{H}_e$, it does not contribute to $1/\tau_K$ since it is invariant under time inversion. This shows an important difference between τ_K and the electron spin relaxation time T_1 (the spin-orbit coupling contributes to T_1).

(4) The whole calculation is valid only to order Γ^2. In higher orders, the situation is complicated by the so-called "Kondo effect."

Gapless Superconductivity

At temperatures slightly below the transition point defined by (8-26), Δ is nonzero, but still small. We can treat it as a perturbation, and compute the excitation energies ϵ_n of the quasiparticles. As usual, knowledge of the wave functions u, v, to first order in Δ, permits the calculation of the eigenvalues ϵ to order Δ^2. Taking u and v from the calculations before Eq. (8-11), we arrive at

$$\epsilon_n = |\xi_n| + |\Delta|^2 \sum_m P \frac{|\langle n|K|m\rangle|^2}{\xi_n + \xi_m} \tag{8-28}$$

(P denotes a principal part, at $\xi_n = -\xi_m$).

If the one-electron Hamiltonian is invariant under time-reversal $[\mathcal{H}_e, K] = 0$ and $\langle n|K|m\rangle$ is 0 unless $\xi_m = \xi_n$. Then the perturbation calculation is *not good:* If ξ_n is small, the energy denominator $2\xi_n$ is small and the correction diverges. This is related to the existence of the BCS energy gap. The same properties appear in all nonergodic systems.

For an ergodic system, $\langle n|K|m\rangle$ is not singular at $\xi_m = \xi_n$ and has nonzero components in a finite band $|\xi_m - \xi_n| \sim \hbar/\tau_K$. Thus the energy denominator in (8-28) remains finite, even if ξ_n (or ξ_m) goes to 0. *The perturbation calculation works.* The expansion parameter is now $\Delta/(\hbar/\tau_K) = \Delta\tau_K/\hbar$ instead of Δ/ξ_n. We can compute ϵ_n explicitly if we make use of (8-24) for the power spectrum of K; performing the integration, we get

$$\epsilon = |\xi| + |\Delta|^2 P \int d\xi' \, g\left(\frac{\xi - \xi'}{\hbar}\right) \frac{1}{\xi + \xi'}$$

$$= |\xi| + \frac{2|\Delta|^2 |\xi|}{(2\xi)^2 + (\hbar/\tau_K)^2} \qquad \left(\frac{\Delta\tau_K}{\hbar} \ll 1\right) \tag{8-29}$$

The form of this dispersion law is shown on Fig. 8-4. For $\xi \gg \hbar/\tau_K$, $\epsilon \cong \xi + \Delta^2/2\xi \cong (\Delta^2 + \xi^2)^{1/2}$, which is the BCS result for conventional superconductors. On the other hand, if $\xi \to 0$, $\epsilon \to \xi[1 + 2(\Delta\tau_K/\hbar)^2]$ tends toward 0 linearly. *There is no gap in the excitation spectrum.* Equation (8-29) can also be expressed in terms of a density of states $N_s(\epsilon)$. To order Δ^2 we have

$$N_s(\epsilon) = N(0)\frac{\partial\xi}{\partial\epsilon} \cong N(0)\left[1 + 2\left(\frac{\Delta\tau_K}{\hbar}\right)^2 \frac{(2\epsilon)^2 - (\hbar/\tau_K)^2}{[(2\epsilon)^2 + (\hbar/\tau_K)^2]^2}\right] \tag{8-30}$$

$N_s(\epsilon)$ is lower than the normal state value N(0) when $\epsilon < \hbar/2\tau_K$, but still finite (Fig. 8-5).

The material has no gap, but it is still a superconductor. For instance, we can compute the screening currents induced by an applied magnetic field H, and we find that they do not vanish—they are of order Δ^2 as usual.

This remarkable "gapless superconductivity" takes place only in the region of small Δ's ($\Delta\tau_K/\hbar < 1$) where our perturbation calculation converges. At larger Δ's we find a gap again. This can be represented in a diagram where we plot along x the strength of the perturbation (measured by $\hbar/\tau_K$) and along y the temperature (Fig. 8-5). In such a plane, we find 3 regions:

N = normal

S_1 = gapless superconductor

S_2 = superconductor with gap

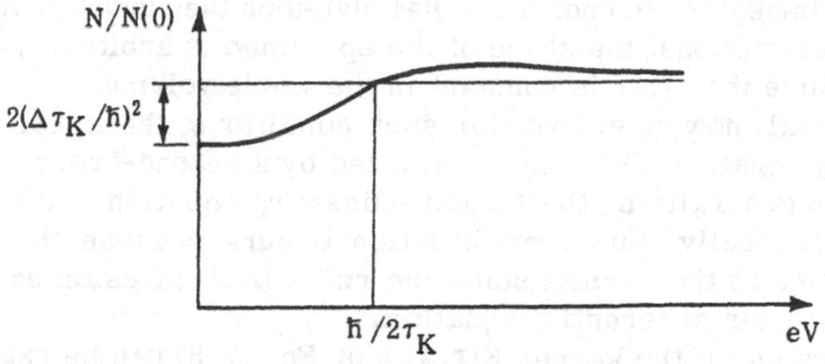

Figure 8-5

Gapless superconductivity: Density of states for Fermi-type excitations as a function of excitation energy. (The curve is valid when the pair potential Δ is small when compared with $\hbar/\tau_K$.) Note that the energy scale is given by $\hbar/\tau_K$ (not by Δ).

The boundary between N and S_1 is given by the universal function of Eq. (8-26′). The boundary between S_1 and S_2 must be computed by more sophisticated techniques (in our language: summing the whole series in powers of Δ).

There is some evidence for the existence of region S_1 in magnetic alloys, from tunneling experiments on InFe (quenched) solid solutions (F. Reif and M. Woolf, 1962). Unfortunately, in the region of greatest interest $[x(\Gamma^2/E_F) \sim k_B T_0]$, the impurity spins precess rapidly in their mutual exchange fields and this complicates the interpretation of all data on magnetic alloys. Colloidal particles or films in a parallel field should also exhibit the region S_1, but, in the present state of the art, their dimensions are not accurately controlled (what an experimentalist calls a 500Å film is in most cases an irregular layer with bumps and holes maybe 100Å or more in amplitude). τ_K depends on the thickness. To display clearly the region S_1, τ_K would have to be fixed to a few per cent, as is clear from Fig. 8-4. This is beyond present technical capability.

More convenient examples of gapless superconductivity will be mentioned in the next paragraph, where we allow $|\Delta|$ to vary in space.

8–3 DIRTY SUPERCONDUCTORS IN HIGH MAGNECTIC FIELDS

From now on, we restrict our attention to "dirty," nonmagnetic alloys. More precisely we assume that the mean free path ℓ is much smaller than $\xi_0 = 0.18(\hbar v_F/k_B T_0)$ and also much smaller than all sample dimensions d. We also assume that the sample is surrounded only by insulators (no normal metal plated on the surface). Apart from these restrictions, the shape of the specimen is arbitrary, and we do *not* assume that $|\Delta|$ is constant in the whole volume.

We shall now show that, for such conditions, the linear self-consistency equation (7-8) can be replaced by a second-order differential equation generalizing the Landau-Ginsburg equation *at all temperatures*. Basically, this simplification occurs because the transport properties in the normal state are ruled by a diffusion equation—a second-order differential equation.

As we know, the kernel $K(\mathbf{r}_1\mathbf{r}_2)$ of Eq. (7-8) can be expressed in terms of the correlation function $f(\mathbf{r}_1\mathbf{r}_2 t)$ defined in Eq. (8-5). In the "dirty" limit, it is not hard to find out what is the differential equation satisfied by f. We consider two successive times t (> 0) and $t+\epsilon$ (ϵ being small) and write down that a particle reaching $\mathbf{r}_1$ at time $t+\epsilon$ has come by diffusion from a point $\mathbf{r}'$ at time t. The average distance between $\mathbf{r}_1$ and $\mathbf{r}'$ is small (of order $\sqrt{\epsilon}$) and thus the corresponding increment in phase $\Delta\phi$ of K is simply

$$\Delta\phi = \frac{2e}{c} \mathbf{A}(\mathbf{r}_1)\cdot(\mathbf{r}_1 - \mathbf{r}')$$

Thus

$$f(\mathbf{r}_1\mathbf{r}_2,\ t+\epsilon) = \int d_3\mathbf{r}'\ f(\mathbf{r}'\mathbf{r}_2, t)\,\mathfrak{D}(\mathbf{r}_1\mathbf{r}'\,|\,\epsilon)\,e^{i\Delta\phi} \tag{8-31}$$

Here $\mathfrak{D}(\mathbf{r}_1\mathbf{r}'|\epsilon)$ is the function describing the spread of a point source at point $\mathbf{r}_1$ after a time ϵ has elapsed. It is a solution of the diffusion equation:

$$\frac{\partial}{\partial\epsilon}\mathfrak{D} = D\nabla'^2\mathfrak{D} \qquad \nabla' = \left(\frac{\partial}{\partial x'}, \frac{\partial}{\partial y'}, \frac{\partial}{\partial z'}\right)$$

together with the initial condition $\lim_{\epsilon\to 0} \mathfrak{D} = \delta(\mathbf{r}_1 - \mathbf{r}')$ and appropriate boundary conditions on the sample surface $(\nabla'\mathfrak{D})_{\text{normal}} = 0$. For an infinite medium, for instance, we have

$$\mathfrak{D}(\mathbf{r}_1\mathbf{r}'\,|\,\epsilon) = \frac{1}{(4\pi\mathfrak{D}\epsilon)^{3/2}}\, e^{-(\mathbf{r}_1-\mathbf{r}')^2/4D\epsilon}$$

Expanding $f(\mathbf{r}'\mathbf{r}_2 t)$ in a Taylor series around $\mathbf{r}' = \mathbf{r}_1$ in (8-31), retaining all terms up to first order in ϵ, and identifying coefficients on both sides, we obtain the required differential equation

$$\frac{\partial f}{\partial t} = D\left(\nabla_1 - \frac{2ie\mathbf{A}}{c}\right)^2 f \qquad (t > 0) \tag{8-32}$$

A similar calculation near the boundary of the specimen leads to the condition

$$\left(\nabla - \frac{2ieA}{c}\right)_n f = 0 \tag{8-33}$$

where n represents the direction of the normal to the surface. Finally, for $t \to 0$, the function f reduces to a δ function suitably normalized:

$$f(\mathbf{r}_1 \mathbf{r}_2, 0) = \mathcal{V}^{-1}\,\delta(\mathbf{r}_2 - \mathbf{r}_1) \qquad \mathcal{V} = \text{sample volume} \tag{8-34}$$

Let us now assume that we have found the eigenfunctions $g_p(\mathbf{r})$ of the linear operator appearing in (8-32); the g's are defined by

$$-D\left(\nabla - \frac{2ieA}{c}\right)^2 g_p(\mathbf{r}) = \frac{\epsilon_p}{\hbar}\, g_p(\mathbf{r}) \tag{8-35}$$

with the boundary condition (8-33). We can then expand the delta function in (8-34) in terms of the g's, obtaining, for any $t > 0$,

$$f(\mathbf{r}_1 \mathbf{r}_2 t) = \mathcal{V}^{-1} \sum_p g_p^*(\mathbf{r}_2)\, g_p(\mathbf{r}_1) \exp(-\epsilon_p t/\hbar) \tag{8-36}$$

We insert (8-36) in (8-1) and perform the integration over t, obtaining

$$K(\mathbf{r}_1 \mathbf{r}_2) = k_B T \sum_\omega 2\pi N(0)V \times \sum_p g_p^*(\mathbf{r}_2)\, g_p(\mathbf{r}_1) \frac{1}{2\hbar|\omega| + \epsilon_p} \tag{8-37}$$

Finally, if we make use of (8-37), our starting equation (8-8) becomes

$$\Delta(\mathbf{r}_1) = N(0)V 2\pi k_B T \sum_\omega \sum_p \frac{1}{2\hbar|\omega| + \epsilon_p}\, g_p(\mathbf{r}_1) \times \int d\mathbf{r}_2\, \Delta(\mathbf{r}_2)\, g_p^*(\mathbf{r}_2) \tag{8-38}$$

The eigenfunctions of this linear integral equation for Δ are simply the g's. For, if we choose $\Delta(\mathbf{r}) = g_q(\mathbf{r})$, the orthogonality of the g's leaves only one term $p = q$ in the sum Σ_p and (8-38) is satisfied provided that

$$1 = N(0)V 2\pi k_B T \sum_\omega \frac{1}{2\hbar|\omega| + \epsilon_q} \tag{8-39}$$

Equation (8-39) has a familiar form, and leads to

$$\epsilon_q = k_B T_0 \,\mathrm{Un}\left(\frac{T}{T_0}\right)$$

where the universal function Un(x) has been discussed earlier. For a fixed field H and a given sample shape, the ϵ_q's are fixed [and can be computed, more or less laboriously, by solving Eq. (8-35)]. What we want to know is the highest temperature T at which (8-38) has a nonzero solution $\Delta(\mathbf{r})$.

Un(x) is a decreasing function of x. Thus we must choose the *lowest eigenvalue* ϵ_0. Thus the recipe to compute a nucleation field for a "dirty" sample of arbitrary shape is the following: We find the *lowest eigenvalue* ϵ_0 of Eq. (8-35) plus the boundary condition corresponding to (8-33). This is then a known function of the field $\epsilon_0(H)$. To obtain the relation between nucleation field and temperature, we write

$$\frac{\epsilon_0(H)}{k_B T_0} = \mathrm{Un}\left(\frac{T}{T_0}\right) \qquad (8\text{-}40)$$

Problem. Derive a formula for the critical field H_{c2} in a dirty superconductor, valid at all temperatures (K. Maki, P. G. de Gennes, and N. Werthamer, 1964).

Solution. Equation (8-35) has the form of a Schrödinger equation for a particle of mass $\hbar/2D$ and charge 2e. The cyclotron frequency for such a particle is

$$\omega_c = \frac{4eD}{c\hbar} H = 4\pi \frac{DH}{\phi_0}$$

where ϕ_0 is the flux quantum. The lowest eigenvalue is

$$\epsilon_0 = \tfrac{1}{2}\hbar\omega_c$$

Writing that $H = H_{c2}$ and $\epsilon_0 = \mathrm{Un}(T/T_0)$, we arrive at the implicit equation

$$\ln\left(\frac{T_0}{T}\right) = \psi(\tfrac{1}{2}) - \psi\left(\frac{1}{2} + \frac{\hbar D H_{c2}}{2\phi_0 k_B T}\right)$$

Discussion of this equation: First, for *T close to* T_0, H_{c2} close to 0, we can make use of the aforementioned expansion of Un(T) and get

$$H_{c2} = \frac{4}{\pi^2}\phi_0 \frac{k_B(T_0 - T)}{\hbar D} \qquad (T \to T_0)$$

It is easily verified that this agrees with the Landau-Ginsburg formula H_{c2} =

$\kappa\sqrt{2}\,H_c$ when we take for κ the Gorkov expression in the dirty superconductor limit (Eq. 7-47), that is,

$$\kappa = 0.75\,\frac{\lambda_L(0)}{\ell}$$

For $T = 0$ on the other hand, we have

$$\frac{2\pi\hbar D H_{c2}}{\phi_0} = k_B T_0 \,\mathrm{Un}(0) = 1.76 k_B T_0 = \Delta_{00}$$

while the bulk critical field is given by $H_c^2/8\pi = \frac{1}{2} N(0)\Delta_{00}^2$, or

$$H_c = \sqrt{\frac{3}{2}}\,\frac{1}{\pi^2}\,\frac{\phi_0}{\xi_0 \lambda_L(0)}$$

This leads to

$$\frac{H_{c2}}{\sqrt{2}H_c} = \frac{\sqrt{3}}{2}\,\frac{\lambda_L(0)}{\ell} = 0.87\,\frac{\lambda_L(0)}{\ell}$$

Thus the ratio $H_{c2}/\sqrt{2}H_c$ increases only by 20 per cent when one goes from $T = T_0$ to $T = 0$. A similar calculation for a plane boundary shows that in a superconductor the ratio H_{c3}/H_{c2} should remain equal to 1.69 at all temperatures. Also, when the field is applied at a finite angle θ from the surface, the angular dependence of $H_{c3}(\theta)/H_{c2}(0)$ has a form independent of temperature. Both these properties appear to be well confirmed by experiment. (In fact, it had been a surprise to notice that in dirty superconductors the Landau-Ginsburg linearized equation applied so well at low temperature.)

A THEOREM ON THE LOCAL DENSITY OF STATES

Just as in Section 8-2, we can, if we wish, study our system at temperatures slightly below the nucleation point T. In particular we can compute the density of states $N_s(\mathbf{r}\epsilon)$ for excitation energy ϵ. It is now a function of the observation point $\mathbf{r}$ since the order parameter $\Delta(\mathbf{r})$ is not constant in space. This function can be measured by tunneling experiments. When N_s is computed to order $|\Delta|^2$, the results are remarkably simple (P. G. de Gennes, 1964).

(1) $N_s(\mathbf{r}\epsilon)$ depends only on the value of $\Delta(\mathbf{r})$ at the observation point $\mathbf{r}$. This local relationship is at first sight very surprising. In the initial equation it is found in fact that $N_s(\mathbf{r}\epsilon)$ depends on the values of $\Delta(\mathbf{r}')$ within a radius $\sim\xi(T)$ from $\mathbf{r}$. But just below the nucleation temperature, we know that the function $\Delta(\mathbf{r})$ has a simple shape. It is proportional to one of the eigenfunctions $g_0(\mathbf{r})$ of Eq. (8-35). This remark permits one to carry out explicitly the required integrations in $N_s(\mathbf{r}\epsilon)$ and the final result involves only $\Delta(\mathbf{r})$.

(2) $N_s(\mathbf{r}\epsilon)$ is given by Eq. (8-30) where Δ is now the local value $\Delta(\mathbf{r})$, and where $\hbar/\tau_K = \epsilon_0$ is related to the temperature by Eq. (8-40). Thus, in the region of validity of the calculation $[\overline{\Delta}\tau_K/\hbar \gtrsim 1$, or (H nucleation − H)/H $\ll 1]$, *our dirty systems are all gapless superconductors.* These predictions have recently been confirmed by tunneling experiments on various alloys in both the H_{c3} and H_{c2} regions (E. Guyon and A. Martinet, 1964). In particular, these experiments show that the energy scale (given by $\hbar/\tau_K = \epsilon_0$) on which the density of states deviates from N(0), is finite even when the order parameter $\Delta(\mathbf{r})$ becomes very small (that is, when H becomes close to the nucleation field). These measurements give in fact an accurate measurement of τ_K.

REFERENCES

Magnetic impurity effects and gapless superconductivity:

Theory:

A. Abrikosov and L. P. Gor'kov, *Zh. Eksperim. i Teor. Fiz.*, **39**, 1781 (1960), Translation *Soviet Phys.—JETP*, **10**, 593 (1960).

Experiment:

M. A. Woolf and F. Reif, *Phys. Rev.*, **137A**, 557 (1965).

Magnetic field effects:

Theory:

K. Maki, *Phys.*, **1**, 21, 127 (1964).

P. G. de Gennes and M. Tinkham, *Phys.*, **1**, 107 (1964).

P. G. de Gennes, *Phys. Condensed Matter*, **3**, 79 (1964).

Experiment:

E. Guyon, A. Martinet, J. Matricon, and P. Pincus, *Phys. Rev.*, **138A**, 746 (1965).

INDEX

Printed in the United States
by Baker & Taylor Publisher Services